T0213475

EAI/Springer Innovations in Communication and Computing

Series Editor

Imrich Chlamtac, European Alliance for Innovation, Ghent, Belgium

The impact of information technologies is creating a new world yet not fully understood. The extent and speed of economic, life style and social changes already perceived in everyday life is hard to estimate without understanding the technological driving forces behind it. This series presents contributed volumes featuring the latest research and development in the various information engineering technologies that play a key role in this process.

The range of topics, focusing primarily on communications and computing engineering include, but are not limited to, wireless networks; mobile communication; design and learning; gaming; interaction; e-health and pervasive healthcare; energy management; smart grids; internet of things; cognitive radio networks; computation; cloud computing; ubiquitous connectivity, and in mode general smart living, smart cities, Internet of Things and more. The series publishes a combination of expanded papers selected from hosted and sponsored European Alliance for Innovation (EAI) conferences that present cutting edge, global research as well as provide new perspectives on traditional related engineering fields. This content, complemented with open calls for contribution of book titles and individual chapters, together maintain Springer's and EAI's high standards of academic excellence. The audience for the books consists of researchers, industry professionals, advanced level students as well as practitioners in related fields of activity include information and communication specialists, security experts, economists, urban planners, doctors, and in general representatives in all those walks of life affected ad contributing to the information revolution.

Indexing: This series is indexed in Scopus, Ei Compendex, and zbMATH.

About EAI

EAI is a grassroots member organization initiated through cooperation between businesses, public, private and government organizations to address the global challenges of Europe's future competitiveness and link the European Research community with its counterparts around the globe. EAI reaches out to hundreds of thousands of individual subscribers on all continents and collaborates with an institutional member base including Fortune 500 companies, government organizations, and educational institutions, provide a free research and innovation platform.

Through its open free membership model EAI promotes a new research and innovation culture based on collaboration, connectivity and recognition of excellence by community.

More information about this series at http://www.springer.com/series/15427

Andrew Banasiewicz

Organizational Learning in the Age of Data

 Springer

Andrew Banasiewicz
School of Science and Engineering
Merrimack College, North Andover, Massachusetts; School of Management
Cambridge College
Boston, Massachusetts, USA

ISSN 2522-8595　　　　　　　　ISSN 2522-8609　(electronic)
EAI/Springer Innovations in Communication and Computing
ISBN 978-3-030-74868-5　　　　ISBN 978-3-030-74866-1　(eBook)
https://doi.org/10.1007/978-3-030-74866-1

This Springer imprint is published by the registered company Springer Nature Switzerland AG
The registered company address is: Gewerbestrasse 11, 6330 Cham, Switzerland

I would like to dedicate this work to those always in my heart and in my mind: my wife Carol, my daughters Alana and Katrina, my son Adam, my mother-in-law Jenny, and the memory of my parents Andrzej and Barbara, my sister Beatka and my brother Damian, and my father-in-law Ernesto.

Preface

Once prominent, the concept of "learning organizations" now lingers in management vaporware obscurity as yet another one-size-fits-all quick-fix consulting flash in the pan. While the idea's unsubstantiated sage nature is perhaps the most visible source of its demise, I believe it was its focus that was the most significant contributor to its ultimate failure. The emphasis of learning organizations is, first and foremost, on organizations, with learning being relegated to a supporting role. Intended or not, the most direct consequence of that framing was subsuming of learning-related considerations under the key organizational considerations of structure, culture, and dynamics. In that sense, the idea of learning organizations was not about "learning" but about "organizations," and so it slowly but irrevocably morphed into a yet another organizational management framework (and not a particularly meaningful one). And so, in the end, the concept of learning organizations ultimately contributed little to advancing organizational thinking, while also failing to contribute to the important yet poorly understood problem of broad base transformation of data into economic value, which is one of the defining challenges of the Age of Data. The rapid proliferation of automated and even autonomous systems is slowly but fundamentally reshaping how work is done and how lives are lived, all of which manifests itself in the relentless march of digitization producing ceaseless flows of vast torrents of data. But the challenge goes far beyond data analytics – the slowly unfolding bio-info-electro-mechanical symbiosis is poised to elevate mankind to new levels of cognitive capabilities, and those transformative changes are redefining what it means to "imagine" and to "create," and ultimately, to learn.

This book is built around the idea of *data-enabled creativity*, which is beginning to enable more and more to "stand on the shoulders of giants," but those giants are no longer just the great minds of yesterday. Machine learning systems can suggest patterns and relationships that may have eluded cognitive resource-constrained and biased human analysts, in addition to which, quantum simulations can vastly expand not just the empirical but, even more importantly, the experiential facets of knowing. As more and more traditional work-related tasks are handled by automated and autonomous systems, the economic value creation oriented contribution of organizational stakeholders, most notably employees, will necessarily need to be re-imagined, and it is the goal of this essay to contribute some food for thought.

My earlier book, *Evidence-Based Decision-Making*, tackled the challenge of objectifying individual-level sensemaking. Set in the context of cognitive reasoning mechanics, the experiential and empirical evidence framework introduced there was designed with the goal of enabling individuals to identify, synthesize, and evaluate all decision-related evidence, in an unbiased manner thoroughly and systematically. The ideas outlined in *Organizational Learning in the Age of Data* aim to build upon that work by considering the task of informed and objective decision-making in the context of human collectives, most notably for-profit and nonprofit organizations, where factors such as group dynamics, organizational culture, and organizational structure influence and shape shared cognition and shared sensemaking processes. But as suggested earlier, it is not just an overview of what is, but rather a look into the future inspired by the desire to prepare for what is likely to be.

The core premise of this book is that the ability to learn is the greatest manifestation of organizational adaptability and, ultimately, the single best predictor of long-term success. When considered from the standpoint of organizational management, in the current digital era, decision inputs can be maddingly varied, overwhelmingly voluminous, and process-intensive, the socio-politico-economic environment is famously volatile and hypercompetitive, and disruptive technological innovation is rampant and unpredictable – to not just succeed, but to survive, organizations have to develop fine-tuned sensing, tracking, and learning capabilities. And given the ever-expanding digitization and digitalization of life, and particularly of work-related aspects of life, the most important aspect of organizational learning capabilities is creative and sound utilization of not just raw data, but also data-enabled automated and autonomous systems.

And yet, at the time when interest in all-things organizational learning should be exploding, it has waned since its short-lived peak in the late 1990s and early 2000s. Why? As noted earlier, it was largely due to the idea of "learning organizations" having been highjacked by management consultants who promptly filled it with all manner of "best in class" unsubstantiated dictums and one-size-fits-all "solutions," which effectively turned a pivotal function into yet another quick fix fad. The vision of organizational learning described in this book is rooted in the belief that while anecdotally supported sage advice and templated decision-making frameworks are both worth considering, it is ultimately the data-enabled creativity that should be the focal point of organizational learning. Framed here as the desire to use data, data analytic tools, and data-enabled technologies in a way that goes beyond what is, data-enabled creativity is not a productized "solution" – it is a combination of a mindset, carefully selected skills and competencies, and a systematic and coherent processes and mechanisms that fuel organizations' ability to not just survive, but to prosper.

Organizational Learning in the Age of Data raises many questions, and answers some, but not all of them. My practitioner-turned-academician experience tells me that some of the ideas outlined here are of the plug-and-play variety, meaning lend themselves to immediate deployment, while some others are conceptual sketches in need of meaningful operationalizations. In all, I see the work discussed here as a start of a conversation in what will, hopefully, become a broadly debated around topics such as: What is the essence of informational literacy and how does that broad capability translate into organizational value? How can organizations make the most of their employees' cognitive assets and data analytic automation? What does it take to be deemed analytically literate in the age of rampant digitization and automation?

Data-enabled automation, most notably the rapid growth and maturation of the Internet of Things, or IoT, is irreversibly changing how work is done and how life is lived. Human reason, which reigned supreme since the times of the great thinkers of antiquity until the rise of modern science in eighteenth century, offered new, and at times better, ways of knowing, but now we are at the precipice of a new era wherein technology is not just a tool that eases human effort, but an independent contributor capable of creating new knowledge, which in turn can propel mankind to new heights of creativity. From the Age of Reason to the Age of Science to the Age of Transcendence, humanity continues to evolve, and what and how we learn simply needs to keep pace. Set in that broad context, *Organizational Learning in the Age of Data* is a product of my practical, hands-on experience as a knowledge worker, and my academic research immersion, aimed at understanding and bringing forth the often-hidden outcome-shaping forces and emerging trends. As such, this book aims to highlight the key already "in play" means and ways of building organizational knowledge base, while at the same time also suggesting opportunities for going beyond the what-is.

Boston, MA, USA   Andrew D. Banasiewicz

Contents

Author Biography

Andrew D. Banasiewicz is a professor of practice and the director of graduate data science and analytics programs at Merrimack College. He also is a professor of business analytics at Cambridge College and the founder of Erudite Analytics, a consulting practice specializing in risk research and estimation. A former senior-level data science and business analytics industry practitioner, where he specialized in development of custom, data-intensive analytic decision support systems, he is now focused on evidence-based decision-making and data-driven learning research and curriculum design. He is the author of five other books and numerous articles and white papers, in addition to also being a frequent speaker at conferences around the world, guest lecturer at domestic and foreign universities, and a fellow of several professional organizations. On a more personal note, Andrew is an avid outdoorsman, scuba diver, and a fitness enthusiast who regularly competes in marathon and Ironman triathlon races.

Chapter 1
Out with the Old – In with the New

1.1 Organizational Decision-Making

The game of chess is widely considered a suitable analogue to organizational strategic thinking, a belief that seems intuitively appealing to most. By and large, that is because organizational managers and chess players alike can rarely predict competitors' moves, yet those of them who possess deeper knowledge of decision patterns, and who are able to correctly read situational features, and also happen to be disciplined yet opportunistic thinkers generally tend to outperform their competitors. Interestingly, studies exploring the underlying dynamics of chess playing behaviours tend to point to skilled players' ability to assess the consequences of potential moves by mentally mapping out the most likely competitor responses, or in a more general sense, the players' superior knowledge of decision patterns, as being the most pronounced indicator of success. Perhaps somewhat less obvious is that accomplished players tend not to commit to a predetermined game plan – rather, they seek to identify situationally arising options that are least vulnerable to competitors' response, and in doing so, they allow their *de facto* strategy to slowly emerge within confines of individual games. Hence in a very general sense, chess masters are disciplined thinkers armed with a robust repository of knowledge of move patterns, the combination of which fuels the deductive aspect of their reasoning. At the same time, their ability to recognize, interpret and respond to situational features can be seen as a manifestation of the inductive side of their reasoning. Not surprisingly, such erudite and balanced persona seems to be a fitting, if not somewhat idealized, portrayal of an effective organizational decision-maker, an individual whose choices are shaped by a robust and evenhanded mix of experience-honed intuition and objective knowledge.

Yet organizational management is, in many regards, quite unlike the game of chess, as there are notable differences between moving of inanimate chess pieces on a chessboard and managing of human collectives working towards shared goals.

© The Author(s), under exclusive license to Springer Nature
Switzerland AG 2021
A. Banasiewicz, *Organizational Learning in the Age of Data*, EAI/Springer
Innovations in Communication and Computing,
https://doi.org/10.1007/978-3-030-74866-1_1

When it comes to the former, the only real consideration is to identify the most competitively advantageous moves; execution, in the sense of the ability to place a particular piece in the desired spot, is for all practical purposes *fait accompli*, because the thinking and the subsequent action involve only a single entity, the player. That is clearly not so in an organizational context where planning and execution are commonly distinct and carried out by different sets of organizational constituents. In that setting, selecting the next 'move' is only the first step – translating that choice into a carefully choreographed sets of activities to be carried out by different parts of the organization is always a complex undertaking, and often riddled with obstacles. Macrofactors such as organizational structure and dynamics, and microfactors such as differences in motivation and capabilities can, individually and collectively, impede execution, effectively muting organizational moves. Moreover, external factors, most notably the general socio-politico-economic volatility and globalization-ushered hypercompetition, exert tremendous pressure to act quickly and decisively. Service quality offers a common and simple illustration: Well-incentivized executives breathlessly expound aspirational visions of their companies' service quality, which are then left to comparatively poorly incentivized (meaning, not nearly as motivated) frontline workers to deliver. And when delivery falls short of lofty expectations, it is often tempting to see that outcome as poor execution of a good strategy, even though the only 'evidence' supporting the presumed efficacy of the said strategy is that it *intuitively* made sense. Stated differently, deductive reasoning may be hindering organizational management.

1.1.1 Hypercompetition and Intuition

The combined effect of globalization, deregulation, privatization, declining technology costs and the resultant technological proliferation, communalization of consumer tastes and preferences, and growing adaption of progressive management practices built around continuous improvement and competitive benchmarking gives rise to rapid and dynamic competition characterized by unsustainable advantage. Commonly referred to as *hypercompetition*, it reduces the lifecycle of competitive advantage as firms imitate and leapfrog each other (cost–quality hypercompetition), attack each other's positions in product or geographic markets (product-market hypercompetition) and use their size and financial advantages (financial hypercompetition) to outgain one another. As a result, organizations' competitive positions are highly changeable, even discontinuously so, which forces firms to develop the ability to respond to competitive moves rapidly and flexibly. In short, the heightened importance of efficiency, speed and responsiveness necessitates that business organizations become better learners.

In a hypercompetitive environment, the only lasting competitive advantage is the ability to generate new sources of temporary advantage. Being the lowest cost or the highest quality provider or having a unique know-how or the deepest pockets is not enough – it is the ability to continue to generate such advantages over and over again that needs to be sustained. That ability, in turn, stems from having developed means to either foresee upcoming disruptions or to cause disruptions, both of which require robust ongoing learning capabilities, which take maximum advantage of wide arrays of objective data, organizationally 'consumed' in the manner that supports purposeful acquisition of competitively advantageous knowledge. At least that is the theory of not only surviving, but even flourishing, in the age of uncertainty and turbulence.

In practice, when confronted with a task of quickly making sense of a large and diverse set of choice related factors, decision-makers often default to relying on sensemaking heuristics, or shortcuts, most notably, *intuition*. Defined as the ability to understand something immediately and without the need for conscious reasoning, those nearly instantaneous conclusions feel very natural, and typically very 'right', but can ultimately turn out to be misguided and ill-conceived, as clearly documented by a robust body of scientific and anecdotal evidence. The underlying causes are both simple and complex; taken together, they could be characterized as cognitive sensemaking inhibitors rooted largely in the physiological mechanics of the brain's innerworkings. Some of the better-known inhibitors include cognitive bias, or sensemaking conclusions that deviate from rational judgement, constrained processing capabilities known as human channel capacity, and brain self-rewiring, or neuroplasticity (or brain plasticity). And yet in spite of voluminous and convincing evidence highlighting the numerous and largely inescapable 'gut feel' pitfalls, the deference to intuition-inspired decision-making shows few signs of dissipating.

1.1.2   To Intuit or to Infer?

Few instincts are as forceful as intuition. Perhaps most vividly illustrated by the evolutionarily fine-tuned automatic physiological flight-or-flight response, a key survival mechanism, intuition can be framed as a manifestation of the subconscious dimension of tacit knowledge (discussed later in this chapter). It is undeniably beneficial to everything from the seemingly mundane, like crossing a busy street, to the most abstract pursuits in the form of creative expression, precisely because it is an actuating mechanism of subconscious knowledge. But, like conscious reasoning, it is also a product of information processing and thus a subject to the earlier mentioned inhibitors; those limitations can become particularly pronounced in the highly evaluative organizational decision-making contexts, as discussed in more depth in *Evidence-Based Decision-Making* [1], which draws attention to dangers of overreliance on intuitive sensemaking.

That very natural-feeling mode of making sense of a problem at hand is often rooted in some initial axiomatic, meaning unquestioned, foundational truths, which may be flawed; moreover, the underlying sensemaking process can also be imperceptibly influenced by a wide array of subconscious reason-warping influences. It is why otherwise well-qualified and competent decision-makers can reach what may ultimately turn out to be empirically unfounded conclusions, as illustrated by the earlier mentioned service quality example. The seemingly well-reasoned strategies can make perfect sense only to fall apart when put into practice, precisely because subjective sensemaking can be so imperfect; that which may seem so seductively logical and obvious can ultimately prove itself to be illusory. Of course, that is not to say that subjective reasoning is somehow destined to lead to failure – after all, the numerous shining examples of brilliant scientific insights of like of Copernicus, Newton, Maxwell, Heisenberg, Planck or Einstein, insights that bucked the entrenched beliefs and altered the course of human thinking, clearly point to the potential power of human reason. In fact, it has been argued that true progress requires norm-breaking, radical thinking that can only come from intuition, but as discussed in later chapters, the unfolding era of data-enabled creativity and bio-info-electro-mechanical symbiosis casts doubt on that argument. The future of learning aside, by their very nature, such rare flashes of genius are really the butterfly effect[1] of reason – they prove the possibility, not the plausibility. Within the confines of organizational management, examples of successful evidence-defying thinking are often celebrated in popular books and dissected in endless academic case studies, yet the far more numerous instances of failed 'creative' reasoning tend to be quickly forgotten. The enduring fascination with stories of success attached to atypical and non-representative choices and behaviours, many of which bucking objective evidence, serves to reinforce the sanctity of intuition. Yet, upon closer examination, it becomes clear that those are just different manifestations of the earlier-mentioned cognitive sensemaking inhibitors, such as the survivorship bias, where atypical, evidence-bucking, examples of managerial reasoning that survived the test of time – most notably because they were immortalized in books and cases – are amplified, while the failed exemplars of maverick management quietly fade away. Out-of-mind, out-of-sight...

The budding evidence-based management movement can be seen as an attempt at resetting of the managerial perspective, allowing creative decision-making to be seen as a product of persistent and systematic gathering, assessment and synthesis of available evidence. More specifically, the 'creative' aspect of decision-making is framed in the context of thoughtful assessment and synthesis of any and all decision-related information, which, as detailed in the *Evidence-Based Decision-Making* book mentioned earlier, encompasses empirical (raw data and research studies) and experiential (expert judgement and industry norms and standards) sources of knowledge. Still, the tendency to favour deductive, which highlights intuition, over inductive, which favours fact-based reasoning, is hard to overcome because it is educationally hard-coded into decision-makers' psyche by years of theory-first educational philosophy. A by-product of the seventeenth century's blossoming of the

---

[1] In chaos theory, a branch of mathematics focusing on the study of seemingly random systems, the butterfly effect captures the idea of systems' sensitive dependence on initial conditions, where a small change can result in disproportionately large effects; the name itself is derived from a metaphorical example of the flapping of the wings of a butterfly in one part of the globe and can, several weeks later, affect the formation of a tornado in another, distant part of the world.

scientific method, the currently dominant hypothetico-deductive educational mindset frames knowledge creation in the context of empirical tests of theoretical conjectures, which can emanate from a wide range of sources including established theories, accepted principles and intuition. Stated differently, organizational decision-makers spend their formative learning years immersed in the idea that new knowledge ultimately stems from prior beliefs, which continually strengthens the reverence of intuition.

Giving credit where it is due, the scientific method has indeed been a marvellous tool; when placed on the roughly 8000-year-old human civilization development timeline, the scientific and technological progress of two last 200 or so years defies rational explanation (though a limited attempt will be made later). The hypothetico-deductive model of knowledge creation has led to unlocking of innumerable mysteries of nature that are, at least in concept, testable, meaning measurable in time and space. However, not all questions can be answered using that approach because not all conjectures are testable (more on that later), thus singularly focusing on deductive reasoning may, at times, lead to myopic thinking. A clear case in point is offered by business education: Management students are commonly immersed in 'principles' of business sub-disciplines such as economics or marketing, are taught to view business firms through the lens of various individual and organizational behaviour 'theories', and are conditioned to make organizational decisions using innumerable 'decision frameworks', ranging from the ubiquitous SWOT (strengths, weaknesses, opportunities, threats) analysis to agile management to business process reengineering and on and on. Devoting the bulk of their educational time and effort to internalizing all sorts of sensemaking templates, they spend comparatively little time learning how to directly, creatively and meaningfully engage with 'raw' organizational management-related evidence, most notably numeric and text, structured and unstructured data. In the end, what should be a journey of discovery paralleling learning to become a skilled chess player, management education is more aptly characterized as indoctrination into the art of pigeonholing, where the primary focus is on learning how to fit observed fact patterns to predetermined solutions. Not surprisingly, it produces far more credentialed zombies than it does innovative thinkers.

Over-reliance on deductive reasoning can also lead to under-utilization of organizational data, which can manifest itself in a variety of ways. Perhaps the most obvious one is that when data are looked at through the prism of prior belief testing, it is easy to dismiss unexpected findings as flukes and thus miss out on opportunities to uncover thought provoking, maybe even eye-opening insights. Simply put, when the goal is to find X, it is easy to dismiss or overlook anything other than X. A somewhat less obvious deductive reasoning pitfall might be the framing of the underlying research question. The logic of the scientific method implicitly assumes that the informational essence of conjectures that are to be tested, which typically emanate from whatever happen to be the accepted view, be it in the form of formal theories or informal beliefs, holds the greatest knowledge potential. In other words, the conjectures to be tested are the 'right' questions to ask. However, in rapidly evolving environmental, sociopolitical, economic and even situational contexts that simply may not always be the case. Trend wrecking change can relatively quickly and quietly lessen the informational value of established beliefs, ultimately rendering empirically testing of conjectures derived from those conceptual roots informationally unproductive, if not outright counterproductive. Hence to yield maximally beneficial outcomes, organizational data need to be seen not only as a source of prior beliefs validation, or conjecture testing, but also as a source of open-ended, exploratory learning. In a more general sense, to succeed, organizations need to approach the task of organizational learning in a manner that is as structured and focused as any other organizational function, such as research and development or marketing.

1.1.3 Learning Organizations Versus Organizational Learning

The recognition of learning as a distinct organizational competency can be traced back to the 1960s and 1970s, but it was not until the 1990s that the topics of organizational learning spurred wider interest among researchers. The resultant rich and varied research streams included several popular books, such as the 1990 work titled *The Fifth Discipline: The Art and Practice of the Learning Organization* [2], which helped to popularize the notion of 'learning organization' among practitioners. Unfortunately, the wider embrace of the idea of learning organizations has led to proliferation of 'tried and true' management solutions in the form of pre-packaged consulting frameworks chock-full of buzzwords and anecdotal stories of inspiring success, and other forms of management lore. Promising quick solutions to even intractable management problems, those templated conceptualizations, often accompanied by unsubstantiated axioms and other self-deceptive dictums became the *de facto* public face of the learning organization concept, which contributed to this important aspect of organizational functioning becoming a yet another passing fad.

Looking past self-anointed prophets, buzzwords and sage anecdotes, the main shortcoming of the learning organization concept is that it is non-informative, because, ultimately, all organizations learn. At their core, organizations are groups of individuals joined together in pursuit of shared goals, from which it follows that as organizational contributors engage in their respective value creation efforts, they continue to add to their tacit knowledge (learning by doing), in addition to which they may also continue to pursue structured learning in the form of formal education. It is in fact difficult to imagine a non-learning organization, and given that, it is more instructive to acknowledge that some amount of organizational learning always takes place, but the efficacy of the organization-specific learning can vary considerably. The notion of learning organizations was somewhat reminiscent of a practicing sports team. While there is a possibility that there are, let us say, soccer or basketball teams that do not practice in preparation for their matches or games, there are more than likely few and far between – by and large, teams that compete practice, just some do it better than others. The important part is to be able to understand what it takes to do it better.

1.1.3.1 Organizational Learning

In a very general sense, learning can be either *inductive*, which entails drawing general conclusions from specific instances, or *deductive*, which does the opposite – it applies general principles to arrive at specific conclusions. An important, and often overlooked, aspect of both inductive and deductive learning is that both equate learning with human endeavours; however, the rise and rapid proliferation of data-producing, capturing and processing electronic infrastructure opened up a parallel, non-human learning avenue, commonly referred to as *machine learning*. Moreover, as readily illustrated by the ever-expanding array of advance machine learning systems, broadly known as *deep learning* technologies, artificial systems are already capable of not only discerning data patterns, but also generalizing from those patterns. And so in at least some organizational decision contexts, it could be argued that inductive and deductive learning characterize not just human, but also machine learning.

What then is organizational learning? A conventional answer might equate it with aggregate acquisition of knowledge by those who directly contribute to the organization's economic output, most notably the organization's employees. Such framing of organizational learning seems reasonable in view of the fact that organizations are essentially groups of individuals joined together in pursuit of common goals. Setting aside, for now, learning related to organizational norms and culture, the human cognition-centric definition of organizational learning does not contemplate decision automation, which is difficult to rationalize in view of rapid proliferation of artificial intelligence (AI). Consider the insurance industry: Many insurance firms now offer online automated quotation systems for certain types of insurance coverage, such as automotive collision; rate-quoting algorithms that power those systems exemplify organizational-level machine learning, as those systems learn from data and make decisions (in the form of insurance quotes they issue) largely without direct human input. It thus follows that should the notion of organizational learning be restricted to human cognition only, the informational, and thus ultimately economic, benefits of such systems would be ignored. The inefficacy of seeing organizational learning through such a narrow lens becomes even more glaring when business organizations that are built entirely around machine learning technologies are considered. To companies such as Grammarly, which offers an AI-enabled writing assistant, Tempus, which develops AI-assisted medical data applications, or DataRobot, which provides data scientists with a platform for building and deploying machine learning models, machine-based learning comprises the informational core.

Given the transformative impact of technology on society (discussed in more detail in the Axial Periods section below), broadening of the traditional conception of organizational learning to also encompass acquisition of knowledge by artificial systems seems warranted. At the same time, however, it is important to also acknowledge that even if the uniquely human social cognition considerations are set aside, currently commercially available AI systems do not (yet?) exhibit what could be called 'creative learning' capabilities. More specifically, although machine learning applications are capable of elementary inductive and deductive learning, those systems are not capable of the kind of learning that spurs imagination-like creativity, in the sense of seeing something new in manifest data patterns, which suggests a need for a more nuanced conception of organizational knowledge acquisition. With that in mind, when considered from the perspective of organizational functioning, learning can be seen as being comprised of three distinct dimensions: adaptive, which supports mostly automatic changes needed to accommodate environmental changes, generative, which produces and disseminates new means of attaining organizational goals, and transformative, which enables strategic shifts as a way of responding to perceived opportunities or threats. When framed in the confines of adaptive, generative and transformative learning mechanisms, the notion of organizational learning can then more easily accommodate both the established, human cognition-centred, and the emergent, machine-based learning dimensions, which suggests a need for a new, more complete conceptualization of the idea of organizational learning.

1.1.4 Time for a New Perspective

When looked at from the standpoint of the most rudimentary organizational need, which is survival, learning capability can be seen as a manifestation of adaptiveness, a trait that enables collectives to fine-tune their behaviours in the form of operating characteristics, and cognitive functioning in the form of choice-related sensemaking. But like individuals, organizations do not exist to merely survive, which suggests that to be truly informative, the notion of organizational learning also needs to account for what could be characterized (following the well-known A. Maslow's hierarchy of human needs) as self-actualization needs, or put more simply, the ability to flourish. Here, learning capability manifests itself in the more elusive idea of creative problem-solving or creative ideation.

Linking together the three sets of learning-related considerations, human cognition vs. machine learning, the adaptive vs. generative vs. transformative learning triality and the duality of survival vs. self-actualization learning purpose, forms an outline of a new perspective on organizational learning, graphically summarized in Fig. 1.1. So framed, organizational learning can then be conceptualized as the acquisition of adaptive, generative and transformative knowledge by means of human cognition and machine learning, and geared towards organizational survival and self-actualization. But as a process, learning does not exist in a vacuum thus to be conceptually and informationally complete, the broader context within which the endeavour of organizational knowledge acquisition is actuated needs to also be taken into account. That context encompasses a myriad of group and individual factors, including organizational structure, culture and group dynamics, in addition to an array of latent psychological and emotional characteristics, and even biological traits, all of which directly or indirectly impact the efficacy of learning efforts.

Still, it takes more than just a good grasp of group and individually determined influences to paint a complete picture of organizational learning mechanics. The

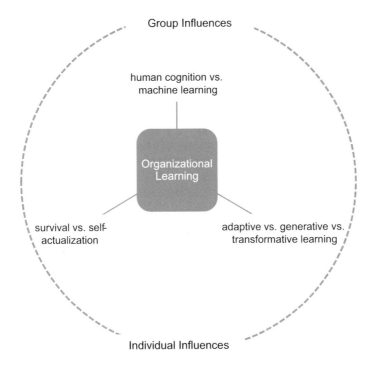

Fig. 1.1 A conceptual model of organizational learning

seemingly ceaseless onslaught of digitization coupled with the rapid expansion of the Internet of Things–driven automation manifesting itself in the proliferation of self-learning algorithmic decision engines and other advanced self-functioning systems are all changing how lives are lived, how work is done and are also beginning to call into question some longstanding aspects of how knowledge is created. In a pragmatic sense, the ability to separate the proverbial chaff from grain, which amounts to finding and institutionalizing decision-guiding insights often hidden in masses of comparatively trivial informational tidbits, is an important prerequisite of effective learning, but the longstanding mechanism for doing so, known as the scientific method, may not match up well with the realities of big data and the demands of organizational decision-making.

1.1.4.1   The Scientific Method in the Age of Data

The circa seventeenth-century emergence of empirical means of knowledge creation has been nothing short of the greatest transformative force in the 8000 or so years old human civilization. Central to empirical learning has been the idea of the 'scientific method', which frames the creation of knowledge in the context of systematically testing of beliefs and assumptions, formally referred to as hypothesis or conjecture testing; in that evaluative context, to be considered 'true', a knowledge claim needs to pass an appropriate empirical test (discussed below). In short, the scientific method ties the creation of new knowledge to analyses of data. Setting aside an obvious observation that the contemporary conception of data, in terms of sources, varieties and volumes, is vastly different from what it was a few decades ago, much less a couple of so centuries ago, as clearly communicated by its name, the scientific method is a tool of science, the goal of which is the discovery and validation of universally true generalizations. While scientific knowledge obviously benefits business and other organizations, as clearly exemplified by medical and pharmaceutical industries, informational needs of organizations are nonetheless frequently more nuanced, as is the case with many situation-specific operational decisions. To the degree to which the core utility of the scientific method, as seen from the perspective of one of its key operational elements in the form of statistical significance testing, is tied to derivation of universal generalizations, its value to discerning contextualized, in the sense of being valid within a given set of circumstances and at a particular point in time, insights becomes suspect. And thus, if a decision-maker is not interested in ascertaining the extent to which some situationally meaningful insights are universally true, does that decision-maker need the statistical significance 'stamp of approval'? To be sure, there could certainly be compelling reasons, but what if there are no such reasons? These questions matter because it is commonplace to apply the logic of the scientific method, or more specifically, to use statistical significance testing, to attest to credibility of data analytic outcomes, whether or not those outcomes are meant to be interpreted as universal generalizations. Here is a simple example: A marketing analyst is looking to determine if there is a material difference between response rates to two separate promotions; under most circumstances, to be considered credible, the said response rate difference needs to be deemed 'statistically significant', even though the comparison might be limited to just the current context (i.e. the analysis does not purport to produce universal generalizations).

In addition to what could be considered to be somewhat philosophical (though clearly important) considerations outlined above, the widely used practice of statistical significance testing becomes even more suspect when data volumes are taken into account. Although as currently used, statistical significance testing logic and mechanics only date back to the earlier part of the twentieth century, those evaluative mechanisms nonetheless predate the modern electronic data infrastructure (e.g. the now common terms 'big data' and 'data science' were coined in the late 1990s) and therein lies the more practical limitation of using those mechanisms to attest to credibility of data analytic results.

Briefly described, statistical significance tests (SST), known to many as F, t or $\chi 2$ tests, are techniques used to differentiate between spurious and material data analytic outcomes, and as such are an important element of the scientific method. While originally developed to support theory building efforts, SSTs have also been used extensively in applied settings, where analysts are also interested in being able to objectively distinguish between spurious and material data analytic results. However, what appears to be communality of goals in fact is not: Building and testing theories call for assessing generalizability of sample-derived estimates (since research studies are usually sample-based), whereas extracting practical decision-guiding insights out of data may just require a reasonable assessment of estimates' validity. In other words, within the confines of significance tests, 'significant' estimates are those that can be attributed to a population of interest, which is critically important to theory building (since 'if-then' type generalizations are the heart of theories), but not necessarily so to practitioners who might just be looking for insights that are applicable to their particular situations only. Moreover, SSTs are also sample size dependent – as the number of records used to compute the test statistic increases, the likelihood of detecting statistical significance increases, more or less in proportion to sample size. That means that a given estimate might be deemed not statistically significant when computed using a small but adequate sample size, such as 200 cases, only to become significant if the underlying sample size was sufficiently increased, everything else being the same. To be clear, that means that the number of data records used to compute the appropriate test statistics, in the form of the earlier mentioned F, t or $\chi 2$ values, can singularly determine the credibility of statistical estimates. If analytic outcomes based on a smaller, but once again statistically adequate, meaning sufficiently large to produce unbiased estimates, are deemed to not be statistically significant, how is one to make sense of a contradictory conclusion stemming from identical-except-for-the-sample-size analysis? And then, what happens when in the age of massive datasets, the determination of statistical significance is routinely made using very large record counts?

Here is what tends to happen: Analysts are often compelled to group statistically significant findings into those that are 'practically significant' and those that are not. While the impulse to do so is understandable, such practice nullifies the benefits of objective estimate validation standards by opening the door to biased result interpretation. As discussed earlier in this chapter, intuition-based evaluation of empirical results is fraught with numerous cognitive and behavioural traps; the core benefit of objective analyses is helping decision-makers to surmount or circumvent those reason-warping effects, thus subjectifying initially objective results runs counter to that goal.

And still, there is more bad news. Being rooted in normal distribution-derived probability estimation, the interpretation of statistical significance cannot be divorced from the underlying symmetric distribution, which is incongruent with many real-life scenarios. More specifically, many real-life phenomena are not normally distributed: banks have far fewer high net worth account holders than low net worth ones, insurance companies incur far more low-cost than high-cost claims, non-profit organizations have far fewer large than small donors, schools tend to have more struggling than high-performing students and so on. There are innumerable examples of measurable states and outcomes that, in aggregate, are not normally distributed; some of those can be mathematically transformed to become approximately normal but others cannot, and when used such persistently non-normal data, the efficacy of statistical significance testing becomes even more suspect.

This brief overview of applicability of the logic and mechanics of statistical significance testing would be incomplete without taking into account some additional, more practical considerations, most notably informational timeliness and desired specificity. Thoughtful, meaning capable of producing valid and reliable insights, use of scientific method does not lend itself to quick, situationally responsive insights. That is primarily because the hypothetico-deductive method is geared towards producing universal generalizations, which, ironically, can run counter to organizational, especially business-related needs. Essentially all business firms, but also non-business organizations, that compete for limited resources are keenly focused on finding ways to get ahead of their competitors, which means that in an informational sense, they seek competitively advantageous, not universal and thus generic, insights. It is intuitively obvious that insights that work equally well for all competitors can only offer parity, not advantage, and so while not getting left behind is certainly an important consideration, getting ahead is the ultimate goal of organizational learning, a sentiment that was captured earlier in the survival vs. self-actualization duality. Though often overlooked, that is an incredibly important distinction between informational demands that characterize (scientific) theory building, and those that characterize making organizational choices in a competitive setting: The former demands well-substantiated and validated claims, while the latter needs timely and situationally tailored insights.

Looking beyond analyses-related methodological considerations, the very essence of learning is changing, which also impacts the idea and the (applied) practice of statistical significance testing. The scientific method rose to prominence because it paved the way to learning about the physical (hard sciences) and man-made (social sciences) worlds. Taken together, that mode of knowledge creation can be characterized as *theoretical* learning; it offers a more systematic and objective parallel to what could be considered the endemic, even primordial *experiential* learning, which entails creating of knowledge by means of immersion or observation. Within that broad context, the core utility of objective data lies in objective validation of theoretically or experientially derived suppositions[2]; when considered within the confines of organizational functioning, that suggests that data can only indirectly shape informational inputs used by organizational decision-makers. With than in mind, the essence of the reasoning laid out in this book is that the idea of organizational learning can greatly enhance the organization's value creation capabilities if, and only if, the traditional, human sensemaking-centric conception of learning is expanded to also include technology-based learning capabilities.

1.1.4.2 Axial Periods

The notion of the 'axial age', first suggested by Karl Jaspers, a philosopher, refers to decisive periods in the maturation of human cognition, in which our sensemaking made very distinct leaps forward. Although there is no universal agreement as to the

---

[2] It is worth noting that even exploratory, i.e. open-ended search for patterns and relationships, data analyses are ultimately geared towards forming hypotheses, which are then to be empirically tested.

number of such periods, the idea itself is compelling because it helps to address what a noted palaeontologist and evolutionary biologist Stephen Jay Gould referred to (in collaboration with Niles Eldredge) as 'punctuated equilibria' – development that is marked by isolated episodes or rapid change between long periods of little or no change. Although Gould and Eldredge's ideas were focused on biological evolution, their reasoning can also be applied to evolution of cognition in the somewhat narrow context of human sensemaking. And in that context, three distinct axial periods can be identified, graphically illustrated in Fig. 1.2.

The period of 600 or so years spanning 800 BCE to about 200 BCE saw the rise of Confucius and Lao-tse in China, Buddha in India, the Hellenic philosophers in Greece and the Hebrew prophets in Palestine, all of whom contributed to lighting of the fire of reason as a way of making sense of the world; rather than fearing nature, man began to try to make sense of it. Nearly 2000 years went by before another sudden intellectual eruption took place, this time giving rise to science as a mean of learning. The period now known as the Scientific Revolution took hold in Europe near the end of the Renaissance, ultimately spawning the intellectual movement known as the Enlightenment, which emphasized objective means of inquiry as a tool of knowledge creation. And now we are standing at the precipice of a yet another cognitive axial period, the Age of Transcendence, where the emerging bio-info-electro-mechanical symbiosis[3] is set to slowly unleash a completely new

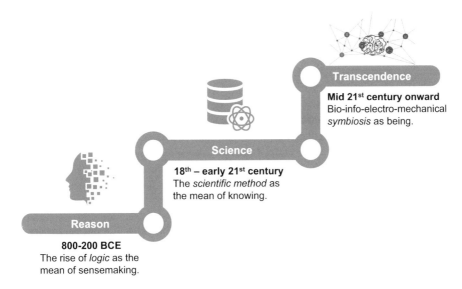

Fig. 1.2 Key cognitive axial periods

---

[3] This term is derived from (published) DARPA (Defense Advanced Research Projects Agency) research, more specifically, from an article by A. Prabhakar, the former director of DARPA, published in January 27, 2017, issue of *Wired* magazine.

ways to learn, which will ultimately allow mankind to transcend our physical, time and space limitations in search of deeper understanding of the nature of reality and our own existence. The same ideas apply to far more 'local', in the manner of speaking, informational pursuits that characterize organizational decision-making, as just because questions that confront organizational managers are not quite as lofty or otherworldly as those regarding, for instance, the conditions that preceded the Big Bang, and it does not mean that those questions should be addressed with less fervour. All of that may sound a bit farfetched, but let us remember that the idea of space or even air travel was equally farfetched not that long ago; looking ahead and imagining that which may seem improbable today is the essence of progress.

What does all that have to do with organizational learning? It compels us to look beyond our traditional conception of what it means to 'learn'. As summarized in Fig. 1.2, the great philosophers of antiquity developed *logical* means of learning by means of structured, rational thought that produced numerous lasting contributions, such as the present-day notions of justice and democracy; building on that great tradition, the pioneers of the Scientific Revolution developed *empirical* means of learning that spun the development of innumerable technological marvels. And now, to paraphrase Isaac Newton's famous words,[4] as we are 'standing on the shoulders of giants', we need to take the next step in our continually developing learning journey by also embracing data-driven *simulational* learning.

1.2   Knowledge and Knowing

From the perspective of machine-like information processing, a motion picture (e.g. a movie) is essentially a series of still frames shown in fast succession, but to a human spectator, it is a story with a distinct, typically implied theme. Senses capture the ultimately discrete audio-visual stimuli, and the brain effortlessly (assuming a well-made production) infers the multitudes of causally and otherwise interconnected meanings necessary to understand and appreciate the underlying plot. At the end of a movie, it is natural to say that we know what it was all about; in fact, we are so comfortable with the idea of knowing that we rarely stop to ask the age-old epistemological[5] question: How do we know that we know? Consider Fig. 1.3.

As graphically summarized above, what constitutes 'knowledge' can be understood in terms of two separate sets of considerations: sources, which capture the 'how' aspect of what we know or believe to be true, and dimensions, which represent the 'what' facet of what we know, in the sense of the type of knowledge. The source of knowledge can take the form of formal learning, often referred to as

---

[4]The full expression, as commonly quoted, which came from Newton's 1675 letter to fellow scientist (and perhaps the great scientist's biggest antagonist) Robert Hooke, reads: 'If I have seen further, it is by standing on the shoulders of giants'.

[5]Epistemology is a branch of philosophy concerned with understanding the nature of knowledge and belief.

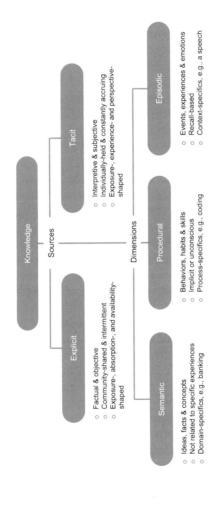

Fig. 1.3 Dimensions of knowledge

explicit knowledge, or informal learning, which is also known as *tacit* knowledge, best exemplified by skills, familiarity with or understanding of topics or endeavours encountered while or working pursuing a hobby. The second and somewhat distinct set of considerations instrumental to dissecting the essence of how we know what we know characterizes the typologically distinct dimensions of knowledge. Here, our individually held 'I know' can be grouped into three broad, general kinds of knowledge: *semantic*, which encapsulates abstract ideas and facts, *procedural*, which captures behavioural abilities to perform specific tasks, and *episodic*, which encompasses an array of hedonic or emotive memories.

Insofar as decision-making is concerned, the differently acquired and differently oriented aspects of knowledge impact each other, so that, for instance, one's explicit knowledge of how to fit regression models will often be somewhat altered by one's practical experience (tacit knowledge) acquired in the course of conducting regression analyses. We could even go as far as suggesting that the ongoing interplay between explicit and tacit knowledge can, within the confines of a specific area of human endeavour, produce different effective topical knowledge. Quite often, individuals might start with the same knowledge of a particular topic – as would tend to be the case with those who took the same course, taught by the same instructor – only to end up at some point in the future with materially different effective knowledge of that topic, because of divergent tacit influences. Perhaps the most compelling example of that phenomenon is offered by advanced degree, e.g. Ph.D., students. Let us say that two similar in age individuals enrolled at the same time in the same statistics doctoral programme. Those two individuals ended up taking the same set of courses, and upon completing their degree requirements graduated at the same time. At the point of receiving their doctoral degrees, both individuals could be assumed to have comparable levels of knowledge of the topic mentioned earlier – regression analysis. Their degrees in hand, they choose to pursue different career paths, with one taking a position as an assistant professor at another academic institution, while the other instead opting to join a consulting organization as a statistical analyst. Over the course of the next decade or so, the academic of the two engaged in theoretical research focused on the use of matrices in linear least square estimation (a subset of mathematical considerations addressing an aspect of regression analysis), while the consultant was primarily focused on developing applied solutions for his corporate clients, such as predictive models to forecast future cost of insurance claims. If their knowledge of regression analysis was to be re-assessed at that point (a decade or so past earning their doctorates), it would be reasonable to expect their respective effective knowledge of regression to be materially different. Why? Because of the interplay of the dissimilarity of their post-graduation work experience (tacit knowledge), and the manner in which their on-the-job learning interacts with their initially comparable explicit knowledge. Ultimately, the two overtly similarly credentialed professionals could have materially different effective knowledge of regression.

The idea of effective topical knowledge is important because that is what a decision-maker draws on when presented with a task or confronted with a choice. At what could be characterized as a 'decision point', effective knowledge represents the totality of what that individual is able to do and believes to be true, within the confines of the problem at hand. The otherwise conceptually distinct sources and dimensions of knowledge – i.e. the sources and dimensions shown in Fig. 1.3 – all converge into that decision point ability (or lack thereof) to perform the task in question or make a rational choice, typically under conditions of uncertainty. Recalling the task of building a regression model, when presented with that task the analyst will need to draw on his or her semantic knowledge of different types of regression techniques (e.g. linear, non-linear, logistic, etc.), his or her familiarity with the appropriate data modelling tools and procedures (e.g. R, SAS or other languages/applications), and, quite possibly, past episodes of performing similar type of work (to pinpoint anticipated challenges, etc.). Stated differently, that analyst will draw on what appears to be a singular reservoir of expertise, but what is really a composite of different facets of know-how that relate to the task at hand.

There is a yet another aspect of knowledge that is important to mention, as it addresses the underpinning philosophical traditions. Much of Western intellectual tradition finds its roots in the Big Three of ancient Greek philosophers: Socrates, Plato and Aristotle. The first, chronologically speaking, of that trio, Socrates, did not write any books, but instead channelled his efforts into publicly asking probing, and at times humiliating questions, which gave rise to the now-famous Socratic Teaching Method.[6] Given the fact that he did not write any books, Socrates' wisdom might have slowly dissipated with the passage of time had it not been for his devoted and talented student, Plato, who wrote down and thus preserved many of his teacher's dialogues. Building on Socrates' teachings, Plato subsequently contributed more formally described abstract, almost other-worldly notions of Forms, Ideals and Ideas (such as Equality or Justice), which were then adapted and further refined by his most outstanding student, Aristotle, by many considered the first empirical scientist. Less 'other-worldly' than Plato in his philosophical approach, Aristotle injected a substantial dose of realism into his teacher's abstract conceptions, laying the foundations for logical theory, or the use of reason in theoretical activity. The many thinkers who followed and carried forward Socrates, Plato and Aristotle's ideas ultimately gave rise to Western intellectual tradition, characterized by the use of reason in knowledge creation, built around of what is now known as the scientific method discussed earlier.

The above summarization of the Western intellectual tradition invites some thought provoking if only tangentially related what-if pondering. What if Socrates did not have Plato as a student? Would the Socratic Teaching Method exist? Would we even know his name, or would his ideas and his memory be long forgotten? It seems reasonable to speculate that there were a number of other, insightful thinkers who had much to offer intellectually, but were not fortunate enough to have talented

---

[6] It is worth noting that many of his contemporaries, especially those in Athens' ruling elite, were significantly less enamoured with his approach, as evidenced by the fact that Socrates was brought to trial (before a jury of 500 of his fellow Athenians) on charges of failing to acknowledge gods recognized by Athens and corrupting the youth; he was not able to sway his jurors and was subsequently sentenced to death.

and dedicated students, willing and able to capture, enshrine in written word and further refine their teachers' ideas. In a very practical sense, one could speculate that those unknown to us thinkers had no first followers,[7] no one to, in a manner of speaking, pick up their torch of wisdom and carry it forward. In that sense, it might be quite possible that the reason we celebrate and embrace Socrates' ideas is as much due to their universal, timeless wisdom, as it is due to a pure stroke of luck that resulted in Plato, a brilliant and committed student, picking up Socrates' wisdom and carrying it forward.

1.3   The Learning with Data Framework

Rapidly expanding automation is not only reshaping how work is done – it is changing the very essence of economic value creation, and by extension, the makeup of core business competencies. Torrents of data spawned by the ever-expanding electronic transactional and communication infrastructure hide enormous informational potential, but tapping into it requires specialized skillset. Not surprisingly, data analytic know-how joins skills such as communication and critical thinking as an essential business competency, which should compel organizations to re-think what constitutes the desired skillset of their employee base and other core organizational value creation contributors (it also gives rise to the notions of 'data analytic literacy' discussed in Chap. 6, and 'informational literacy' discussed in Chap. 9).

In the bygone era of some data being used to support some of the organizational decisions in some contexts, it was sufficient to think of data analytic capability as a standalone department staffed by appropriately qualified specialists. Those days are no more. With data-generating electronic transaction processing and communication infrastructure permeating virtually all aspects of organizational functioning, coupled with the ever-expanding varieties of externally available add-on data, it is hard to now think of a data-less decision context or situation. It thus follows that the earlier discussed organizational knowledge creation has to entail systematic, ongoing utilization of the readily available, insights-rich data. Stated differently, any organization that wants to remain not only competitive but viable going forward has to embrace the idea of a broad base rudimentary level of data analytic know-how. In an organizational setting, being able to engage in basic manipulation and analysis of data and having adequate foundation of explicit knowledge of elementary concepts of statistics and probability is rapidly becoming as fundamental as the ability to think critically and to communicate; in fact, as discussed later (Chap. 9), those three distinct competencies can be seen as subsets of the more broadly framed informational literacy.

---

[7] Credit for this idea should be given to Derek Sivers and his 2010 TED (technology, entertainment, design) talk.

But since modern organizational functioning also entails a growing use of data analytic automation, ongoing organizational learning also calls for data-technological competence, as exemplified by the ability to use advanced simulation and virtualization technologies. As captured by the idea of data-enabled creativity discussed in Chap. 7, the combination of rich, detailed data and advanced machine learning applications is beginning to enable planning and general organizational sensemaking-related simulations and virtualizations – essentially, advanced 'what-if' scenario exploration and evaluation tools that can vastly expand the current conception of tacit knowledge and knowledge-building learning.

Just considering only the first two decades of the twenty-first century, the pace of information technology progress and proliferation has been nothing short of astounding, but the readiness of organizations to transform their informational investments into discernible economic benefits has been spotty, largely because of ineffective informational resource utilization. 'Learning how to learn' is an idea that receives relatively little attention in organizational settings, but in the era of distributed data access, it is a necessary prerequisite to productive and purposeful utilization of readily available, but not always readily apparent, informational resources.

1.3.1   Learning Inputs and Outcomes

In the most rudimentary sense, learning can be seen as a process of consuming inputs, in the form of various stimuli, with the goal of generating outputs, in the form of knowledge. Broadly defined, learning process inputs can be either episodic, which are situationally and context dependent, or ongoing, which are recurring but otherwise invariant. Learning process outcomes, on the other hand, can take the form of newly created incremental knowledge, or updates to existing knowledge, as graphically summarized in Fig. 1.4.

Fig. 1.4 Learning inputs and outcomes

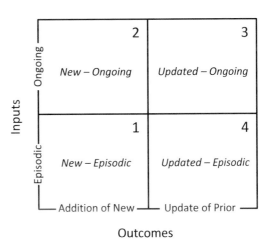

The resultant 2×2 organizational learning input-outcome classification yields four distinct learning scenarios: new knowledge produced episodically (quadrant 1), new knowledge produced on ongoing basis (quadrant 2), ongoing update of prior knowledge (quadrant 3) and episodic update of prior knowledge (quadrant 4). Jointly, those four dimensions of organizational learning capture the 'what' aspect of learning in the form of distinct types of knowledge assets derived from different informational sources and learning modalities, such as theoretical understanding of a new phenomenon of interest derived from the most recent empirical research findings (quadrant 1) or the most up-to-date frequency of a particular type of insurance claims (quadrant 3). Essential to properly framing and contextualizing those distinct types of knowledge assets is a more in-depth detailing of the 'how' aspect of organizational learning, with particular emphasis on learning modalities.

1.3.2 Learning Modalities

Building on the preceding discussion, the basic tenets of broadly framed organizational learning suggest that the ability to reason, conceptualized as the power of the mind to think and understand by a process of logic, is most emblematic of human learning, while the ability to identify patterns in vast quantities of data is most descriptive of machine learning. That core human vs. machine distinction is captured in two meta-categories of organizational learning: *reason-based*, which embodies individual and collective cognitive and behavioural knowledge, and *technology-based*, which embodies the efficacy of the broadly defined informational infrastructure to autonomously perform specific, typically routine, organizational tasks. Each of the two meta-categories can be further subdivided into more operationally meaningful categories of *tacit* and *explicit*, and *computational* and *simulational* learning, for reason- and technology-based, respectively. Figure 1.5 offers a graphical representation of the resultant MultiModal Organizational Learning (MMOL) typology.

Although outlined in Fig. 1.5 as four distinct manifestations of organizational learning, experiential, theoretical, computational and simulational modes can also be thought of as progressively more sophisticated means of knowledge creation. From early man gazing at the stars and trying to make sense of natural phenomena (experiential learning), to philosophers and scientists who came later trying to discern the underlying laws of nature (theoretical learning), to modern data- and technology-enabled investigators identifying new and testing presumed relationships (computational learning), to the now-emerging mechanisms of simulating reality as a mean of learning that transcends experience and physical reality (simulational learning). And although experiential, theoretical, computational and simulational learning can be seen as progressive and more sophisticated means of learning, it is important to think of those modalities in the context of expansion rather than replacement. More specifically, each of the four distinct modalities contributes incrementally to the overall learning, which is to say they complement

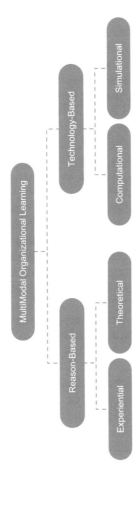

Fig. 1.5 The MMOL typology of organizational learning

rather than replace one another, in the same way that planes, automobiles, bicycles and simply walking all offer different means of traversing distance.

1.3.2.1 Reason-Based Learning

At an individual level, learning can be broadly characterized as acquisition of new or reinforcement existing knowledge. The underlying process begins with awareness-arousing stimulus being encoded into short-term memory in one of two forms: iconic, or visual, and echoic, or auditory. The process of learning is then initiated: It starts with the formation of new neural connections, which is followed by consolidation, or strengthening and storing of remembrances as long-term memories, with distinct clusters of neurons being responsible for holding different types of knowledge. Any subsequent retrieval of memories from long-term to active memory brings about re-consolidation, or strengthening of the stored knowledge, often referred to as remembering.

In a more abstract sense, learning can be characterized as adding new or modifying information already stored in memory based on new input or experiences. It is an active process involving sensory input to the brain coupled with extraction of meaning from sensory input; it is also a fluid process, in the sense each subsequent experience prompts the brain to (subconsciously) reorganize stored information, effectively reconstituting its contents through a repetitive updating procedure known as brain plasticity. Though generally resulting in improvements to existing knowledge, brain plasticity can nonetheless bring about undesirable outcomes, most notably in the form continuous re-casting of memories – in fact, that is one of the reasons eyewitness accounts become less and less reliable with the passage of time. All considered, the widely used characterization of learning as the 'acquisition of knowledge' oversimplifies what actually happens when new information is added into the existing informational mix – rather than being simply 'filed away' and stored in isolation, any newly acquired information is instead integrated into a complex web of existing knowledge.

Within the confines of reason-based learning, the most elementary knowledge acquisition mechanism entails immersion in, or observation of, a process or a phenomenon, commonly referred to as experiential learning. Subjective and situational, this mode of learning can be seen as a product of a curious mind driven to understand the nature of a particular experience, and it is built around systematic examination of sensory experiences, particularly those obtained by means of direct observation or hands-on participation. Being entirely shaped by person-specific factors, experiential learning implicitly dismisses existence of innate – i.e. generalizable – ideas, resulting in knowledge that is entirely defined by individual learners. That mode of learning is particularly important in the context of specific tasks, such as underwriting of risk or managing retail customer loyalty programmes.

Complementing the experiential learning dimension of reason-based learning is theoretical learning, which is focused primarily on common knowledge, or innate ideas that transcend individual experiences. It entails developing an understanding of universally true and commonly accepted abstract formulations and explanations, as exemplified by the axioms and rules of mathematics or the laws of nature. That mode of learning typically plays a critical role in the attainment of professional competence, as evidenced by numerous professional certification requirements.

Both the experiential and theoretical facets of reason-based learning are impacted, and to some degree shaped by a number of distinct physiological and psychological factors. The importance of expressly considering those largely subconscious learning influencers is particularly pronounced in organizational settings because those factors are most directly responsible for learning outcome variability. More on point, those factors can offer at least a partial explanation for why two otherwise similar individuals exposed to comparable, or even the same, learning inputs and contexts may end up with materially different levels of knowledge.

Physiological Factors: In a very general sense, knowledge can be thought of as a library – a collection of systematic, procedural and episodic remembrances acquired via explicit and tacit learning. However, as suggested by the notion of brain plasticity, unlike physical libraries, neural networks-stored 'collections' are subject to ongoing reshaping, triggered by the process of assimilating of new memories. The resultant continuous re-writing of old memories means that an individual-level effective topical knowledge is ever-changing (which is why eyewitness accounts become less and less reliable with the passage of time), and that the ongoing interpretation and re-interpretation of knowledge can exert a profound impact on individuals' perception and judgement.

While the ongoing reshaping of knowledge affects the validity and reliability of individuals' knowledge, cognitive bias impacts the manner in which stored information is used. Reasoning distortions such as availability heuristic (a tendency to overestimate the importance of available information) or confirmation bias (favouring of information that confirms one's pre-existing beliefs) attest to the many ways subconscious information processing mechanics can warp the manner in which overtly objective information shapes individual-level sensemaking. To make matters worse, unlike machines that 'remember' all information stored in them equally well at all times, the brain's persistent self-rewiring renders older, not sufficiently reinforced memories progressively 'fuzzier' and more difficult to retrieve. As a result, human recall tends to be incomplete and selective.

Moreover, the amount of information human brain can cognitively process in attention at any given time is limited due to a phenomenon known as human channel capacity. Research suggests that, on average, a person can actively consider approximately 7 ± 2 of discrete pieces of information. When coupled with the ongoing reshaping of previous learnings (brain plasticity) and the possibly distorted nature of perception (cognitive bias), channel capacity brings to light cognitively biological human reasoning limitations.

Psychological Factors: Looking beyond the brain mechanics-related reasoning limitations, reason-based learning is also impacted by numerous attitudinal factors, most notably related to emotions and motivation. For instance, more positive experiences tend to manifest themselves in more complete recollections than negative events, and those events that occurred more recently appear to be more significant or thus likely to recur. Moreover, desire to perform better has been shown to lead to deeper learning, even when time spent on learning, as well as learners' gender and ability were controlled for, highlighting the importance of intrinsic motivation to learning. While commonly considered in the context of individual-level characteristics, emotion and motivation also have important group-level analogues, outlined next.

An even more profound, in an organizational setting, psychological factor manifests itself as group dynamics. Contradicting conventional wisdom which suggests that groups make better decisions than individuals, research in areas of social cognition and social psychology instead suggests that the efficacy of group-based decisions cannot be assumed to outperform the efficacy of choices made by individuals; in fact, it is the combination of cognitive (individual), social (group) and situational (expressly neither individual nor group) factors that jointly determine the efficacy of decisions. The widely embraced higher levels of confidence attributed to group decisions may at times even be misguided because of a phenomenon known as groupthink, which is a dysfunctional pattern of thought and interaction characterized by closed-mindedness, uniformity expectations and biased information search, all of which manifesting itself in strong preference for information that supports the group's view (it is how potentially brilliant breakthrough ideas die in committees).

A yet another psychological influencer of reason-based learning is group conflict. As suggested by social exchange theory, which views the stability of group interactions through a theoretical lens of negotiated exchange between parties, individual group members are ultimately driven by the desire to maximize their benefits, thus conflict tends to arise when group dynamics take on more competitive than collaborative character. Keeping in mind that the realization of group decision-making potential requires full contributory participation on the part of all individual group members, within-group competition reduces the willingness of individuals to contribute their best to the group effort. Not only can that activate individuals' fears of being exploited, as well as heighten their desire to exploit others, it can compel individuals to become more focused on standing out in comparison with others. That in turn can activate tendencies to evaluate one's own information more favourably than that others', and also to evaluate more positively any information that is consistent with one's initial preferences.

1.3.2.2 Technology-Based Learning

The growing sophistication and proliferation of self-learning technologies, commonly referred to as artificial intelligence (AI), is beginning to challenge the traditional, human-centric conception of organizational learning. Machine learning, a sub-category of AI that focuses on endowing computers with the ability to learn without being expressly programmed, discerns patterns from available data, accumulates and synthesizes the resultant knowledge, and then executes specific tasks using self-discerned decision logic. In fact, as implied in the term 'artificial intelligence', AI systems are expressly designed to mimic the functioning of the human brain, as illustrated by one of the more commonly used artificial learning approaches represented by a family of algorithms known as neural networks. Unimpeded by human limitations in the form of cognitive bias, fatigue or channel capacity, and taking advantage of practically limitless computational resources, AI is pushing the broadly defined ability to learn beyond the traditional limitations of human-centric information processing. And in some contexts, most notably when performing routine, repetitive tasks, AI-based decision engines can in fact outperform humans, not because they can 'see' more, but primarily because those systems can rapidly, tirelessly and objectively funnel vast quantities of data into decision alternatives that exhibit the highest probability of desired outcomes.

It is important to emphasize that technology-based learning is a complement, not a replacement, for human learning. When decisions are characterized as repetitive and structured, and the decision-making environment is characterized as stable, technology-based learning can offer incremental value to the organization by enabling more exhaustive and objective utilization of the available data. Conversely, there are many decision situations in which technology-based learning offers considerably less value, as is the case when available historical data have limited explanatory or predictive value, and/or decision environment is highly volatile. That said, organizations typically face a mix of repetitive-structured-stable and *ad hoc* decisions, suggesting that technology-based learning should be considered an important aspect of the overall organizational learning strategy.

Also comprised of two complementing dimensions, as depicted in Fig. 1.5, technology-based learning can take the form of computational or simulational learning. While overtly quite similar in the sense that both learning modalities are built on the foundation of analysis of raw data using sophisticated data analytic tools and techniques, computational learning is primarily focused on the 'what-is' dimension of knowledge creation, whereas simulational learning offers an inroad into more speculative 'what-if' dimension of data analytic knowledge. More concretely, the former takes the form of informational summarization and pattern identification, while the latter is built around anticipatory, forward-looking data-based simulations of future outcomes of interest. Utility-wise, computational learning is invaluable to guiding recurring, routine decisions characterized by high degrees of longitudinal stability, as exemplified by managing insurance claims, while simulational learning is essential to infusing objectivity into non-routine decisions, as exemplified by emergence of disruptive technologies.

Simulational learning can be thought of as machine equivalent of human reason-based theoretical learning. More specifically, it enables constructed reality-based knowledge creation, or discovery of universal generalizations within artificial representations of the world, broadly referred to as *virtual reality*, perhaps best exemplified by astrophysical research delving into the birth of our physical universe. Virtual reality-enabled learning makes possible generation of previously inaccessible insights (e.g. conditions that existed shortly after the Big Bang) because it enables the creation of possible but not-yet-observed situations, and virtually limit-less what-if type of scenario planning.

While obviously unaffected by physiological and psychological factors, technology-based learning is not a panacea either. Ironically, one of the most profound limitations of that facet of organizational learning is also the reason for its very existence: the challenge posed by overabundance of data.

In the most rudimentary sense, data can be conceptualized as a mix of signal, which is potentially informative, and noise, which is generally non-informative. Hence, one of the core aspects of data utilization is to separate signal from noise, a task that becomes increasingly more challenging as the volume and variety of available data expand. Walmart, the world's largest retailer, handles more than a million customer transactions per hour; in aggregate, as of 2020, the total volume of business-to-business and business-to-consumer transactions surpassed 450 billion per day. And many of those transacting consumers, considerably more than 5 billion as of 2020, are calling, texting, tweeting and browsing on mobile devices, all of which add informationally rich pre- and post-purchase details. That, of course, is just one, comparatively narrow, facet of what is now commonly known as big data. Still, whatever the source or type of data a given organization has available to it, it is safe to assume that the bulk of it represents a product of passive recording of an ever-growing array of states and events, which means that finding the few organizational decision-related insights typically entails analytically sifting through vast quantities of non-informative noise.

The problem of overabundance is certainly not limited to computational and simulational learning, but within the confines of technology-based learning, epistemology or the essence of validity and reliability of what is considered 'knowledge', poses a particularly formidable challenge. Lacking the face validity or 'does it make sense' aspect of the reason-based learning, technology-based learning has to rely on generalizable decision heuristics to enable the automated algorithms to independently and consistently differentiate between material and spurious conclusions. Consider a common scenario: A computer algorithm sifting through data in search for material patterns pinpoints a recurring association between X and Y – once identified, the association is 'learned' and subsequently used as a driver of the algorithm-enabled decision rule. However, there is typically a non-trivial possibility that what manifested itself as a recurring association between X and Y is erroneous, due to both X and Y being influenced by unaccounted for (i.e. not captured in the available data) factor Z, effectively rendering the presumed association illusory. Moreover, even if the X-Y association is unaffected by the unaccounted for factor Z, the widely used statistical significance testing may produce falsely positive conclusions. That is due to the earlier discussed sample size dependence of significance tests: the now almost inescapably large number of records used to compute test parameters can quite easily inflate the perceived (i.e. statistical) significance of magnitudinally trivial effects, such as correlation coefficients.

1.3.3   Data Analytic Literacy

Although in principle data analytic know-how can be conceived as a continuum, in practice, differences in the depth of analytic proficiency can be well approximated by three distinct categories of entry-, mid- and expert-level skills. The core distinction among those three levels of ability can be expressed in terms of the interplay between tacit, or theoretical, and explicit, or applied, knowledge, as graphically depicted in Fig. 1.6.

Analytic literacy can be seen as a foundational level of tacit and experiential data analytic knowledge, reflecting essential ability to manipulate (i.e. extract, review, correct) data and conduct simple, typically descriptive analyses; analytic competency is attained when a high degree of tacit data analytic knowledge is added to the earlier attained analytic literacy, which typically reflects knowledge of more advanced data analytic methodologies (i.e. multivariate statistics and machine learning methods), and finally, analytic proficiency is attained by adding a meaningful amount of practical experience of working with complex data using advanced data analytic techniques. There are numerous benefits to viewing data analytic know-how through such tiered prism, but in general, they reflect the commonization of data analytics. The ubiquity of data and the sense of urgency felt by organizations to turn data into an economic engine, or at least into a driver of organizational deci-

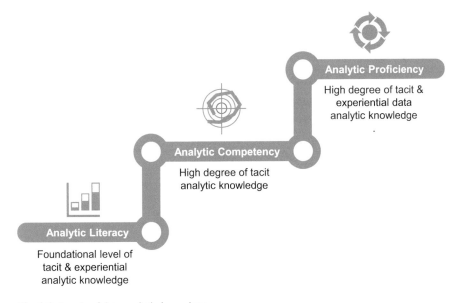

Fig. 1.6 Levels of data analytic know-how

sions, calls for broad base dissemination of data analytic responsibilities – in other words, it calls for infusing of data analytics into professional and occupational areas that previously did not formally encompass systematic analyses of data. And while such decentralization of data analytics is appealing, it is a risky undertaking given the generally poor level of (formal) educational data analytic preparation. Explicit and or at least somewhat formalized credentialing manifesting itself in operationally distinct and meaningful capability tiers suggested in Fig. 1.6 offers a tried-and-true[8] risk mitigation strategy. Oftentimes it is difficult, if not outright impossible, to discern methodological and computational validity of insights derived from data analytic outcomes, typically because at least some of the required technical details are not readily available, or users of those outcomes might lack the requisite technical knowledge. And lastly, such practice is well established in other organizational functioning-related contexts such as accounting or actuarial science.

As summarized in Fig. 1.6, the analytic literacy → competency → proficiency journey manifests itself in varying degrees of tacit and explicit knowledge, but those two distinct dimensions of data analytic know-how are themselves products of the interplay among data analytic reasoning, familiarity with data analytic methodologies and knowledge of computational means, which within the confines of data analytics encompass appropriate programming languages, such as R or Python, and/ or applications, such as SAS or SPSS. Figure 1.7 captures those three distinct elements of data analytic know-how, along with their interdependencies.

Within the realm of data analytics, the ability to reason, framed here as the 'know-why' dimension of data analytic skillset, encapsulates the purpose of data analysis. The overabundance of data can obscure the true informational need, instead drawing attention to the possibility of generating arrays of analytic outcomes not because those outcomes address specific informational needs, but because the available data lend themselves to easy summarization. Hence, the essential part of sound organizational learning with data is analytic discipline, which amounts to staying focused on analytic outcomes that address stated informational needs. Here, the expression that 'less is more' seems particularly appropriate.

The remaining two dimensions – the methodological 'know-what' and computational 'know-how' – comprise what is commonly seen as the core data analytic skillset. The former encompasses a diverse body of mostly explicit knowledge encompassing understanding of data structures, data due diligence and data feature engineering, and familiarity with statistical and machine learning approaches to data analysis. The latter is focused on more narrowly scoped data manipulation capabilities, which typically range from data extraction and processing to execution of specific data analytic procedures; in applied organizational settings, it tends to skew more towards experience-based tacit knowledge.

While each of the three analytic skillset dimensions encompasses a distinct set of competencies, as depicted in Fig. 1.7, the overall skillset constitutes a self-contained system. Each part interacts with, and in fact needs, the other parts – it is not possible to engage in productive applied analyses of data without adequate command of the

---

[8] The Society of Actuaries, which is a global professional organization for actuaries, uses a series of rigorous exams to bestow two Associate-level designations (Associate of Society of Actuaries and Chartered Enterprise Risk Analyst), and its highest designation, Fellow of the Society of Actuaries.

Fig. 1.7 The elements of
analytic skillset

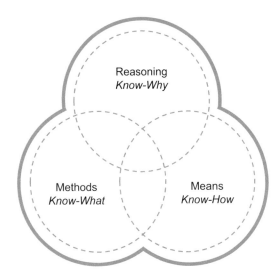

'why', 'what' and 'how' facets. Stated differently, when considered individually, robust understanding of the logic of learning with data, deep theoretical knowledge of statistics and machine learning, and strong programming skills are all certainly valuable in their own right, but are singularly insufficient in the context of data analytic skills. That is precisely why the new domain of data science emerged over the last couple of decades – effective learning with data is ultimately an endeavour that draws from multiple established disciplines, most notably from computer science and statistics, but also from neuroscience, psychology, management and even philosophy. In contrast to the past when large-scale analyses of data were seen as special purpose vehicles, in the Age of Data, large-scale data exploration is not only an everyday undertaking, but more and more a necessary component of everyday organizational functioning.

1.3.4   A Microperspective: Analytics as Search for Answers

The once common quip of organizations feeling 'data-rich but information-poor' spurred an onslaught of standardized reporting tools which, interestingly, is now leaving more and more of the same organizations feeling 'information-rich but knowledge-poor'. Among the key contributors to that trend is the often-faulty data analytic mindset: Trying to keep up with the ever-growing torrents of data, business (and other) organizations tend to lose sight of the critical distinction between *information* and *knowledge,* where the former represents summarized data (i.e. a quarterly sales report), while the latter manifests itself as decision-guiding insights (e.g. drivers of quarterly sales volatility). It is common for information to be produced simply because it can be produced, meaning available data compel creation of summary reports and other generic informational products, from which it follows that information may or may not be useful to organizational decision-makers (and given the vastness of data available nowadays, common sense suggests, and practical experience confirms that lots of superfluous information is being created). Knowledge, in contrast to information, is a manifestation of purposeful analyses of data aimed at producing answers to specific questions. In other words, the creation of knowledge is compelled by the need to know, and to the degree to which data that relate to the question at hand are available, those data are analysed with a well-defined informational goal in mind. At the same time, other data are set aside, not because they are non-informative, but because their informational content does not contribute to the informational goal at hand. Implied by the so-framed information vs. knowledge distinction is that summarization of raw data lends itself to tool and process standardization while answering emergent business questions does not, which helps to explain the preponderance of information and the comparative paucity of decision-guiding knowledge in many organizational settings.

Given the above and considering that it is natural, even expected, for organizations to seek efficiencies, it is not at all surprising that many efficiency-seeking organizations may indeed feel information-rich but knowledge-poor. When mining data for insights is approached more as an exercise in time- and resource-economical data wrangling and summarization than a search for new knowledge, the few nuggets of insights will likely be drowned in volumes of, perhaps interesting, but rarely truly informative, tidbits, which will ultimately result in little organizational learning. Still, to be a source of value, data analytics-produced knowledge needs to be both timely and not prohibitively resource-intensive; in other words, while time and resource efficiency should not be the targets, they need to be seen as constraints. All considered, actuating the desire to use objective data as a source of knowledge, at an organization-wide level, requires explicit guidelines, and the systematic step-by-step process schematically summarized in Fig. 1.8 was designed to offer such guidelines. It frames data analytics as a mean of answering business questions using data, or stated differently, competency-expressed capability to translate data into decision-guiding knowledge.

An implicit but important assumption on which the process of answering business questions with data is based is that the goal of business analytics (which is also likely true in non-commercial domains) is to address manifest informational needs, rather than to summarize or otherwise process the totality of available data. Given the well-established conception of data as mixture of informative (also referred to as 'signal' on account of the idea's communication theory roots) and non-informative (also known as 'noise') components, such implicitly selective analytic reasoning seems theoretically warranted and practically appealing. It is important to note, however, that the process outlined here is geared towards addressing known questions, which is only one facet of data analytics; there are numerous situations, discussed later, which warrant open-ended, exploratory analyses of data. Turning back to the data-driven knowledge creation process summarized in Fig. 1.8, the logic embedded in that process aims to balance organizations' need for efficiency with assurance of data analytic discipline. Rich varieties of available data can trap

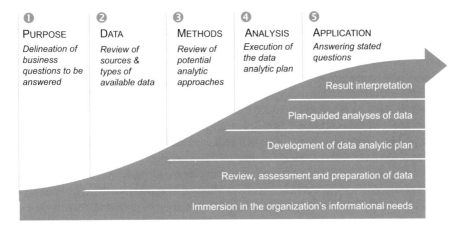

Fig. 1.8 Data-driven knowledge creation

analytic efforts in the habit of generating large volumes of 'interesting' but ultimately non-informative outcomes; at the same time, it is important to acknowledge
that unbounded, purely exploratory data immersion could yield eye-opening
insights. With that in mind, effective utilization of available data resources calls for
discipline of purpose: If the goal at hand is to answer clearly delineated business
questions, analyses ought to be focused on, and guided by, those questions; if, on
the other hand, the goal is to explore patterns and relationships without a clearly
defined informational goal, analyses need to be structured accordingly. The general
approach captured in Fig. 1.8 is expressly focused on the former, without overlook-
ing or downplaying the potential value of the latter.

In terms of the specifics of the data-driven knowledge creation process, the first
step is to clearly delineate business questions, or more generically, questions that
emanate from organizational management-related choices. The idea here is that purposeful analyses are predicated upon outcome clarity, which can be phrased in
terms of identification (e.g. what is the demographic profile of my most loyal customers?) or confirmation (e.g. are retirees more likely to own an RV than those still
working full-time?). Once informational needs have been clearly stated, data to be
used need to be identified, reviewed and prepared for analyses. The scope, depth and
the general level of effort required to do so can vary considerably depending on the
combination of organizational context and informational needs, but recalling the
earlier discussion of the difference between information and knowledge, it is important for the data review and preparation process to expressly link data choices with
the questions to be answered using those data (which can entail potentially extensive data feature engineering efforts). The next step is to develop an analytic plan,
the overt purpose of which to guide the sequencing and the specifics of ensuing data
analytic work, while also facilitating a holistic assessment of the knowledge creation initiative from the standpoint of informational goals, data inputs, data processing and analysis throughputs, and decision-guiding outputs. The process of

developing a formal analytic plan should be, as much as possible, a collaborative effort bringing analysts and end users (of data analytic results) together in a way allowing analysts the opportunity to develop a more intimate understanding of the essence of business questions at hand, while also giving users of data analytic outcomes more refined expectations.

Using the analytic plan as a guide, the next step of the process is to conduct the plan-delineated analyses and produce the initial data analytic outcomes. Quite commonly, those initial outcomes may need a non-trivial amount of post-analytic refinement and interpretation, initially to assess the results' validity and reliability, and then to discern the non-technical framing of the often esoterically worded statistical results. The last step in the process of answering business question with data is to do just that – to answer the stated business questions. Depending on the type of analyses that were utilized, that could mean a simple restating of data analytic outcomes, which would likely be the case with the first of the two earlier examples (i.e. what is the demographic profile of my most loyal customers), or it could entail drawing inferences from data analytic outcomes, as would likely be the case with the second of the two earlier examples (i.e. are retirees more likely to own an RV than those still working full-time).

The generally worded outline of the data-driven knowledge creation process implicitly assumes a considerable degree of technical data analytic knowledge, ranging from familiarity with a wide range of data review and preparation considerations and techniques, estimation and inference concepts, and statistical and machine learning techniques and applications, as captured in the earlier discussed notion of analytic literacy. The inherent multidisciplinarity of data analytics often means that that even those who have the requisite technical foundation may lack the clarity of how computational programming and data analytic skills are to be shaped into a consistent learning mechanism. The next section offers a high-level overview of the recommended 'learning how to learn with data' approach.

1.3.4.1   Data-Driven Knowledge Creation Skilling

The core belief imbedded into the learning with data reasoning is that technical knowledge of data manipulation and analysis is a mean to an end, where the end takes the form of decision-guiding insights. Given that, the most fundamental aspect of becoming a proficient (recall the three-tiered learning with data capability progression summarized in Fig. 1.6) data learner is captured in the interplay between 'conceptual understanding' and 'functional capabilities', where, as summarized in Fig. 1.9, the 'what' and 'why' rationale is merged with 'how-to' abilities. That may seem obvious, yet the traditionally fragmented, even siloed data analytic learning does not naturally produce that type of system-level understanding and functional capabilities. The often-piecemeal building of statistical, computational programming and related know-how implicitly burdens learners with putting those disparate pieces together into a coherent knowledge creation process, as individual domain-specific (e.g. statistics) elements of data analytic know-how tend to be more like

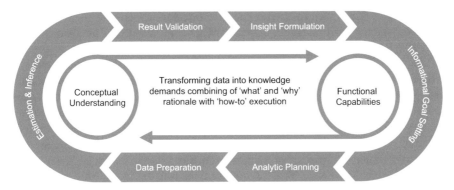

Fig. 1.9 The skill dimension of data-driven knowledge creation

fragments of related but ultimately different stories than chapters in a singularly themed book. To a large degree, the entrenched business education philosophy is to blame here, as it favours teaching in a piecemeal fashion, 'about' data, statistics, management, etc., over 'how to function' as a data analyst or a manager.[9]

The ability to answer business questions with data is encapsulated in the progression graphically depicted in Fig. 1.9. Informational Goal Setting → Analytic Planning → Data Preparation → Estimation & Inference → Result Validation → Insight Formulation frames the logic of the data-driven knowledge creation process discussed earlier in the context of discrete sets of skills. The intent behind causally linking of the distinct data analytics capability-shaping competencies is to highlight the importance of each individual skillset to the overall capability, while also emphasizing the importance of the interplay between conceptual understanding (i.e. what purpose does it serve?) and functional capabilities (i.e. how do I do that?). And lastly, the overall data analytic process shown in Fig. 1.9 is meant to be iterative because learning is an ongoing process where new knowledge is continuously added in perpetuity, at least in principle.

In terms of the individual components, Informational Goal Setting is, in keeping with the earlier discussion, the first step in the learning with data process, and it may seem deceptively simple. At face value, delineation of business questions that are to be the focal points of ensuing analyses is often taken to be nothing more than restating the challenges facing the organization, but the hidden difficulty here is that not all of those challenges can be meaningfully addressed with objective, empirical data. To restate the well-known expression, 'not everything that is worth knowing is knowable (and not everything that is knowable is worth knowing)'; important questions such as 'how will the newly introduced regulatory regime impact our business?' or 'how many new products will our competitors roll out this year?' may be better answered through pooling of expert projections than projecting of historical

---

[9] I delve considerably deeper into the details and ramifications of the academic–practice gap in *Evidence-Based Decision-Making* (New York: Routledge, 2019).

data trends.[10] It all goes to saying that informational goals have to be carefully curated, which requires a sound understanding of the essence of the question and available data resources.

Analytic Planning, the next core element, serves a dual purpose: Firstly, it attempts to match end users' anticipated informational benefits with analysts' data analytic process through bridging of the expertise gap. For example, within the confines of risk analytics, end users of data analytic outcomes tend to have deep knowledge of business domains, such as insurance underwriting, and comparatively limited knowledge of data analytic methodologies, while analysts tend to have comparatively limited knowledge of business domains and, obviously, high degree of methodological knowledge. That disparity is at the core of the still common distrust of advanced data analytic tools, such as likelihood and severity of adverse outcomes-predicting models, on the part of experienced business professionals,[11] and analytic plan collaboration offers an opportunity to close that gap. Secondly, explicit analytic planning offers analysts an opportunity to carefully examine their own data processing and analysis choices. It is common for information workers to have a great deal of familiarity with some methods and quite a bit less with others, or have deep experience with one data source and much less with others, which may skew their data processing and analysis choices, ultimately leading to suboptimal outcomes. Although explicit analytic planning may not be able to fully correct for uneven knowledge and experience, the process will more than likely create an opportunity to become at least aware of some habitual, but not necessarily ideal, analytic choices.

The next two elements of the process encompass distinct-but-related sets of skills: Data Preparation and Estimation & Inference. Both domains are generally seen as core data analytic competencies, and both entail a combination of conceptual and applied knowledge of data structures, manipulation and feature engineering (Data Preparation), and broadly scoped statistical as well as automated means of analysing data (Estimation & Inference). In a sense, those are the most 'tangible' or defined elements of learning with data, encompassing computational programming skills using some mix of open-source programming languages, such as R or Python, domain-specific languages such as SQL, comprehensive statistical software suites such as SAS or SPSS, and data visualization applications such as Tableau or Microsoft Power BI, in addition to knowledge of statistical and machine learning data analytic methodologies.[12] Typically, data preparation and analysis skillset

---

[10] I discuss that topic at length, as a part of the Empirical & Experiential Evidence framework, in *Evidence-Based Decision-Making*, New York: Routledge, 2019.

[11] It is worth noting that there are numerous well-documented cognitive biases, such as base rate fallacy, neglect of probability bias, overconfidence effect or confirmation bias (all discussed in more detail in the *Evidence-Based Decision-Making* book mentioned earlier), that shed light on psychology of distrust of advanced data analytics.

[12] In practice, the line demarking statistical and machine learning methods is somewhat blurry because some techniques, such as linear or logistic regression, are often included under both 'statistics' and 'machine learning' umbrellas; as used in this book, statistical methods are those based

require appropriate university preparation, which typically means training in statistics or computer science, but it is common for those with training in hard sciences or engineering to be able to comparatively quickly adapt their knowledge and skills to become high-functioning business analysts or data scientists.

Result Validation, the next step in the process, should be considered from the perspective of two distinct manifestations of validity: face and inferential. The former is the 'does it make sense' type of confirmation, while the latter entails a more systematic determination of the extent to which results of analyses utilizing (typically) imperfect data and probabilistic methods of inquiry produced results that can be trusted. To be dependable, face validation requires appropriate, as in stemming from similar situations, experiential knowledge, while inferential validation is more deeply rooted in appropriate methodological knowledge.

The final step, Insight Formulation, entails translation of the often-esoteric data analytic outcomes into meaningful, to the intended user audience, knowledge. Within the confines of learning with data, it boils down to answering the business questions delineated in the Informational Goal Setting part of the process. It is both an obvious and frequently overlooked step, as data analytic initiatives by their very nature lend themselves to greater emphasis on data analytic processes, and comparatively less emphasis on applicability of results.

1.3.4.2 Learning Outcome Assessment

While the development and deployment of robust learning systems and supporting mechanisms tend to be the focal point of organizational learning-oriented efforts, those learning enablements alone are insufficient to assure the desired outcomes. Especially within the confines of reason-based learning, the earlier discussed cognitive bias, channel capacity, group dynamics as well as emotional and motivational factors all can potentially impede the degree of learning that takes place within organizations. Moreover, the relative newness of technology-based learning also has important implications for efficacy of the overall organizational learning efforts, as it often pits human experience against machine algorithms. All in all, assessing the efficacy of organizational learning efforts should be considered a core element of the broader organizational learning efforts.

As suggested by the Hawthorne effect, when faced with formal learning assessment, learners are likely to engage more deeply in the learning process. However, to meet the stated objectives of data-driven learning, which focus on assuring the overall organization-wide assimilation and utilization of applicable knowledge, assessment needs to be situationally meaningful, with particular emphasis on the type and character of knowledge-based competencies. Given the diversity and the depth of organizational skillsets, that degree of learning outcome assessment customization

---

on defined mathematical distributions (such as the well-known standard normal distribution that forms the basis for linear regression), and machine learning methods are techniques based on principles of mathematical optimization.

needs to reflect not only the type of knowledge, but also the type or the depth of learning. For example, as 'doers', data scientists should be expected to acquire functional, i.e. theoretical plus experiential, knowledge of new machine learning algorithms, while those managing data science teams, as 'overseers', may only need to acquire adequate theoretical understanding of those algorithms.

1.4 Organizational Learning Transformation

Recognizing the transformative impact of data, organizations strive to make more data more readily available to more users by a variety of means, including embracing self-service business intelligence solutions; however, many quickly recognize that the bulk of prospective data users do not have the necessary data manipulation, assessment and analysis skills. It is clear that data and data-related technologies are evolving and proliferating much faster than the overall workforce's readiness to embrace data-centric work, and merely making tools available to those who are apprehensive about 'jumping into data' is unlikely to bring about the desired organizational learning transformation.

The lack of the requisite know-how is but one of numerous impediments to broad base utilization of organizational data assets. Evidence-based decision-making also requires some degree of silencing of subjective intuition, which is difficult, even unnatural. A part of what makes that challenging is that perception- and evaluation-warping cognitive biases are all inescapable consequences of human brain's wiring, thus considerable cognitive effort is required to overcome those natural mental impulses. At the same time, there is nothing natural about trusting data-derived conclusions, especially when those conclusions contradict one's intuition.

Lastly, making systemic organizational learning an operational reality also calls for structural 're-wiring' of organizations, especially those that reached operational maturity in the pre-data-everywhere period. Given that the main objective of organizational structure is to ensure that the chosen strategy is effectively executed, the creation of learning-friendly environment is at least partly tied to the embrace of evidence-based rather than managerial hierarchy-centric decision-making structure. Just adding tools and systems to an authoritarian organization where decisions are driven by subjective evaluations is unlikely to yield meaningful benefits, even if knowledge is created and disseminated throughout the organization. The ultimate goal of organizational learning is not just to become wiser, but to act more wisely.

References

1. Banasiewicz, A. D. (2019). *Evidence-based decision-making: How to leverage available data and avoid cognitive biases* (1st ed.). New York: Routledge.
2. Senge, P. M. (1990). *The fifth discipline: The art and practice of the learning organization.* New York: Doubleday/Currency.

Chapter 2
Democratization of Organizations

2.1 Organizations as Human Collectives

At their core, organizations are groups of people joined together in pursuit of shared goals. As it is intuitively obvious from that broad definition, the reason humans tend to 'join together' is because doing so helps with the attainment of goals, which in turn can be seen as manifestations of a wide range of human needs. Within the confines of the well-known Maslow's hierarchy of needs, those can range from the most fundamental, namely self-preservation, to those that could be considered refined or evolved, most notably self-actualization. It thus follows that organizations are formed to facilitate the pursuit of a variety of objectives, and so their influence on lives of individuals can be expected to grow as those individuals' needs and goal expand. In fact, throughout human history, organizations have been steadily expanding into more spheres of life, while also growing in size and complexity. And so as aptly summarized by a noted organizational scholar Amitai Etzioni, '…our society is organizational society'. Most individuals are born in hospitals, educated in schools and employed by business, governmental or other organizations; those individual then elect to join social, religious or professional organizations, some for the duration of their lives. From birth through death, at work or at play, life in modern societies is increasingly conducted in organizational settings, so much so that in view of many sociologists, organizations are now the dominant institutions of modern societies.

2.1.1   A Brief History

Starting from the premise that organization, as in coming together to accomplish a set of goals, is essential for human activities that provide the basic physical needs such as food and clothing, the history of human collectives is the history of the evolution of the methods by which humans structure the activities necessary to our survival. Starting with prehistoric times, the early men began to come together primarily to increase their food supply, to develop better clothing and to build better shelters. The transition to what is now described as organized society, manifesting itself in the rise of cities and states, began approximately 6000 years ago, and it is believed that it was in response to factors including geographical features (which could explain the rise of the early civilizations in Mesopotamia and Egypt, where unstable water supply compelled larger-scale irrigation efforts), distribution of natural resources and security. With more stable food supply came population growth and the rise of central authority through feudalization and the emergence of organized religion. With cities becoming large centres of settlement, more people became disconnected from subsistence activities and began to carry out function in production, exchange, administration and religious ceremonies, which ultimately spurred the rise of occupational and educational organizations.

2.1.1.1   Business Organizations

The transition from the opportunism of hunting and gathering of early horticultural societies marks the onset of economic opportunism and the rise of business organizations. Some 3000 years ago, around the eighth century BCE, in India, early organizations, called shreni, emerged and were the first known business firms that could independently enter into contracts or own property. Some 2000 years later, around 960 CE, China's Song dynasty saw the advent of printing presses and the first paper money; shortly thereafter, the first partnerships and joint stock companies resembling modern, publicly owned business organizations began to emerge. Starting in the early 1500s, government-backed firms, such as the Dutch East India Company and British East India Company, began building global trading empires and floating stocks and bonds on newly emerging stock exchanges. The birth of modern science in the latter 1700s can be seen as a slowly actuating trigger of the large-scale[1] mechanization that began to unfold in the late 1800s, a period known today as the Industrial Revolution. The rise of industrialization also shaped the modern for-profit, externally owned, professionally managed and staffed by progressively more specialized workforce business organization; the increasing sophistication of the means of production in turn gave rise to the early conception of organizational

---

[1] There are numerous examples of chronologically much earlier localized mechanization – for instance, the Persians (present-day Iran) were using wind power grain mills and water pumps as early as 500–900 CE.

learning, manifesting itself in the rise of new domains of knowledge, such as industrial engineering and management science.

When considered from the perspective of organizational learning, it is appropriate to see the Industrial Revolution as a composite of three distinct periods, separated by roughly 50-year timespans: The Steam & Oil period was trigged by the rise of initially steam- and soon after oil-powered technologies; the Electric period followed the discovery of electricity, which was then replaced by the Electronic period triggered by the emergence of electronics. (The relatively recent emergence of interconnected networks, the rise of quantum computing and artificial intelligence, and the now emerging bio-info-electro-mechanical symbiosis are all signalling the arrival of a yet new industrial development era, discussed in more depth in Chap. 4.) Each of those periods spawned new types of business organizations, even giving rise to whole new industries, ultimately translating into persistent reframing of organizational value creation-related skillset. Setting aside period-native business firms (i.e. those that came to be during a particular industrial period), the ability of older, established organizations to adapt to new techno-business realities has been, and will likely continue to be, tied to the efficacy of organizational learning.

The importance of organizational learning is, of course, not limited to just business firms; non-commercial organizations, such as professional associations, foundations or charitable organizations, all need to be able to adapt to changing social, environmental and other circumstances. And while the general tenets of learning might not materially vary across the different types of organizations, the manner in which individual organizations can implement those tenets is often tied to organizational contexts. There are numerous ways of making sense of the overwhelming diversity of organizational types, the most rudimentary of which is focused on the very general purpose, which could either be the pursuit of economic profit or advancement of social and related goals which points towards two mutually exclusive categories of commercial and non-commercial entities.

2.1.2   Commercial Organizations

For-profit or commercial organizations exist primarily to generate profit, that is, to earn more money than they spend; their owners can decide to keep all profits, spend some or all of it on the business itself or share all or some of it with employees through the use of various types of compensation plans, such as profit sharing. As legal entities, for-profit companies can be organized as corporations,[2] partnerships, limited liability companies (LLC) or sole proprietorships; the owners of a

---

[2] Under the US tax law, there are C and S corporations: The former pay taxes on their income, and income received by their owners, typically in the form of dividends, is also subject to income taxes (hence the often heard 'double taxation' quip); S corporations do not pay income taxes, but instead their owners, who report the corporation's income of personal income, pay income taxes.

corporation are called 'shareholders[3]', the owners of a partnership are called 'partners', the owners of an LLC are called 'members' and the owner (by definition, there is only one) of a sole proprietorship is called, well, the owner. There are a number of legal- and taxation-related differences separating those four organizational forms, but those details fall outside of the scope of this overview, that said, one difference that is important here is that ownership in corporations and partnerships can be publicly traded, but ownership in LLCs and sole proprietorships cannot.[4] However, just because the underlying legal structure allows a company to sell its ownership interest to the investing public, it does not mean that all firms do. Consequently, corporations and partnerships can be either public or private – the former are companies whose shares, also known as stocks, can be bought by the public on a formal stock exchange (or over-the-counter), and the latter are companies whose shares cannot be freely bought by the public; LLCs not structured as public partnerships and sole proprietorships, which are owned and run by one person (and thus there is no legal distinction between the owner and the business entity), can only be private.

An important aspect of commercial companies' operations is their leadership, which is typically comprised of two distinct elements: operational management and supervisory oversight. The former entails active control over day-to-day activities of a firm, and it typically rests with executive officers, defined here as anyone with decision-making authority in an organization; in large business organizations, those positions tend to be staffed by professional managers. The latter is the domain of boards of directors, which are bodies charged with jointly overseeing activities of the executive leadership. It should be noted that in the United States, for-profit companies are only required to form board of directors if they are publicly traded, but many larger private (i.e. not publicly traded) companies also have formal boards of directors. In contrast to executive managers, corporate directors tend to be elected by shareholders in publicly traded companies (and appointed in private ones) for the purpose of supervising activities of corporate executives; on average, a for-profit company's board will have about nine members. Although ideally independent of one another, the two organizational management bodies tend to overlap due to a common practice of the highest-ranking executive officers also serving as members

---

[3] A quick personal perspective in regard to that customary framing of corporate shareholders as 'owners'. It is my view that large corporations with many thousands of individual and institutional shareholders do not in fact have owners, just speculative investors. The reason for that is that, under property law, ownership entails exclusive rights in the form of possession, control and disposition or transfer, but those rights do not fully extend to corporate shareholders – in fact, the only unabated ownership-related right of corporate shareholders is the right to sell or transfer their shares. The argument that there are fairly obvious practical reasons for that abridgement of the rights of ownership is unconvincing to me because if two out of the three fundamental rights of ownership do not apply to shareholders, then neither does the label of ownership.

[4] It should be noted that the very flexible tax structure of limited liability companies makes it possible for those entities to structure themselves as publicly traded partnerships and have their ownership interest traded on securities exchanges.

of the board of directors, which clearly blurs the distinction between operational management and oversight (more on that later).

Looking beyond legal structure and leadership, as economic entities, business firms have customers, owners, quite commonly non-owner managers and employees. In addition, all public business firms, by virtue of being publicly traded, and the vast majority of private (because under the US tax law, the optimal financial structure entails a mix of equity and debt financing since the interest on the latter is tax deductible) business firms also participate in capital markets and are thus subject to at least some regulatory oversight.[5] Consequently, the traditional view of business firms casts those organizations as distinct economic entities operated for the benefit of their owners and aiming to fulfil the needs of their customers.

However, that once-dominant view is giving way to the enterprise perspective, which frames business firms as social institutions operated for the benefit of multiple stakeholders, not just those with an economic interest in it. What emerges is a fairly complex picture, where numerous stakeholder groups exert some degree of direct or indirect influence on organizational choices via different modes of interaction, while being driven by different motivating factors. For example, business owners are motivated by economic gains, while regulators are focused on assuring adherence to applicable laws and regulations, a distinction that translates into materially different informational and thus learning demands. All considered, when viewed from the enterprise perspective, business organizations are networks of distinct constituents, and the management of those networks demands well-thought-out ongoing organizational learning mechanisms. More specifically, within the confines of for-profit organizations, learning plays an important functional role as an enabler of organizations' efficiency and responsiveness. What does it mean for an organization to be 'responsive'? When put in the context of an organization's interactions with, for instance, its employees or regulators, a responsive organization is one that *knows* its current employee-related and regulatory obligations, and has the means of staying abreast of, which means *learning* about, emerging regulatory and employment-related developments. Similarly, when considered in the context of the organization's interactions with its customers, a responsive organization is one that knows its customers' needs and has the ability to stay abreast of any changes in those needs.

---

[5] It is a little-known and often surprising fact that in the United States alone there are more than 500 separate federal agencies (a lot more when state agencies are added into the mix) that monitor and/or enforce a myriad of existing rules, in addition to the some 10,000 or so new rules, decrees or pronouncements that are added annually; that is why large diversified organizations need to maintain sizable departments whose whole purpose is regulatory compliance assurance.

2.1.3 Non-Commercial Organizations

In contrast to for-profit commercial firms, which are differentiated primarily based on their business focus and the underlying legal structure, non-commercial organizations can be grouped together based on the type of their underlying tax exemption, and their broadly defined advocacy focus. In terms of tax exemption status, the US Internal Revenue Service recognizes two main types of non-profits: non-profit organizations (NPO) and not-for-profit organizations (NFPO); the former serve the public via goods and services, as exemplified by various social advocacy groups such as American Civil Liberties Union (ACLU) or Greenpeace, while the latter may serve just a group of members, as is the case with professional associations such as the American Bar Association. In terms of organizational focus or purpose, charitable organizations comprise the largest segment (out of the total of about 1.5 million non-profits, as of 2019, in the United States, nearly 1 million are registered charities), which includes religious, educational, social, scientific and literary organizations. The second largest non-commercial segment is made up of foundations (more than 100,000 in the United States), which are typically focused on sponsoring, typically by means of funding, events, programmes and research aimed at building awareness or education. Trade and professional associations (around 63,000 in the United States) round off the top three; that category encompasses the ubiquitous chambers of commerce and business leagues, as well as special interest groups such as the American Farm Bureau or the American Association of Retired Persons (AARP). Looking beyond those main groupings of non-commercial organizations, there are many other types including social welfare organizations, veterans associations, social or recreational clubs and fraternal societies; also included are manifestly business organizations endowed with non-profit status which includes retirement fund associations, credit unions, mutual insurance companies, cooperative hospital service organizations or cemetery companies.

While non-profit and not-for-profit organizations share some communalities with for-profit business firms, namely, both are subject to regulatory oversight and are managed by generally similar leadership structures (i.e. operational management + board of directors, although non-profit boards tend to be a lot larger), there are a number of distinct differences. Perhaps the most visible of all is the purpose. For-profit commercial companies exist to develop products and service aimed at meeting the needs of the marketplace (comprised of individual consumers and organizations) and by doing so to generate profits for their owners, whereas non-commercial organizations exist to advance specific causes, without aiming to earn a profit. The second key difference is the source of capital that is required to operate. Commercial entities' primary source of capital is investment, which is speculative allocation of money by investors with the expectation of future profits, whereas non-profit and not-for-profit organizations' primary source of capital are donations, which are (typically, but not always monetary) gifts given in support of causes furthered by organizations. Another key difference is the nature of leadership. For-profit firms tend to be hierarchical, and the operational decision-making power is typically concentrated in the hands of executive leadership, while in non-profit organizations, the bulk of the decision-making power rests with the board of directors. A yet another common source of difference is organizational culture. Although social responsibility, sustainability and related social goals have been gaining visibility in the corporate world of commercial enterprises, by their very nature, those companies are primarily focused on financial outcomes, while the very essence of non-profit and not-for-profit organizations is to focus on furthering their focal social, scientific, cultural and other general or special interest agendas.

While differences between commercial and non-commercial organizations are indicative of somewhat different informational needs, both types of organizations operate within the general confines of their distinct organizational ecosystems. It follows that in order to remain viable going concerns and continue to advance their goals, non-commercial entities also need to be responsive to the voices and needs of their constituents, most notably their donors and the populations they serve, and also need to be efficient in their use of the resources, in order to make meaningful progress towards their stated organizational goals. In other words, they have a somewhat differently oriented but just as compelling a need to develop robust organizational learning capabilities.

2.1.4 Organizational Dynamics

Collaborative decision-making is a seductive idea. It hints of inclusion and equality, and it circumvents the many challenges inflicting individual sensemaking, including cognitive bias precipitated distortions, choice evaluating limitations brought about by channel capacity and the unreliable nature of recall, largely attributed to brain plasticity. In addition, it offers a forum for critical reviews of individual ideas, which in turn creates opportunities for individuals to refine their ideas; all told, it is not surprising that conventional wisdom argues that groups tend to make better decisions than individuals. And while that line of reasoning is quite convincing, there is little-to-no research evidence to back it up.

In fact, studies in social cognition and psychology suggest that cognitive and situational influences are generally the strongest determinant of the quality of decision-making, and that when evaluated in the context of objective standards, decisions made by groups do not consistently outshine decisions made by individuals. More on point, research results suggest that collective decision-making tends to heighten decision confidence, but there is no consistent uptick in decision quality. Why? Because just as individual-level sensemaking is affected by numerous brain mechanics shortcomings discussed earlier, group sensemaking is subject to different, but potentially equally profound, reason-warping influences. Perhaps the most visible of those is biased information search, which manifests itself as strong preference for information that supports the group's view; it could be thought of as a group analogue to confirmation bias or bandwagon effect, two somewhat related individual-level cognitive biases. Often harder to discern is groupthink, a dysfunctional pattern of thought and interaction during group decision-making characterized by closed-mindedness, uniformity expectations and group bias. It can be easy to confuse groupthink with a generally positive phenomenon of group cohesion, although the latter usually manifests itself in group members being drawn closer together by socioemotional similarity and communality of goals, whereas the former leads to reduction in cognitive diversity. All considered, it is important to recognize that while groups have the potential to enhance the capacity, objectivity and creativity of organizational reasoning, care must be taken to at least limit the impact of group dynamics-warping influences, in a way that protects the organization's cognitive diversity.

An aspect of cognitive diversity that is of particular importance to organizational learning is critical thinking, a broadly framed reasoning skillset that manifests itself as the ability to rigorously evaluate an issue of interest. A critical thinker is one who is able to examine the underlying assumptions and discern hidden meanings or values, and then has the capacity (in the sense of cognitive abilities and reasoning and analytic tools) to assess the weight of evidence in reaching well-reasoned, supported and balanced conclusions. While framed using attributes shared by all normal functioning organizational constituents, critical thinking is inescapably individual because it calls upon individual-specific mix of explicit and tacit knowledge, which is what gives rise to individually unique perspectives, and, in a group setting, to cognitive diversity. Hence, the earlier mentioned adverse impacts of groupthink and biased information search, the two key manifestations of dysfunctional group dynamics, are ultimately tied to hindering individual group members' critical thinking.

A yet another important, organizational decision-making-related aspect of group dynamics is group conflict. As suggested by social exchange theory, which views the stability of group interactions through a theoretical lens of negotiated exchange between parties, individual group members are ultimately driven by the desire to maximize their benefits. When group interactions become more competitive, then collaborative conflicts are likely to arise, adversely impacting contributory participation of individual group members; simply put, within-group competition reduces the willingness of individuals to contribute their best to the group effort. In traditional organizational settings (discussed in the next section) that employ hierarchical structures, the overtly collaborative interactions can be covertly competitive as individual group members who eye, openly or not, advancement may perceive group sharing as detrimental to their self-interest. Ironically, the immediately obvious framing of group conflict as disagreement may actually be a positive manifestation of group dynamics as it may be indicative of cognitive diversity, whereas the not-so-immediately obvious framing of group conflict as competition may be a sign of group dysfunction.

The mechanics of competition-fuelled adversarial group dynamics are fairly well understood. Actual or perceived competition is the trigger that activates individuals' fears of being exploited, while also heightening those individuals' desire to exploit others. So primed group members tend to become more focused on standing out in comparison with others, which in turn heightens their instinct to evaluate their own information more favourably than information of others', a phenomenon known as ownership bias; moreover, individual group members also become more inclined to evaluate more positively any information that is consistent with their initial prefer- ences, a phenomenon known as preference effect. It is worth noting that those self- preservation instincts are at least somewhat involuntary, to the degree to which preference-consistent information is usually more salient and thus more accessible from memory. Reasoned or involuntary, the distinct possibility adversarial group dynamics casts doubt on the wisdom of emphasizing team collaboration in organi- zational settings characterized by pyramid-like reward structures. That realization is at the root of the emergence of more community-minded organizational structures such as flatarchies and holacracies, discussed in the next section.

2.2   Organizations as Commercial and Social Units

In order for the abstract idea of 'organization' to become benefits-producing entity, those expected benefits need to be clearly formulated as organizational mission and the appropriate mission-aligned structure needs to be erected. The former, as an expression of the organization's purpose, is typically framed in the context of serv- ing different types of customer needs for commercial organizations, and the pursuit of specific social or professional goals for non-commercial organizations. Organizational purpose is often captured in a formal mission statement, and it is usually evident in the organization's activities, but discerning of organizational structure is considerably more complex because it hinges on addressing several dis- tinct considerations. The most important of those are organizational governance, departmentalization, inclusive of job/function design and operating structure. It is worth noting that the expression of operating structure in the form of organizational chart is rather commonly used as a tangible depiction of organizational structure, which is at the very least misleading as operating structure is only a component part of organizational structure; it is the combination of the three aforementioned ele- ments that paint a complete picture of organizational structure.

2.2.1   Organizational Governance

One of the most foundational aspects of organizations is the design and mechanics of their governance, or the manner in which policies, rules and actions are set and administered. Officially defined as a formal system of rules, processes and practices aimed at controlling and directing the manner in which organizations pursue their stated objective, organizational governance can be broken down into two distinct components of operational management and supervisory oversight. The first of those two dimensions implies active control over day-to-day activities of an organi- zation, and it typically rests with executive officers, defined here as anyone with decision-making authority in an organization. In large commercial and non-commercial organizations alike, those positions tend to be staffed by professional managers, hired by boards of directors and entrusted with carrying out of broad strategic priorities set by those boards. The second dimension of governance – supervisory oversight – is the domain of boards of directors, which are bodies charged with jointly overseeing activities of the executive leadership.[6] In contrast to executive managers, who are typically appointed by boards of directors, organizational directors can be elected (a common practice in public corporations where shareholders elect board members) or appointed (a comparatively common practice among non-profits where directors tend to be appointed by third parties or serve in *ex-officio* manner based on office they hold) for the purpose of supervising activities of executive management.

The rather abstract ideas comprising organizational governance need a tangible 'skeleton' to be implemented, which is the role of the remaining two facets of organizational structure: departmentalization and functional job design, which lay out the way an organization structures its jobs to coordinate work, and operating structure, which encompasses chain of command and span of control. In the organizational governance sense, the former spells out lines of authority, while the latter addresses the specifics of reporting relationships, the combination of which further details the manner in which an organization coordinates its work.

2.2.2 Operating Structure

Commercial and non-commercial organizations alike can be either centralized, which is where decision-making authority is restricted to higher levels of management, or decentralized, which is where decision-making is distributed throughout organizational hierarchy. The former is preferred when conflicting goals of individual operating units create a need for a uniform policy, while the latter is deemed more effective when conflicting strategies, uncertainty or complexity require local adaptability; Fig. 2.1 shows an abstract graphic representation of both. The choice between those two different structural designs is commonly captured in organizational charts, which offer visual depictions of departmentalization and chains of authority; together, those two component parts offer a summary representation of how different organizations coordinate work. While in theory any arrangement

---

[6]An intent that is surprisingly often overlooked, if not outright ignored in commercial companies, where it is quite common for a single individual to hold the post of the chief executive officer and the chairman of the board of directors. That particular practice has the effect of inescapably concentrating chief rulemaking and execution powers in one person, which in turn nullifies the very point of corporate checks and balances. The generally weak and unconvincing arguments that are offered as justification of that obvious disregard of the core principles of good corporate governance do not alleviate the potential abuse of power, which is underscored by the recurring problem of corporate governance breakdowns, with Enron, WoldCom and Tyco scandals offering some of the more egregious examples.

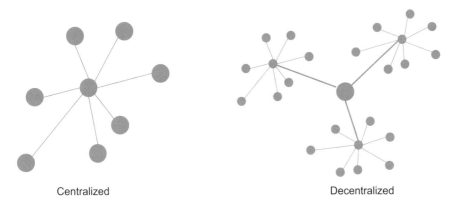

Centralized   Decentralized

Fig. 2.1 Centralized versus decentralized structures

within the broad centralized vs. decentralized design templates is possible, in practice there are several established arrangements that tend to be used ranging from rigid, vertically integrated, hierarchical, autocratic structures, to relatively boundaryless, empowered, networked organizations designed to be more commercially and socially agile.

2.2.2.1   Centralized Structures

The traditional, hierarchical structural arrangements can take on one of the three forms: functional, divisional or matrix. The functional structure divides work and employees by specialization, which emphasizes standardization of organizational processes and thus translates into relatively narrow job design; individual departments are constructed around established functions such as marketing, production, human resource management, etc. In functional structures, employees report directly to managers within their functional areas who in turn report to a chief officer of the organization, and the senior management centrally coordinates interactions among specialized departments. The advantages of the functional structure include development of expertise in functional areas, competency-based employee-job matching and the overall simplicity of design. The disadvantages include dividing of organizations into 'silos', difficulty of sparking cross-functional cooperation and resistance to change. Under most circumstances, the functional structure works best for organizations that remain centralized because there are few shared concerns or objectives between functional areas, and the organization can take advantage of economies of scale stemming from centralized purchasing functions, and the learning curve effects that emanate from specialization.

Often seen is an alternative to the functional structure, the divisional structure typically divides work and employees in accordance with product, market or geography. For example, a manufacturer of athletic apparel could organize itself by product line such as clothing, shoes and accessories, or by market such as retail stores and online sales, or by region such as Northeast, Midwest, etc. The advantages of the divisional structure include heightened focus on core competencies, greater cross-functional (e.g. marketing and production) coordination and greater ability to responding to changing marketplace needs because of more decentralized decision-making. The disadvantages include potentially lower efficiency, duplication of effort (e.g. clothing, shoes and accessories with separate marketing and accounting functions) and possible customer cannibalization stemming from different divisions competing for the same customers. Balancing the good and the bad, divisional structure might be beneficial when the company's markets are volatile, assuming the cross-divisional competition for organizational resources and customers can be minimized, and the diseconomies of scale do not reach dangerous levels.

However, when competition among divisions becomes significant, or the organization is not adapting quickly enough to the changing marketplace, or when diseconomies of scale continue to increase, the organization may require a more sophisticated matrix structure. The matrix structure attempts to combine the core elements of functional and divisional structures in a way that leverages each of the two structures' advantages while trying to sidestep their disadvantages. In practice, that tends to lead to split chain of command arrangements, where an employee can report to two managers who are jointly responsible for that employee's performance. Under that type of an arrangement, one manager would typically come from a functional area, such as marketing, and the other from a business unit related to a product, market or geography. The advantages of that structure include heightened possibility of forming functional and divisional collaborations, reduction of resource sharing costs and better view of the totality of individual products, in the sense of their design, marketing, distribution, etc. The disadvantages include fuzzy and confusing lines of responsibility, which may undermine efforts to enforce fair and equitable treatment practices, addition of likely significant, management and coordination burden, and the possibility of placing conflicting demands on employees. Overall, matrix structures offer greater visibility into the overall value creation processes of organizations, which in turn supports more effective governance; though complex, those arrangements are particularly well-suited for heavily project-driven organizations, such as construction companies.

2.2.2.2   Decentralized Structures

The changing nature of work detailed in Chap. 1 is giving rise to new organizational structures, most of which eschew the traditional autocratic hierarchies in favour of more progressive arrangements. Those new organizational structures aim to eliminate, or at least dramatically reduce, impediments to creative problem-solving by means of collapsing the traditional 'rank-and-order' hierarchies and replacing rigid arrangements with flexible ones. Among the most visible manifestations of the philosophy that underpins those non-traditional organizational structures is that teams replace departments, the organization and suppliers work as closely together as parts of one company and everyone, which includes managers and employees alike, participates in the decision-making process.

Unlike the earlier discussed hierarchical organizational designs which can be seen as 'either-or', meaning, functional, divisional or matrix forms are distinct and mutually exclusive alternatives, non-hierarchical organizational designs can be envisioned along a continuum of the degree of flatness. In some of the most extreme cases, there are no job titles, no seniority, no managers or no executives – everyone is seen as equal, and employees are generally left to self-manage. Though it may seem utopian, Valve, a gaming company, made it work. There, nobody has a job title, and nobody directs anybody else's work; instead, all employees at Valve join whatever project they want to work on. While that lack of structure seems to work for Valve, the same approach was more of a mixed bag for Zappos, an online shoe and clothing retailer. When the company, back in 2015, offered severance packages to all employees for whom Zappos' style of self-management, known as holacracy (discussed in more detail below), did not work, of the 18% of employees who took the severance, 6% cited holacracy as the reason. It turns out that the very structure, or more correctly lack thereof, that for some translated into making most of their talents and the welcomed opportunity to gain participation in organizational governance, for others produced role ambiguity and lack of career progression clarity. Given that, an important question arises: Is such total lack of hierarchy a reasonable goal for established organizations seeking to evolve into some post-bureaucratic, more socially responsive form? W.L. Gore & Associates, a materials science company, started out as a flatarchy, a term now commonly used to describe non-hierarchical organizations, but as it grew (the company currently employs close to 10,000 people), it found that it needed to add some structure. Similarly, Zappos as a retailer employs a wider, more general skillset oriented, cross-section of workers than a video game design company like Valve, which suggests that self-governance-based organizational structures (or really, lack thereof) may be more feasible for organizations comprised mostly of ultimately self-directing specialists.

All considered, non-hierarchical organizational designs share a core similarity in the form of some degree of flatness, thus as a group, those organizational designs exhibit a set of common-to-all advantages and disadvantages. Starting with the former, they seem better positioned to leverage all employees' skills and competencies and are also better positioned to foster collaborative work practices; in addition, they should also be, at least in theory, more in tune with shifts in the marketplace. In terms of disadvantages, the committee-and-consensus-driven decision-making can contribute to a lack of a singular, common-to-all vision, and it can mire the organization in time-consuming group decision processes.

However, flatarchies are not one-size-fits-all designs; as noted earlier, there are numerous designs falling along the 'degree of flatness' continuum, each exhibiting a somewhat unique mix of idiosyncratic strengths and weaknesses. Holacracy could be considered the flattest of the self-management organizational designs, and thus can be seen as the most radical. Within that organizational design, all employees are free to work on what they do best, and the only tangible structure, as considered from a traditional command-and-control perspective, is based on 'circles' roughly resembling departments; in other words, individual employees are accountable to circles that are most closely aligned with employees' projects. This is not to say that there is no structure – Zappos, for instances, makes use of 'lead link' roles which perform some of the traditional leadership functions. The difference, however, is that while in a traditional structure the assignment and the nature of work both tend to be entrusted to managers' judgement, which may or may not be informed by adequate evidence, in a holacracy, roles are flexibly assigned and the assignment of work is driven by the requirements of what is to be accomplished, as determined by appropriate circles, and facilitated by appropriate lead links.

Holacracy, however, is just one of the numerous embodiments of the idea of a flatarchy; another, relatively recent structure-flattening idea is podularity, which is an organizational design structure rooted in the premise that small, autonomous full-service teams, called 'pods', can replace large traditional departments, ultimately spurring greater productivity. Jeff Bezos' (the founder of Amazon) rule at Amazon perhaps says it best: If it takes more than two pizzas to feed a team, the team is moving towards unnecessary levels of bureaucracy (and thus its productivity is declining). It is also argued (by others) that the earlier discussed dysfunctional group dynamics, most notably groupthink and biased information search, tend to become more pronounced as the size of work group grows. To become podular, organizations need to reimagine the idea of a cross-functional team: Rather than being focused on representativeness, which tends to pile on members for often bureaucratic reasons, organizations need to focus on creative efficiency, seen as the ability to solve the problem at hand not only in a novel way, but also as quickly as possible. To do so, platoon-sized 'steering committees' should be replaced with more numerous, four-to-six member supercells, which have been shown to remain more innovative and more individually invested than the large, often unwieldy groups.

There are still other hierarchy-flattening organizational design ideas seeking to reduce bureaucracy, albeit by taking a broader view of the 'whats' and 'hows' of the company's value creation process. Here, three distinct designs have been proposed: hollow, modular (not to be confused with podularity) and virtual. Hollow organizational structures divide work and employees by core and non-core competencies, and, typically, core processes are maintained internally while non-core processes are outsourced. Under most circumstances, hollow structures favour firms operating in highly (price and otherwise) competitive industries, where robust outsourcing options are also available. A fairly obvious advantages of that organizational structure include lower overhead and the freedom to focus on the company's core business, while the equally obvious disadvantages centre on the loss of control over potentially important functions (non-core does not mean not important), and greater dependence on outside suppliers who may expectedly raise prices or engage in other adverse behaviours.

The modular organizational structure differs from hollow organizations in that it is more manufacturing oriented. That difference aside, similarly to hollow designs, modular structures entail outsourcing of distinct sets of product components, a practice that is quite common with manufacturers of complex equipment, such as airplanes, automobiles or computer equipment. Not surprisingly, some of the same advantages and disadvantages that characterize hollow structure are also true of modular designs: cost efficiencies and the heightened ability to focus on core competencies are the design's core advantages, whereas the loss of control and greater dependency on outsiders are its most visible disadvantages. The final, and in some sense the ultimate, embodiments of the idea of a flatarchy are virtual organizations, which are better seen as expression of cooperation between companies, institutions or individuals delivering a product or a service. Although the resultant 'companies' tend to take the form of contractual arrangements binding together otherwise distinct entities, the market-facing entities usually present themselves as unified organizations. There are numerous advantages to that organizational structure which include greater agility, extremely low organizational overhead and the ability to put together an 'all-star' type mix of contributors; there are also numerous disadvantages, which include extremely low organizational identity (i.e. the degree of identification with a virtual organization), generally low levels of trust among individual participants (who may even be competitors, as standalone entities) and heightened communication burden.

2.2.3 Organizational Culture

Famously elusive, organizational culture is a broad summary notion that encompasses that somewhat hard to verbalize sense of what it is like to be a part of an organization. It is the totality of formal and informal rules, routines, rituals and beliefs that tend to be shared and maintained by members of an organization, and that can be used to characterize and to distinguish one organization from others. As seen from the perspective of an incoming organizational member, culture manifests itself as a set of coercive and normative processes that take the form of formally proclaimed philosophies, often enshrined as organizational values, as well as routine behaviours, norms and rules of conduct. Lastly, given that it combines formal and informal elements, organizational culture is usually communicated both tacitly and explicitly, the latter often intended to reduce ambiguity.

The key to discerning organizational culture is understanding the source of it. In most business organizations, culture emanates from their founders or leaders, from demonstrated success and at times from new members. Founders and leaders usually sow the seeds of organizational culture through the process of implementing their personal beliefs, values and assumptions regarding strategies, competition and the environment; those beliefs, values and assumptions precipitate patterns of thinking and behaviour that are then learned and adopted by organizational members. If and when the resultant business strategies and philosophies lead to success, they are then embraced as cherished organizational values. Over time, it is common for business organizations to enshrine the key elements of their culture as 'core organizational values', prominently displayed and eagerly passed onto new organizational members.

   The explicitly communicated aspects of organizational culture are primarily cognitive, and tend to encompass everyday organizational rules, such as the dress code,
which are typically unambiguous. The tacitly communicated aspects, on the other
hand, are predominantly symbolic and thus highly interpretive. Cherished (or at
least subscribed to) organizational beliefs and values usually constitute the core of
symbolic elements of culture – under most circumstances, organizational stakeholders need to make sense of those noticeably more ambiguous elements of organizational culture on their own. Not surprisingly, misinterpretation of the tacitly
communicated, symbolic elements of organizational culture is a common occurrence, especially in environments in which organizational culture can be broadly
characterized as emergent, which is to say it can be characterized as organic, participative and adaptive. Organizations can, and sometime do, provide explicit cultural
frames of reference in hopes of reducing cultural ambiguity, although that degree of
organizational introspective is rare among younger, less mature organizations,
which is where the risk of cultural ambiguity is the highest.
   When considered from the perspective of an uninvolved onlooker, an organization's culture can be seen as a mean of communicating its self-identity to internal
and external stakeholders; in a less direct way, it can also be seen as an indication of
the manner in which the organization reacts to broad societal, legal and environmental forces. As such, an organization's culture can be framed in the duelling
'either-or' context of being either participative or non-participative, adaptive or
inelastic, and organic or synthetic. The resultant composite gives rise to two common cultural archetypes: participative – adaptive – organic, and non-participative –
inelastic – synthetic. It is tempting to see the first of the two cultural profiles as
being somewhat 'better', in the sense of being preferred, because labels are evaluative in nature and thus prone to being value-laden – such global conclusion, however, might be unwarranted, as illustrated by the earlier example of 6% of Zappos'
employees stating that the primary reason for their acceptance of the company-
offered severance package, and thus leaving the company, was the company's organizational structure, holacracy. In the view of those dissatisfied employees, Zappos'
embrace of that particular organizational structure created a culture that simply did
not work for them, most notably because it resulted in role ambiguity and lack of
clearly communicated professional growth opportunities. Stated differently, the
choice of management structure brought with it culture that was highly participative, adaptive and organic, which in turn created an environment that was welcomed
by most (as, after all, 82% of Zappos' employees opted out of the offered buyout),
but not by all.
   In a more general sense, embracing participative culture is considerably more
viable for a small start-up than a large, geographically and operationally dispersed
organization; moreover, participative culture might also be more desirable for a
small firm focused on growth than for a large, established organization where control is of more concern. On a more individual level, participative culture may not
suit all personalities – in a way of contrast, non-participative culture tends to offer
clearly structured and laid out roles and career pathways, and just as some may shy
away from such an environment, others find it enticing. Young, entrepreneurial

organizations tend to have naturally participative – adaptive – organic cultures largely because they may not yet have found their truly defining values and beliefs; those organizations usually attract a somewhat different mix of employees than established, successful organizations whose culture tends to be non-participative – inelastic – synthetic as a reflection of having found their unique defining values and beliefs.

At least in part, the reason that the non-participative – inelastic – synthetic cultural type may seem unappealing on its face value is that it is difficult to get around the somewhat automatic conclusion that it implies a monochromatic entity made up of like-behaving, like-thinking human-robots. In fact, when taken to the extreme, it conjures up images of drones marching in unison under the command of some over-lord. Even abstracting away from that somewhat pejorative, science fiction-inspired perspective, non-participative organizational culture seems to suggest over-emphasis on efficiency and the relative de-emphasis of creativity, though as exemplified by the famed 3M culture of innovation, that may be an overgeneralization. 3M Corporation (once known as Minnesota Mining and Manufacturing), a multinational conglomerate, steadily grew, for more than 100 years, to become one of the most celebrated product innovators in the world, currently offering more than 55,000 different products, in a large part because of its non-participative – inelastic – synthetic culture built around innovation. Built around five core values,[7] 3M's culture of innovation has been widely replicated, particularly its emphasis on setting aside a fraction, 15% to be exact, of employees' work time to work on their individual ideas. In fact, one of the company's best known products, the Post-It notes, was invented during that 15% time (as were Google's Gmail and Google Earth products since Google embraced the same idea, even increasing the innovation time allowance to 20% of employees' worktime). Naturally, those 'best in class' examples can be seen as outliers, but they nonetheless underscore the fact that both participative and non-participative cultures can produce outstanding results, thus it is important to focus more on how well the culture of a particular organization is suited to its long-term mission and its near-term goals, while setting aside biased beliefs.

Still, those normative considerations notwithstanding, the reality of technological and environmental turbulence, globalization and hypercompetition are not only changing how work is done but are also reshaping the often longstanding cultural norms. Ranging from fundamental organizational reconfiguration, ushered by the emergence of less hierarchical organizational structure models, to the steady rise of alternative working models, perhaps best exemplified by telecommuting (which, due to the COVID-19 global pandemic, saw an earlier unimagined mass acceptance), the very notion of organizational cultural norm is being called into question. Perhaps that is why, instinctively, many may find the rigidity of a fixed, longstanding norms so passé, while at the same time seeing the participative variant as a

---

[7] Feel the customer pain, empower employees, dedicate time to innovation, develop collaborative platforms and attract good talent.

refreshingly new, post-industrial alternative, especially when those instincts are propped up, the stereotypes of will-less drones contrasted with creative and engaged but independent thinking contributors. And so it follows that to capture the essence of a given organization's culture in a way that leads to sound assessment of the extent to which organization-specific rules and values impact its ability to learn, one must look past the exaggerated and the over-idealized cultural clichés. Also, given that culture 'exists' within the confines of the organization's structure, it is essential to also consider the interplay between the organization's culture and its structure, which is addressed next.

2.2.4 The Interaction Effect

On the one hand, in order to discern the true essence of organization-specific structural and cultural characteristics, it is important to immerse oneself in each individually, but on the other hand, understanding of the broader organizational ecosystem calls for careful examination of the structure-culture interaction effect. In that sense, it is the culture-structure interaction in conjunction with the reflexivity of organizational members that determine the shape of key organizational functions, or the manner in which organizations interact with internal and external stakeholders. Stated differently, in practice it makes little sense to try to make sense of the organization's culture without also addressing its structure, and vice versa. Considering that, for the most part, organizational culture manifests itself in terms of relatively abstract, general rules, beliefs and rituals, it stands to reason that the same set of cultural norms might 'feel' differently in centralized and in decentralized organizations. The reason for that is that the former tend to exhibit greater uniformity in how cultural norms are interpreted and applied throughout the organization, while the latter are more prone to allowing for a greater degree of localization, be it by geography or by division, of the same cultural norms. And so to an objective onlooker armed with the ability to holistically examine the totality of a given organization's culture, the application of the same cultural norms across different organizational structures could likely produce noticeably different cultural 'looks and feels'.

It is, however, more than just appearances, and at least to some degree, it could be argued that organizational structure begets organizational culture. By monopolizing the core elements of decision-making, centralized organizations effectively discourage greater embrace of, and immersion with, the spirit of cultural norms, which can effectively erect barriers to the earlier mentioned cognitive diversity, ultimately leading to a more constrictive, 'our way' type of culture. Moreover, recalling the three distinct dimensions of organizational learning discussed in Chap. 1 – adaptive, generative and transformative – centrally controlled organizations tend to exert profound, though not always intended influence on organizational learning. More specifically, a bureaucratic organizational structure may allow some degree of adaptive learning, which encompasses reactive changes primarily brought about by internal pressures, so long as it does not come into conflict with existing norms, policies and practices, but it leaves little room for the other two forms of organizational learning to flourish. More specifically, the rigidity of centralized organizations' structures tends to effectively discourage generative (acquisition of new skills and knowledge) and transformative (strategic shifts in skills and knowledge) learning on a larger, organization-wide scale. As famously advocated by Max Weber,[8] a sociologist, bureaucracy offers an effective means of controlling organizations (which is why it played an important role in the rise of large modern business enterprises), but its typically inelastic decision-making character tends to reduce broadly conceived organizational learning to mechanistic responses to environmental demands, and all but suppresses higher-order, critical thinking.

2.3 Changing Organizational Dynamics

While the slow but steady embrace of flatarchies is perhaps the most visible manifestation of democratization of organizations, a more impactful though much harder to observe evolutionary change is the gradual shift away from seeing organizations as entities that exist primarily for the benefit of their stakeholders, and towards seeing organizations as entities that exist for the benefit of society at large. In other words, the idea of democratization of organizations encompasses not just the manner in which organizations are structured and governed, but also the very purpose for their existence. In accordance with that view, organizations do not just exist to make money for their shareholders (commercial entities), establish and maintain professional standards (not-for-profit entities), or offer social assistance to those in need (non-profit entities) – they exist to further the development and the wellbeing of society as a whole, which means that not just non-commercial, but also commercial, organizations need to be socially conscious. In practice, in order to become more than aspirational slogans, those ideals need to manifest themselves in mindsets of organizational governing boards, which set organizational priorities and strategic objectives. However, as it tends to be the case with so many (sadly, too many) aspects of social functioning, that broadening of organizational scope is slowly taking hold not because of some sort of mass enlightenment, but because of series of conflicts and resolutions that slowly propel organizations along that evolutionary pathway. Thus to make sense of the underlying mechanics of organizational evolution, it is instructive to take a closer look at the key influencers of institutional boards' functioning, most notably the duty of care and the business judgement rule, as well as the key sources of stakeholder conflict, in particular heightened stakeholder activism, and organizational governance failures.

---

[8] Maximilian Karl Emil Weber (1864–1920) was an influential German sociologist, philosopher and political economist whose ideas had a profound impact on the shaping of Western socioeconomic institutions.

2.3.1 The Duty of Care and Business Judgement Rule

Organizational governance and information technology management share one peculiar characteristic: both tend to receive little attention until problems emerge. It should come as no surprise that governance-related problems, for both commercial and non-commercial organizations alike, are most commonly tied to financial irregularities. For commercial businesses, those routinely take a form of inaccurate or incomplete financial disclosures, 'creative' treatment of matters such as revenue or expense recognition, or insufficient financial controls[9]; for non-profit organizations, governance problems tend to be more reflective of organization-specific mission, as exemplified by the Planned Parenthood profit debate (the organization was accused of selling aborted fetuses for profit), the Sierra Club's funding concerns (acceptance of more than $26 million from Chesapeake Energy, a major energy company) or the Red Cross relief fund usage questions (the nearly $500 million raised to aid with Haiti's earthquake aftermath). Needless to say, documented or suspected organizational governance missteps can take a wide array of forms, some serious, others comparatively minor manifest; when considered in aggregate, their root causes can be traced to two important corporate law concepts: the duty of care, and business judgement rule.

The concept of duty of care encapsulates obligations of organizational decision-makers, most notably executive managers, and members of the board of directors (commonly referred to as officers and directors, respectively), to adhere to a standard of reasonable care when discharging their duties. In a very general sense, it simply means that those who are entrusted with acting on behalf of business and non-business organizations are expected to act in the best interest of the organizations they directly manage, as executive managers, or oversee, as members of boards of directors. The very open-ended notion of 'acting in the best interest of the organization' manifests itself in two somewhat distinct requirements: the duty to act, and decision-making due diligence.

The duty to act compels board members to be actively involved in the affairs of the organization; it amounts to saying, 'It is unacceptable to be a figurehead passively watching, or not, the organization-pertinent events unfold – one must actively engage in activities aimed at helping to positively affect the wellbeing of the

---

[9] In the United States, the Securities Act of 1933 and the Securities Exchange Act of 1934 spell out specific informational disclosure requirements for publicly traded companies, which ultimately give rise to what is known as 'executive risk', an umbrella term for threats that emanate from incorrect, misleading, incomplete or not timely disclosure of legally required financial information. Shareholders who believe that the company in which they hold stock did not fully, accurately and timely disclose pertinent information can sue (typically as a group, called 'class', as in 'class action') the directors and officers of the company for damages; on average, about 300 to 400 such lawsuits are filed in the United States annually, and to protect themselves against the cost of what is known as 'securities litigation', which can be considerable (the median settlement cost is about $10 million, and the top 10 most expensive ones are all in excess of $1 billion), virtually all publicly traded firms purchase what is commonly known as 'directors and officers, or D&O, insurance'.

organization'. Stated differently, the power and the trust placed in the hands of managers and board members do not come with an option to act – they come with an obligation to act.

Decision-making due diligence requirement further clarifies demands placed on organizational directors and officers by positing that making decisions without thoughtfully considering available and applicable information may also constitute dereliction of decision-makers' duties. In other words, the duty to act compels executive managers and directors to make thoughtful decisions for well-considered reasons. This is not to say that the duty to act demands clairvoyance, in the sense of avoiding what might ultimately turn out to be disadvantageous courses of action; it simply requires that decision-makers carefully consider available evidence when deploying or otherwise engaging organizational resources.

The demands placed on organizational directors and officers are somewhat balanced by what could be framed as a measure of decision-making immunity known as the business judgement rule.[10] In a very general sense, it can be seen as a manifestation of the idea that since, in principle, however diligent and well-considered, actions of organizational decision-makers are ultimately speculative, which means that any decision may, in retrospect, turn out to unsound. Thus, so long as organizational decision-makers' actions were taken in good faith (i.e. can be characterized as sincere in intention), and did not conflict with any of the applicable laws, rules or regulations, directors and officers cannot be held liable for outcomes of their decisions. As such, the business judgement rule shields well-meaning board members who are misinformed, misguided or simply mistaken from legal liability, meaning it protects them from being sued in court, except in rare cases of egregiously flawed judgement (the litmus test for which is usually based on a legal 'reasonable person' standard). Such extreme cases aside, the business judgement rule's cloak of immunity disallows any legal second-guessing of executive managers' and board members' choices, even if those choices ultimately led to catastrophic outcomes, such as bankruptcy, so long as board members can be shown to have acted in good faith.

Industrial history is full of examples of once prosperous – even dominant – but now extinct companies including retailers, such as Sears, Marshall Field's or Circuit City, banks, such as Washington Mutual, Lehman Brothers or Bear Stearns, or manufacturers, such as American Motor Company (AMC), Polaroid or Cambridge Sound. Non-profit failures are generally less publicized, but in many regards can be even more disappointing as was the case with the demise of the Federal Employment Guidance Services, a New York City–based human services organization, or NC New Schools, an education organization based in North Carolina's Research Triangle. Much has been written about poor strategic choices, even decision-making inaptitude that ultimately precipitated those and numerous other organizational

---

[10] The 'business' part of the name should not be taken to suggest that it only applies to for-profit organizations; although by their very designation non-profit organizations have manifestly fewer business-like operating aspects, non-profits are nonetheless economic entities that engage in a variety of financial transactions in the form of purchases, credit, etc.

failures, but the business judgement rule protected their well-meaning but manifestly incompetent decision-makers from legal and economic liability.

Though on the one hand the duty of care and the business judgement rule can be seen as a yin and yang of organizational functioning, both of those behaviour-framing notions ultimately compel executive and governing decision-makers to immerse themselves in organizational functioning. That said, such historically narrow application of those concepts to just a select group of organizational officers and directors, which even in large organizations comprises a relatively small subset of organizational stakeholders, demands reinterpretation when considered in the context of the earlier discussed flatarchies, where decision-making authority can be distributed among a wide cross-section of organizational stakeholders. If the duty to act expectations are expanded to include any organizational stakeholder who is in a position to commit organizational resources, there needs to be a corresponding broadening, and perhaps also deepening of available and accessible 'ways of knowing'. In other words, it is not enough to just erect a more participative organizational structure and culture – in order for such broadening of organizational control access to yield tangible organizational benefits, and to keep tap on the numerous individual- and group-level cognitive and behavioural dysfunctions (e.g. cognitive bias, groupthink, etc.), there also needs to be sufficient informational support infrastructure.

2.3.2   Stakeholder Activism

Though only tangentially related to organizational learning, an important aspect of the changing landscape of organizational dynamics is the growing visibility of shareholder activism, a broad category of concerted efforts by equity holders (i.e. shareholders) to exert direct influence on organizational decisions.[11] As one of many frictional facets of organization-stakeholders interactions, the contemporary practice of shareholder activism can be linked to the emerge of the new institutionalist view of organizations, which argues that organizations ought to be seen as political and economic entities that seek stakeholder acceptance and broader legitimacy. Emanating from that rationale is the belief that given their need for acceptance and legitimacy, organizations should be responsive, if not accountable, to social and economic pressures, and in fact such pressures could be seen as necessary means of communicating environmental demands to organizations. In that sense, shareholder

---

[11] In a very general sense, shareholders have several delineated rights which include the right to participate in a company's profit, the right to buy new shares, the right to vote (in a company's annual or general meeting), the right to sue for wrongful acts (as illustrated by the earlier example of shareholder class action litigation) and the right to influence management, but that right is indirect as it is effectively limited to the election of company's board of directors – the goal of shareholder activism is to directly influence a company's executive management, which is not a delineated shareholder right.

activism can be framed as an organic complement to the formal and indirect share-holder communication mechanism, looking to go beyond the enumerated share-holder rights (see footnote 11).

The roots of contemporary direct stakeholder action can be traced back to the 1890s and early 1900s Populist and Progressive movements which represented growing anger at the influence of well-financed special interest groups on govern-mental policies, and government's inability to deal effectively with social problems of the day. Focused on more decisive and more immediate impact, shareholder activism can take either adversarial or supportive paths. The former commonly entails application of coercive pressure by means of activities such as lobbying efforts, protests, demonstrations, boycotts and lawsuits, all aimed at sanctioning specific policies or practices; the latter commonly takes the form of praising of poli-cies or behaviours of interest, as is the case with publishing listings of the most environmentally friendly, sustainable or socially responsible organizations. It is worth noting that organizational activism is no longer a domain of select groups of like-minded individuals or special interest groups. Fortune magazine's annual list of the '100 Best Companies to Work for in America' is based on data gathered each year and addressing organizational attributes such as pay and benefits, life balance, opportunities, job security and pride in the company; as such, it can be seen as a broader embodiment of the idea of activism, as can the annual ranking of the '100 Best Companies for Working Mothers', published annually by the Working Mother magazine.

2.4   Changing Organizational Environment

It is difficult to notice, and even more difficult to appreciate, immersive change. In ways that are sometimes forceful and other times nearly imperceptible, the early years of the twenty-first century have been shaped by a seemingly endless stream of socio-politico-economic transformations. From Internet-enabled resurgence of pop-ulism coupled with rising social consciousness and the strengthening of the drive towards individual self-determination, to the rapid proliferation of electronic inter-connectivity and the resultant emergence of big data as a key source of competitive advantage and economic value creation, the established norms and ideals are being replaced by new visions and interpretations. Set in that fervent context, the tradi-tional conception of what organizations are and how they function is also changing.

2.4.1   Now That Everyone Has Voice, Is Anyone Listening?

Social media is ubiquitous, a claim that is perhaps best illustrated by Facebook, which, as of the latter part of 2020, had over 2.7 billion active users, or more than a third of the world population. With so many individuals and organizations looking to social media platforms as a communication outlet, fatigue and apathy seem unavoidable. Recognizing the growing difficulty of cutting through the ever more cluttered social communication media, researchers and organizations have been trying to find creative ways to counteract the growing social media fatigue. For example, a recent *Journal of Marketing Research* study focused on the impact of Twitter and Instagram messaging found a strong association between inclusion of visual images, particularly high-quality, professionally shot pictures, and user engagement. A different example illustrates the benefit of quick, opportunistic usage of social media via a technique known as 'improvised marketing interventions', which is the composition and execution of a real-time marketing communications proximal to an external event. Using that approach during the power outage that disrupted Super Bowl XLVII in 2013, quick-witted Oreo marketers tweeted, 'Power out? No problem', along with a brightly lit image of a solitary Oreo cookie containing a caption reading 'You can still dunk in the dark'. This clever, opportunistic action produced over 15,000 retweets within the next 8 hours, creating significant publicity for Oreo at minimal expense (especially considering that around the same time, Super Bowl television advertising rates were about $4 million per a single 30-second spot).

However, it is not just the communication density that poses a problem to productive use of social media – it is also the informational content itself, as seen from the perspective of validity. For instance, the current (2020/2021) COVID-19 pandemic has been widely discussed and dissected on social media, but evidence is mixed with regard to whether this medium, on balance, jeopardizes or promotes public health. As a whole, internet-based communication media have been variously described as the source of a toxic 'infodemic', or as a valuable tool, in large part because no accepted evaluation norms or standards are readily available. Still, given the open character of social media, in the sense of lack of any form of peer review provisions or mechanisms, the posted content inescapably requires reader discernment, which itself is prone to a host of subjective biases and interpretations, not to mention varying degrees to fact-check any posted content.

Those considerations raise important, organizational functioning-related questions. Are the traditionally hierarchical, command-and-control organizations able to effectively utilize, both as a source as well as a conveyor of information, communication platforms where anyone can communicate anything at any time? Are those organizations able to respond in a timely manner to threats and opportunities afforded by the new, organic media? What is the most effective strategy for adapting to the 24/7, open-to-all communication regime?

2.4.2 De-Departmentalization of Innovation

If data are ubiquitous, and if the informational content of data is now the primary source of innovation, should the manifestly creative organizational thinking continue to be departmentalized, as in a formal R&D function, or should it be disseminated throughout organizational contributors? Nowadays, organizations almost routinely turn to solution crowdsourcing platforms, such as Kaggle, for new data analytics-driven solution, why not do the same in other areas?

For more than a century, the departmentalized specialization model pioneered by F.W. Taylor[12] has been the staple of organizational design because it offers an elegant solution to problems of control and efficiency maximization. Using Taylor's design, the 'innovation' function tends to be contained in a standalone department, primarily because it is implicitly seen as just a different organizational function that can be made more efficient by leveraging learning curve and economies of scale effects.[13] Yet some of the most breakthrough innovations came from 'skunkworks'[14] type efforts, which tend to be loosely organized, if organized at all, and characterized by a high degree of problem-solving autonomy, as well as also being unhampered by bureaucracy. The earlier discussed 3M's Post-It Notes as well as Google's Gmail and Google Earth products offer some of the better-known examples of such non-departmentalized innovation. To the degree to which, in the Age of Data, a core part of value-creating organizational innovation is tied to creative use of data, the ubiquity of data coupled with the inherent individuality of data analytic knowledge and abilities strongly suggest that organizational creativity ought to be seen as an organic, decentralized, all-organizational-stakeholders-involving effort.

2.4.2.1 Centralized – Decentralized – Networked

The earlier discussed (in the context of organizational structure) and here abstractly expressed 'to centralize or not' idea is operationally somewhat more nuanced when considered from the perspective of organizing for innovation. In the organizational structure design sense, it entails not just two but three distinct organizational design schemas: centralized, decentralized and networked, as graphically depicted in Fig. 2.2.

Within the narrow context of innovation, each of the three structures abstractly outlined in Fig. 2.2 has a different mix of advantages and disadvantages. The centralized innovation design is commonly seen as outdated and not in keeping with changing nature of work, in addition to also being adversely impacted by the inescapable bureaucracy of its command-and-control structure. Still, by seeing

---

[12] A mechanical engineer by education and trade, Frederick Winslow Taylor (1856–1915) is widely regarded as the father of scientific management.

[13] A learning curve is a generalized graphical representation, approximated by the so-called s-curve (sigmoid), which captures the relationship between proficiency and experience; i.e., the more someone performs a given task, the better they get at it (until a plateau is reached). The notion of economies of scale captures the cost advantages that can be obtained as cost per unit of output decline (primarily due to fixed costs being spread over progressively more units) with increasing scale.

[14] The name derived, most probably, from Lockheed Martin's (an aerospace and defence company) Advanced Development Programs, officially nicknamed Skunk Works and best known for their development of the now legendary U-2 and SR-71 Blackbird spy aircraft.

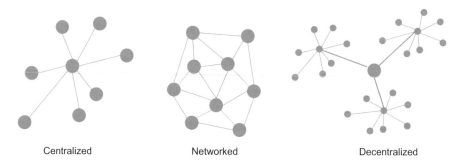

Centralized   Networked   Decentralized

Fig. 2.2 Organizing for innovation

innovation of planned rather than a spontaneous undertaking, the centralized design has some distinct strengths that stem from an express recognition of innovation as a distinct organization function, one that is supported by its own dedicated staff, budgets and other resources. As such, it is particularly well-suited for situations where innovation tends to take the form of continuous improvement rather than an outright 'out of the box' type of pursuit, perhaps best exemplified by the automotive industry. Consider automobile as a functional product: Since its inception more than a century ago, its basic functional outline changed relatively little – it was and still is a wheeled carriage powered by its own engine and (currently still) steered by a driver. Yet due to the ongoing process of continuous improvement, the same basic functional product has been slowly evolving in terms of its core functionalities such mechanical capabilities, reliability, comfort, efficiency and a host of other attributes; even what could be seen as major evolutionary shifts, such as the introduction of automatic transmission, power-assisted steering or antilock brakes, were still just incremental improvements over earlier systems. Interestingly, since such continuous improvement-focused innovation rarely inspires books or makes headlines, it tends to go unnoticed and, in a sense, unappreciated. Even more importantly, however, in order to produce automobile-like continuous improvement, the underlying innovation efforts need the type of purposeful, focused and innovation-sustaining 'engine' that comes with the dedicated staffing and investment of centralized research and development function.

At the same time, one of the core drawbacks of centralized and departmentalized innovation model is that becoming good at incremental improvement-focused innovation often also means being more and more invested in 'what is' and thus less likely to see, or even embrace, 'what could be'. The automotive industry again offers an instructive illustration: The now widespread drive towards electrification did not pick up steam, no pun intended, until new industry entrants, who were unencumbered by massive investments in the old, fossil fuel-powered technologies, made a compelling case. It is quite remarkable in view of the incredible scientific and technological advances of the past century that essentially the same, in terms of its basic design principles, internal combustion engine technology has been the backbone of the automotive industry throughout its roughly 150-year history. The underlying reason also best captures the value of decentralized and networked innovation models, which, colloquially speaking, are less about perspiration and more about inspiration.

Recognizing the difficulty of inspiring 'out-of-the-box' thinking in the often groupthink-permeated centrally departmentalized innovation function, together with the importance of pursuing new, mould-breaking ideas (sometimes characterized as discontinuous innovation), some forward-thinking large centrally controlled organizations have been creating semi-autonomous units tasked with the pursuit of radical, disruptive ideas and technologies. The other two innovation structures depicted in Fig. 2.2 – decentralized and networked – emphasize some of the key aspects of the open-ended nature of discontinuous innovation. The decentralized innovation design mindset is perhaps best illustrated by a multiunit organization, such as a business firm comprised of multiple strategic business units, or SBUs, where each SBU is engaged in a discernible and distinct commercial activity, as exemplified by General Electric's (GE) Power, Healthcare, Capital, Aviation, Digital and Renewable Energy units. In that simple but illustrative context, the overt dissimilarity of anything ranging from production inputs to value creation know-how to final outputs just about necessitates framing of the task of innovating within the confines of each distinct SBU. In a sense, however, such a natural organizational structure–innovation design fit is largely confined to diversified conglomerates like GE; it is important to keep in mind, however, that number-wise, such broadly diversified organizations are vastly outnumbered by companies with far narrower business scopes. And not all more narrowly focused business organizations are necessarily small – Amazon, an online retailer, and Walmart, a (primarily) brick-and-mortar retailer are both far larger (in terms of revenue and market capitalization) than GE. And that is where the value of distributed data-enabled creativity, and the larger idea of widely distributed organizational learning become particularly pronounced: In its truest form, innovation decentralization is a manifestation of distributed and organic data-enabled organizational learning. Organizationally empowering and informationally enabling of, in principle, the entire stakeholder ecosystem ought to be the aspirational goal of the decentralized organizational innovation design model. However, the framing of the idea of 'organizational ecosystem' needs to take into account organizational composition, which means that within the confines of a diversified, GE-like, conglomerate, it could entail individual SBUs, or the entire organization for a specialized company such as Walmart.

When considered in the context of the idealized framing of the decentralized innovation design model, the third variant – networked design – might sound quite similar to the decentralized model, but there are distinct differences, graphically summarized in Fig. 2.2. Whereas the decentralized model is built around distributed innovation 'units', which could be formal departments or somewhat informal teams, the networked design is maximally organic, in the sense that it does not entail any pre-determined groupings of contributors, instead relying on spontaneous, opportunity- and need-driven collaborations. Given its lack of pre-arranged structure, networked design is typically associated with progressive non-hierarchical flatarchies, such as the earlier discussed holacracy and podularity, which emphasize broadly shared collaboration and self-governance; such completely de-departmentalized system of interconnected collaborators is somewhat reminiscent of academic research cooperation. Each 'nod' in the network (see Fig. 2.2) can be seen as an independent, meaning self-directed and self-governed, pursuit of a common idea; while perhaps the least efficient, it is also least effected by creativity-hamstringing effects of groupthink and other limiting factors that are hard to avoid in formal departmental structures.

2.4.3 The Rise of the Gig Economy

It is official – musicians no longer are the only ones with 'gigs'. What was once a cottage industry of temporary, short-term staffing has now matured into a major segment of the economy where short-term contracts and freelance work are viable alternatives to permanent employment. Organizations like the adaptability of workers' mix of skillset to the needs of the moment, and those providing their services value the flexibility that often accompanies. Uber and Airbnb are now multibillion dollar companies and are built entirely on services provided by armies of contracted suppliers; their impact on the economic ecosystem is both complex and profound. Consider the case of New York City's taxi medallions, which are tightly controlled, limited resource that give the owner the right to pick up passengers, specifically street hails. The value of non-corporate (i.e. held by individual drivers) medallion reached its peak at over $1,000,000 in 2013, but just five years later it dropped by less than a fifth of its original value, because of the Uber and other ride sharing services. While such direct impact metrics are not readily available to capture the impact of Airbnb on the hospitality industry, the spike in industry consolidation, perhaps best exemplified by Marriott International's recent buyout of Starwood Hotel & Resorts Worldwide to create the world's largest hotel company, offers compelling albeit indirect evidence.

   The important aspect of those trends is that the rise of the so-called gig economy has implications that span a wide spectrum, ranging from how the view of economic employment is beginning to change, to how the emergence of aggressive new organizations focused expressly on that economic segment is impacting numerous established segments of the economy. Once again, multiple, dynamic forces impact organizations in ways that rigid hierarchical structures may simply not be able to respond to in a timely and adequate fashion. That conclusion is equally applicable to identifying and reacting to potential threats, as well as seizing emerging opportunities, particularly in the sense of organic ideas that could arise in skunkworks-like environment. Lastly, rigid organizational structures may be viewed unfavourably by prospective new employees, especially the now up-and-coming Generation Z (i.e. those born between 1995 and 2010). According to research by McKinsey & Company, a consultancy, they are cause-oriented, hypercognitive, feel comfortable seeing the world through multiple perspectives and are particularly focused on authenticity and freedom of expression. As a broad group, Generation Z paints a picture of organizational constituents who are more loyal to a particular cause than to an organization, a characteristic that aligns a lot more closely with the basic ideals of holacracy-like flatarchies than with the command-and-control organizational structures.

Another aspect of the growing generational value and attitude differences that already have had a profound impact on the functioning of commercial and non-commercial organizations is the idea of work-life balance. The frequently heard declaration stating that someone 'works to live' rather than 'lives to work' offers perhaps the best way of capturing those sentiments. At the risk of over-generalizing, the older generations – Baby Boomers (born between 1940 and 1959), Generation X (born between 1960 and 1979) and, perhaps to somewhat lesser extent Generation Y (also known as Millennials, born between 1980 and 1994) – placed premium on professional success and economic security, which boded well for traditional command-and-control organizations. Explicitly or implicitly offering higher socio-economic status as an incentive, organizations were able to direct and manage large pools of capable and willing organizational contributors, but that may be more challenging with the up-and-coming Generation Z who, as noted earlier, tend to be less focused on socioeconomic success and advancement.

2.5 Inspiration or Perspiration?

Although nearly a century has passed since the pioneering work of F.W. Taylor, the father of scientific management, management is quite often still viewed (and more importantly, practiced) more as an art than science. Implicitly favouring subjective gut feelings or flashes of inspired insight over objective evidence, organizational hierarchies continue to put their fate in the hands of select few all-powerful autocrats. And once again, arguments that organizations can be both hierarchical and evidence-driven are unconvincing because true evidence-based management renders the bulk of organizational bureaucracy superfluous, just as automation renders manual operation unnecessary. Decision-making hierarchy inescapably infuses subjectivity, which in turn opens Pandora's box of cognitive bias and other reason-warping distortions[15]; it is often one of the key reasons why organizations that 'should know better' (i.e. have ample data and means of using it) make disastrous decisions. All and all, it is hard to escape the conclusion that true embrace of fact-based organizational decision-making calls for flat organizational structures, but letting go of command-and-control is turning out to be a challenge.

To some degree, it is hard for organizations to look past autocratic management structures because the notions of 'organization' and 'leadership' seem inseparable, as the history of organizations is also the history of organizational leadership (even

---

[15] A more in-depth discussion of those influences, and the manner in which they warp subjective sensemaking is offered in *Evidence-Based Decision-Making* (New York: Routledge, 2019).

the idea of democracy, as currently practiced, is ultimately geared towards selection of leaders, not some sort of mass governance). Does that mean that organizations, as human collectives, have to be led, in the command-and-control sense? It is tempting to point to the earlier discussed 6% of Zappos' employees who took the offered severance and left the company because of the company's embrace of holacracy and using that as 'evidence' to cast doubt on the appeal of non-hierarchical organizational structures. However, it is important to not lose sight of the fact than twice as many (a total of 18% of Zappos' workforce who took the severance package) left for a number of other reasons, and even more importantly, the overwhelming majority, or 82%, did not leave. No management style fits everyone, but the high degree of self-determination offered by non-hierarchical organizational structures appears to be better aligned with the earlier discussed demographic trends (i.e. the Generation Z's notably different values and preferences), and by more actively engaging larger cross-sections of organizational stakeholders in organizational decision-making, it fosters fuller utilization of talents and capabilities of those contributors. This reasoning parallels the manner in which the state of scientific knowledge is advanced. Multiple, independent researchers contribute to the overall knowledge discovery efforts, some through conceptual and others through empirical means, and the resultant outcomes are shaped by the totality of those independent, distributed efforts. And while a handful of generational geniuses indeed infuse periodic flashes of inspired brilliance, it the focused and disciplined work of many other contributors that brings about what we call scientific progress, and it all happens without any overarching command-and-control structure.

It is also hard to envision decision-making without some measure of intuition, because according to cognitive psychologists and neuroscientists, human evolutionary development encoded us with instinctive reliance on innate sensemaking. Obviously, intuition-based decision-making worked remarkably well within the confines of grand evolutionary processes – why could not it work equally well in organizational management? Indeed, there are times when instinctive decision-making can indeed be quite effective, but that is usually so when the organizational and the broader environmental climates are calm and stable (of course, it could also be a manifestation of a stroke of good luck). On the other hand, there are ample reasons, such as the ever-present hypercompetition and the widespread socioeconomic volatility, to believe that for most organizations, business and non-business alike, the future will be anything-but-stable (and counting on good luck is probably not a viable strategy), as the emerging consensus view is that socioeconomic turbulence is not an aberration, but the new face of normal. What does all that mean for organizational decision-making? The frequency with which decisions need to be made will continue to increase, the time-to-decision will continue to shrink and decision-related informational demands will continue to grow. Extrapolating those trends onto operational practicality, it is difficult to envision how pyramid-like organizational structures can effectively discharge their managerial duties under such conditions, which also implies that it will become progressively more difficult to justify the overhead cost of hierarchical bureaucracy. There is simply too much to be learned in too short a time to entrust only a relatively small subset of organizational stakeholders with essentially all decision-making authority, and to bog down the decision-making processes in bureaucracy. To remain viable, organizations need to flatten their structures, more widely distribute decision-making authority and foster sound organizational learning practices and mechanisms; putting their future in the hands of a small handful of executive managers and directors, who for reasons that defy rational explanation are endowed with the presumption of seeing further than others, is akin to tying retirement to winning lottery.

In a more prescriptive sense, the future of organizations lies in more inclusive and immersive engagement, where the key organizational stakeholders each contribute to the best of their ability and are motivated by meaningful participation in organizational decision-making (and sharing in the fruit of organizational success). The traditional, hierarchical organizational model is a throwback to the industrial society, where there were 'thinkers' and 'makers', and where the former set goals and laid out plans and the latter simply executed. It is also rooted in myopic, short-term (as reflected in the still prevalent emphasis on quarterly earnings), self-serving philosophy of maximizing shareholder value. In post-industrial society, technology is rapidly replacing the function of 'makers', and organizations are awash in vast quantities of data which, at least in theory, can be used by all organizational stakeholders to contribute to creation of economic and social value, as organizations aim to maximize broader and longer-term social benefits.

The preceding reasoning is not meant to suggest that organizations should become rudderless ships, where everyone does everything and thus no one has the specific responsibility of steering. To start with, the goal of large-scale information gathering and processing is to reduce or altogether eliminate tedious, non-thinking and non-creative tasks, and by doing so to free up time and resources for a more constructive use. Recall the earlier (Chap. 1) drawn distinction among three distinct manifestations of organizational learning: adaptive, which supports mostly automatic changes needed to accommodate environmental changes, generative, which produces and disseminates new means of attaining organizational goals, and transformative, which enables strategic shifts as a way of responding to perceived opportunities or threats. Those learning dimensions can be arranged on a continuum ranging from tactical to strategic, with adaptive being highly tactical, transformative highly strategic and generative falling somewhere in between; an organization could flatten its structure, become highly adaptive and responsive and maintain the still important elements of role assignment (e.g. objective setting and planning vs. operational execution) by using those three broad learning modalities as the basis for its organizational structure. Here, frontline employees, suppliers and other stakeholders who are most involved or familiar with day-to-day operations would focus on adaptive learning, which means deep immersion with transactional and customer/client data; they would then be entrusted with operational decisions. At the other end of the spectrum, strategic planning-focused stakeholders would engage in ongoing, deep immersion with organization specific as well as external to organization trends, patterns and projections as the key element of their ongoing transformative learning; they would then be entrusted with setting organizational objectives and mapping out general plans for reaching those goals. And lastly, the enablers, a broad swath of stakeholders that provide support to both the customer/client facing stake-holders and those engaged in strategic planning, would themselves be focused on generative learning as a source of new ideas continuously improve the efficacy and efficiency of the organization's tactical and strategic decision-making.

Chapter 3
Organizational Dynamics and Individual Cognition

3.1 Informational Overabundance

The organizational learning typology outlined in Chap. 1 illustrated the distinct modalities of knowledge creation, in addition to which – somewhat more tacitly – it also hinted at an ever-expanding informational richness, even potential informational overabundance. The problem of 'too much' can become particularly pronounced at a decision point, where it can manifest itself in two distinct challenges: informational volume-related overload, and informational diversity-related ambiguity, typically brought about by the interplay between conflicting decision inputs. Considering that organizations are, at their core, human collectives pursuing shared goals, building of operationally meaningful understanding of the mechanics of organizational knowledge creation and utilization demands an in-depth examination of individual-level and group-wide facilitators and inhibitors of information assimilation.

A natural point of departure is proper framing of the general notion of *learning*. First of all, learning is typically seen as an individual-level quality or attribute; hence, the 'standard' characterization of that quality is framed in the context of an individual (implicitly a person in this analysis, but in a more general sense, a living organism). Within that broad context, learning is commonly conceptualized as the process of adding new or modifying already existing, in one's memory, information. It should be noted, however, that the characterization of 'adding to the already existing knowledge' should not be taken to mean that new information is simply layered atop of old information; instead, newly acquired information is dynamically integrated into a complex web of existing knowledge (typically during rest, hence the importance of sleep to learning) through involuntary, ongoing brain self-rewiring process known as brain plasticity. Consequently, it is more appropriate to describe individual-level learning as an ongoing expansion and evolution of one's knowledge base; while worthwhile in its own right, such redefinition of the general conception

© The Author(s), under exclusive license to Springer Nature
Switzerland AG 2021
A. Banasiewicz, *Organizational Learning in the Age of Data*, EAI/Springer
Innovations in Communication and Computing,
https://doi.org/10.1007/978-3-030-74866-1_3

of learning has important implications for organizational learning, which will be explored later in this chapter.

Looking beyond the general conception of learning, individual-level acquisition of knowledge can take one of two forms: experiential or theoretical (see Fig. 1.4 in Chap. 1). The former manifests itself in immersion in, or observation of, processes or phenomena; subjective and situational, it is tied to sensory experiences, particularly those obtained by means of direct observation or hands-on participation. Given its reliance on subjective interpretation of observed reality, experiential learning tends to be highly personalized, from which it follows that manifestly similar experiences can result in noticeably different knowledge. That is generally not the case with theoretical learning, which entails acquisition of individual perspective-transcending innate ideas. In contrast to subjective experiences, acquisition of theoretical knowledge is focused on assimilation of universal truths, perhaps best exemplified by rules and axioms of mathematics, and thus can be expected to, at least notionally, result in essentially the same knowledge across individuals.

As hinted above, the universality of principles (of mathematics or any other domain of knowledge) should not be confused with individual-level understanding of those principles. Just as like experiences can yield somewhat individually differentiated experiential knowledge, exposure to universal truths can result in varying degrees of comprehension of those principles. Stated differently, as readily evidenced by seemingly endless examples of formal learning outcome assessment variability, individual-level differences in learning capacity and motivation persistently lead to interpersonal variability in theoretical knowledge. At the core of much debated individual differences in ability and motivation is the comparatively rarely examined aspect of cognitive functioning known as *thinking*. Broadly conceptualized as a mental, typically purpose-driven manipulation of ideas and associations, thinking is as essential to theoretical learning as computing devices are to technology-based learning. Moreover, it is an inescapably individual characteristic that impacts theoretical learning in a manner that is evocative of the way in which individual perception affects experiential learning.

When considered from the typological perspective, thinking can fall into one of two types: critical or creative. Critical thinking is commonly framed as a reasoned, structured and reflective problem-solving process, while creative thinking is usually characterized as unconstrained, unstructured and even unorthodox manipulation of ideas. In a very general sense, the former produces carefully reasoned conclusions by means of objectively discernible logic, while the latter tends to 'flash' inspired ideas in a manner that often transcends coherent explanation. Implied in those brief descriptions is noteworthy difference in *metacognition*, or awareness of one's thought processes. Under most circumstances, critical thinking entails greater degrees of metacognition which translates into a greater ease of externalization and thus sharing of the resultant insights, which has important implications for group learning.

When the idea of externalization of individually conceived insights is considered in an organizational setting, it becomes important to account for the potential impact of group dynamics. In particular, the degree of organizational involvement, framed here as cognitive immersion in organizational functioning, comes to the forefront. According to a 2016 Gallup poll, only about 13% of employees around the world perform their work enthusiastically, which suggests systemic organizational involvement problems. That conclusion seems to be tacitly supported by the widespread and enduring reliance on all manners of quick fix management solutions, ranging from flawed anecdotes to one-size-fits-all management frameworks. It seems plausible that disengaged employees who just want to 'get it done' are prone to embracing ready-to-use truisms as easy pathways to, well, just getting it done. And if that line of reasoning holds up to scrutiny, then organizations need to, somehow, make work more personally relevant, more interesting, in order to foster culture of persistent organizational problem solving.

However, there are sound reasons to pause before embracing such a sweeping conclusion. Top among those reasons is that while the employee engagement line of reasoning might be compelling, stepping right into such a seemingly obvious conclusion is reminiscent of the very traps that compel otherwise capable professionals to put their faith into empirically unsubstantiated truisms. A couple of top-of-mind considerations may help to illuminate that point: First, the notion of employees 'performing work enthusiastically' is highly conjectural as it purports to capture an inherently elusive aspect of organizational behaviour; it is easy to imagine how different conceptualizations and operationalizations of that notion can yield starkly different conclusions. In short, one may find exactly what one is looking for, not necessarily because it is there in an objectively verifiable sense, but because of subjective definitional and measurement choices (a tendency sometimes characterized as theory-laden research). Second, aggregate metrics, as exemplified by the 13% work engagement estimate, tend to mask important nuances such as work type (e.g. professional vs. staff) or organization type (e.g. for-profit vs. non-profit), not to mention the potentially critical impact of organizational dynamics in the form of organizational structure and/or culture. In other words, reliance on global generalizations poses a considerable risk of grossly misrepresenting individual contexts. Taken together, just those two top-of-mind considerations suggest that organizational managers would be well advised to not jump to the conclusion that the vast majority of employees are disinterested in their work, but might instead consider investing more thought and diligence into assessing their specific workforce's depth of organizational thinking involvement. Doing so, however, demands careful examination of the very essence of organizational learning-related functioning.

3.2   Institutional Influences

Although commercial and non-commercial organizations can vary considerably across a wide array of factors ranging from the overall mission to operational specifics, there are a handful of characteristics that are exhibited for all formal organizations. All need to choose an organizational form, such as for-profit corporation or non-profit 501(c), in addition to settling on a defined structure that spells out roles, responsibilities and rules of conduct, as a set of mechanisms supporting and coordinating stakeholder activities aimed at accomplishing organizational goals. In addition to those structural choices, all organizations also embrace, not always in a fully considered manner, somewhat less tangible 'ways of being', typically framed as organizational culture and dynamics.

3.2.1 Organizational Form

Chapter 2 offered a broad overview of key differences between commercial and non-commercial organizations, in addition to also exploring the core aspects of organizations' legal (e.g. corporation vs. partnership vs. limited liability company, or non-profit vs. not-for-profit) and organizational (centralized vs. decentralized) characteristics, all of which contributed to a broadly scoped review of dominant organizational forms. There is, however, more to consider in order to develop a better appreciation of organizational form, especially within the realm of business entities, which can be also characterized in terms of focus or size. More specifically, a business entity can be characterized as either the so-called pure play firm, if it focuses on specific types of products or services, or a conglomerate, if it is comprised of what could be seen as multiple pure play units bound together under the umbrella of a singular legal and financial entity. Size-wise, business organizations are commonly ranked based on either revenue, which is the basis for the well-known Fortune Magazine's ranking of companies (e.g. the Fortune 500), or market capitalization, or the aggregate value of a publicly traded company's outstanding shares; as it regards to the latter, one commonly used categorization schema groups business companies into nano, micro, small, mid, large and mega market capitalization segments.

It is a widely held belief that management of a large or mega market capitalization conglomerates presents noticeably different challenges than management of small pure play companies. In fact, even ardent advocates of flatarchies acknowledge the practical implausibility of effectively managing far-flung conglomerate holacracies; although probably to a lesser extent, some of the same challenges are also seen as being true of large non-commercial organizations. That line of reasoning, however, implicitly links the idea of 'effective management' with the notion of 'control', which harkens back to the more than a century-old practice of scientific management pioneered by Frederick Taylor, and the even older Max Weber's conception of bureaucracy. Built around a distinction between knowledgeable managers and (comparatively) less knowledgeable workers, so-framed idea of effective management was rooted in the need for control because control was seen as the core enabler of the ability of knowledgeable taskmasters (managers) to ensure that the less knowledgeable workers structured their work in output-maximizing manner. That Industrial Era reasoning is, however, largely at odds with the emerging reality of data-enabled creativity (discussed at length in later chapters). No longer just cogs in giant industrial wheels, modern-day information workers are often as knowledgeable, and at times even more knowledgeable than their managers, which is reflected in the now-common managerial self-characterization as enablers and facilitators rather than task-masters. Taking that line of reasoning a step further, the contemporary framing of effective management ought to be rooted in the ability to provide clearly spelled out objectives, supportive and rich (organizational) informational environment, and facilitative assistance in overcoming situational and other obstacles. The ceaseless march of automation will continue to claim more and more Industrial Era-like production, which in turn will continue to reshape the essence of organizational members' contributions to the value creation process – the resultant fundamentally changing nature of work has direct and material impact on the rationale that underpins how organizations structure and organize themselves.

In a conceptual sense, the idea of organizational form is a summary concept that captures the core organizational attributes in a way that supports meaningful mapping that gives shape to organizational distinctiveness. Commercial and non-commercial entities tend to present themselves to their constituents as singular wholes, even though some are in fact clusters of otherwise distinct entities held together by the glue of an overarching legal and control structure; while such organizational packaging makes good sense from the perspective of the investing public or regulatory oversight, it is inappropriate when considered from the standpoint of organizational learning. To meet the earlier noted challenge of offering supportive and rich informational environment, multiunit (e.g. business companies comprised of distinct SBUs) organizations ought to be seen as sets of effectively standalone entities.

3.2.1.1 Policies and Procedures

While organizational form offers the structural shell, or more specifically the legal and economic framework, explicit policies and procedures spell out the character of organizational functioning. To a large degree, policies and procedures are applications of various form-implied organizational design choices, thus are typically best understood in the context of formal–informal continuum. On the one hand, formal policies and procedures are binding rules, typically rooted either in externally defined laws and regulations, perhaps best exemplified by the US employment non-discrimination laws, or internal rules, such as organization-specific conflict of interest provisions. On the other hand, informal policies and procedures tend to encompass a broader, possibly more tacitly framed norms and best practices, many of which can be seen as manifestations of organizational living, often characterized as organizational culture.

The generally easy to discern formal policies and procedures and the somewhat more elusive informal rules of conduct shape organizational behaviour in a way that can be facilitative or impeding of organizational learning. It is in that broad context that the command-and-control features of centralized hierarchical organizations are on full display, and it is also the context in which organizational ambiguities of non-hierarchical organizations also become evident. Not surprisingly, careful delineation of both formal and informal policies and regulations is an important contributor to crafting an effective organizational learning strategy.

3.2.2 Group Dynamics

Irving Janis, a psychologist, coined the notion of *groupthink* in the early 1970s in an effort to put a face on what he saw as a group decision-making impairment stemming from group cohesion. Now generally considered a manifestation of undesirable facet of group dynamics, groupthink can be characterized as a dysfunctional pattern of thought and interaction exhibiting itself in closed-mindedness, uniformity expectations and biased use of information. It can ultimately lead to coerced adherence to the group's consensus view, effectively muting unique insights of potentially uniquely insightful individuals.

While there are numerous root causes of groupthink, and more broadly framed deleterious group dynamics in general, two of the more visible and persistent contributors are organizational structure and organizational culture, both of which were discussed at length in the previous chapter. The former encapsulates the arrangement of authority, communications, rights and duties of distinct organizational stakeholders, determining how the roles, power and responsibilities are assigned, controlled and coordinated, and how information flows between the different organizational decision-making units; those arrangements can be described as either hierarchical, also known as tall (centralized organizations), or non-hierarchical, or commonly known as flat (decentralized organizations). In contrast to the relatively easily discernible structure, organizational culture is considerably more elusive. It encapsulates the often hard to put into words ways of organizational living that are shared and maintained, and as such tend to reflect values and priorities that are more enduring than those embodied in organizational designs; it is often thought of as an organization's 'personality', which means that while important to organizational stability, it can also contribute to excessive philosophical assertiveness that can impede adaptiveness and innovation.

While in principle groupthink can arise in just about any organizational context because it is an attribute of an individual group rather than an entire organization, a combination of rigidly hierarchical structure and assertive culture greatly increases both the likelihood as well as the severity of that adverse dynamic by tacitly enforcing adherence to the 'accepted view'. Still, it is important to emphasize that conformance to established norms and standards is not inherently problematic, just as non-conformance is not inherently desirable. While in some ways a traditionally hierarchical organization, 3M persistently stuck to its 'culture of innovation' for about a century now, and the results are nothing short of enviable.[1] At the same time, there are ample instances where non-conformist thinking can lead to demonstrably bad outcomes, as so visibly illustrated by multibillion-dollar fines levied on financial sector firms by regulators as a result of some of those firms' 'maverick thinking'. It follows that there is more to consider than just the interaction between the organization's structure and its culture – other organizational characteristics also need to be taken into account to better understand the very essence of organizational functioning.

3.2.2.1 Organizational DNA

Just like the biological notion of ecosystem has gained widespread usage as a way of capturing the totality of organizational constituents, another biological concept, DNA (deoxyribonucleic acid), which encapsulates the fundamental and distinct physical characteristics of nearly all living organisms, has also gained popularity as a way of capturing the essence of organizations. Given the general familiarity with the concept of DNA, the use of the DNA metaphor in an organizational context offers the means of capturing and communicating complexities of organizational 'personalities' that might otherwise be difficult to succinctly and meaningfully summarized.

To start with, the interaction between organizational culture and its structure, coupled with reflexivity of organizational members determine the shape of key organizational functions, or the manner in which organizations interact with internal and external stakeholders. Under some circumstances, structure may influence culture – for instance, the embrace of hierarchical structures and thus concentrating the bulk of the decision-making power in the hands of relatively few individuals can be expected to give rise to the so-called 'rank and file' culture, often characterized by low levels of cognitive diversity and generally limited innovation and creative problem-solving initiative. When put in the context of the three dimensions of organizational learning – adaptive, generative and transformative – rigid bureaucracies bolster adaptive learning in the form of reactive responses to internal pressures, while at the same time they structurally discourage generative (acquisition of new skills and knowledge) and transformative (strategic shifts in skills and knowledge) forms of learning. In short, command-and-control-oriented organizational structures tend to inspire cultures of mechanistic responses, ripe for groupthink-type dysfunctions and inimical towards higher-order, critical thinking.

It would, of course, be a mistake to paint hierarchical, centralized organizations as inherently flawed. As persuasively argued by Max Weber more than a century ago, bureaucracy is a highly efficient and effective way of organizing human activity, and so for a large, established organization that is more concerned with

---

[1] 3M is widely regarded as one of the most innovative companies in the world, employing 93,000 people globally and boasting a staggering array of more than 60,000 products used in homes, business, schools, hospitals and countless other settings.

efficiency and control than organic creativity, it could be seen as a rational structure. That conclusion, however, rests on some non-trivial assumptions, most notably, that organizational employees, and other stakeholders who contribute to the organization's mission, are sufficiently motivated, and their contributions can be measured with adequate precision. But as argued through this book, the nature of work is changing, primarily due to automation and digitization, as are many of the core societal and demographic factors, as illustrated by the comparatively non-traditional work-related attitudes of the currently rising Generation Z (those born between 1995 and 2010). Those changing societal dynamics, most notably work-related values, needs and expectations, coupled with other transformative sociotechnological shifts have profound consequences on the hallmark characteristics of hierarchical bureaucracies: control and efficiency. When workers of the past were tasked with producing widgets, either manually or using mechanical tools, and those workers' incentives centred on economics, primarily in the form of wages and job security, 'control' and 'efficiency' had a straightforward meaning – to meet the stated production goals. But when widgets are being produced using automated, largely self-functioning systems, and workers are charged with assuring effective operation of those systems, the essence of what constitutes control and efficiency becomes considerably less obvious. And those notions become even less clear when more complex (than basic economic needs) motivational factors, such as attainment and maintenance of work–life balance or contributing to a meaningful, greater social goal, need to be taken into account. When the machine-like Weberian vision of an effective and efficient organization is mostly 'staffed' (in the production sense) by, well, machines, is there still a need for command-and-control bureaucracies, or do such bureaucracies become redundant if not outright counterproductive?

As the nature of how organizations function, in the sense of pursuing their stated goals, and the values, needs and expectations of those who contribute to achievement of those goals continue to evolve, it becomes more and more pressing to view organizations through the lens of living organism-like organizational DNA. When considered as standalone attributes, organizational structure, culture, broadly defined informational capabilities (e.g. performance measurement and tracking, knowledge transfer and utilization, etc.), and motivations of core organizational stakeholders, most notably, employees, shareholders and/or donors, regulators are singularly not sufficiently informative. In other words, seeing an organization as being, for instance, hierarchical or flat, or characterized by organic or synthetic culture may say little about the overall organizational functioning because it is the interaction among distinct organizational attributes that determines the essence of organizations. But being able to decode organizational DNA requires not just a sense of the interaction between the core manifestations of group dynamics, such as structure culture; it also calls for a summary assessment of numerous individual-level characteristics.

3.3 Individual-Level Factors

Framed as mental and purposeful manipulation of ideas and associations taking place along critical and creative dimensions, individual-level *thinking* can be seen as a key contributor to aggregate organizational cognitive functioning. Another distinct but closely related cognitive functioning contributor is *reasoning*, broadly defined as the construction of arguments to derive and justify conclusions. Paralleling the critical-creative thinking duality, reasoning can use either deductive or inductive logic, where the former applies general principles to specific instances, and the latter aims to draw generalizable conclusions from specific instances. It is important to note that the term 'logic' is used here to denote a process rather than an implied alignment with the critical dimension of thinking. In fact, at its core, reasoning actually entails drawing of inferences using either critical (logic) or creative (imagination) thinking processes – as noted by R. Gerard Ward, 'reason can answer questions, but imagination has to ask them'.

When considered in the context of earlier discussed group dynamics, individual thinking and reasoning gives rise to *social cognition*, a manner in which individuals process and use information about others in group contexts. More specifically, it encompasses the totality of mental processes related to the perception, understanding and implementation of linguistic, auditory, visual and physical cues that communicate emotional and interpersonal information. Typically, individuals tend to be consciously aware of some of those impressions or representations, but the prevailing view among social psychologists is that the bulk of social cognition-related processes are subconscious. That means, for instance, that differences between one person's perspective and that of others are usually explicitly discernible to him or her, but when he or she engages in group interactions, they often assess others' perspectives implicitly, which can lead to misinterpretations. It is important to note that, especially in group organizational settings, such as a meeting in which important matters are discussed, 'how' a perspective is communicated, in the sense of non-verbal cues such as facial expressions or the tone of voice, can play nearly as important a role as 'what' is being communicated, in the sense of expressly stated perspectives.

The aggregate organizational learning process can thus be likened to an orchestra, where multiple instrumentalists contribute, in accordance with their instrument type and role, to produce the desired musical outcome. To appreciate how individual-level thinking, reasoning and social cognitive processes contribute to the aggregate organizational know-how requires taking a closer look at the mechanics of how learning occurs at an individual level, and how cognitive, behavioural and emotive factors shape the learning processes and outcomes.

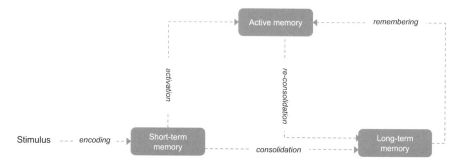

Fig. 3.1 The mechanics of learning and remembering. (Reprinted with permission from Evidence-Based Decision-Making, Banasiewicz, A., Routledge, 2019)

3.3.1 The Mechanics of Learning and Remembering

As currently understood, memories are formed by developing new neuronal connections, known as synapses, through a general process that encompasses encoding, consolidation, storage and recall, commonly referred to as remembering; Fig. 3.1 captures the overall process.

As graphically depicted above, a stimulus is first encoded in short-term memory, into which sensory information enters as either visual or auditory (the 'short' in short-term memory is indeed brief; on average, visual memories only last about 1 second, while auditory memories last about 4–5 seconds). The ensuring learning process begins with the formation of new neuronal connections, followed by consolidation, where preformed remembrances are strengthened and then stored in specific part of the brain as long-term memories. Subsequent retrieval of earlier stored information from long-term to active memory results in reconsolidation, or strengthening, which is commonly known as remembering.

In a more abstract sense, learning can be characterized as modifying information already stored in memory based on new input or experiences. It involves sensory input to the brain (which occurs automatically), followed by extraction of meaning from sensory input which is then placed into short-term, or temporary, memory, from which it may, if reinforced, be transferred into long-term, or permanent memory. An important aspect of that process is its fluidity, which is a key difference between organic, or human, and artificial, or machine-based storage and retrieval systems. Each subsequent experience cyclically prompts the brain to (subconsciously) reorganize stored information, effectively reconstituting its contents through a repetitive updating procedure known as *brain plasticity* (that involuntary process is why eyewitness accounts become less and less reliable as time passes). That brain self-rewiring is generally viewed as advantageous since improvements are made repeatedly to existing information, but it can have adverse consequences as well, most notably when memories of the past rather than being maintained and protected are amended, sometimes beyond recognition. Hence as noted earlier, the widely used characterization of learning as the 'acquisition of knowledge' oversimplifies what actually happens when new information is added into the existing informational mix – rather than being 'filed away' and stored in isolation, any newly acquired information is instead integrated into a complex web of existing knowledge, resulting not only in 'more' but also somewhat 'different' knowledge base.

3.3.1.1 Mental Models

One of the more ethereal aspects of thinking, as in sensemaking, are subjective prior conceptions, often referred to as *mental models*, which are abstract representations of the real word that aid the key aspects of cognitive functioning, particularly reasoning about problems or situations not directly encountered. As is the case with all individual perception and experience-based generalizations, mental models are subject to flawed inferences and bias, and, as implied by their name, are not objectively discernible or reviewable. That said, while mental models as such are hard to ascertain, their impact on sensemaking can be estimated using the combination of Bayesian and frequentist probabilistic inference (this is one instance where the two, otherwise competing estimation approaches can work collaboratively). Starting from the premise that the former, which aims to interpret objective evidence through the lens of subjective prior beliefs, can be used as an expression of mental model-influenced inference, whereas the latter, which sees probability as an algorithmic extrapolation of the frequency of past events, can serve as an objective baseline, a generalizable subjectivity estimate can be derived. Given that, it is possible to estimate the extent of subjectivity of individual mental model representations by relating individual perspectives (Bayesian probability) to a baseline represented by a group-wide perspective composite (frequentist probability). The resultant pooled individual vs. baseline perspective variability can be interpreted as the aggregate measure of perspective divergence, summarized here as the *cognitive diversity quotient (CDQ)*. Computation-wise, leveraging the widely accepted computational logic of statistical variance, CDQ can be estimated as follows:

$$CDQ = \sqrt{\frac{\sum_{i=1}^{} \left(Perspective_i - Baseline\right)^2}{Number\ of\ Perspectives}}$$

Outcome-wise, the cognitive diversity quotient is a standardized coefficient, which means its magnitude can be compared across different group opinion situations. Therefore in situations that lend themselves to such analyses, the use of CDQ can infuse elements of analytic logic into the often ill-structured mechanics of organizational consensus building. Moreover, CDQ can also provide important insights when used longitudinally, meaning it can be used to track and evaluate changes in situation- and decision-point-specific consensus across time. Also worth mentioning is that the use of such an objective assessment method with an inherently individual perspective-ridden evaluations of problems or a situations can help with addressing the seemingly inescapable polarity of organizational diversity. More specifically, it is fairly common to come across cherished organizational attributes that at the same time can also act as obstacles to efficient organizational decision-making; colloquially stated, even good things may need to be managed to continue to produce good outcomes. And lastly, being able to address perspective differences in what could be seen as a more democratic manner (in computing of the CDQ score, all opinions are treated the same) may diminish the threat of groupthink, and by doing so, it may foster greater cognitive diversity.

3.3.2   Cognitive, Behavioural and Emotive Factors

When massive open online courses (MOOCs), self-paced, pre-recorded learning modules aimed at unlimited participation and open access via the web, were first introduced around 2006 and began to quickly garner widespread interest, many in higher education circles became worried about the future of traditional, face-to-face educational institutions. Nearly decade and a half later, MOOCs are certainly still a part of the higher education landscape but are no longer seen as a threat by traditional universities, many of which developed their own, albeit typically limited MOOC offerings. What changed? While numerous potential explanations can be cited here, the one that is of particular interest from the standpoint of organizational learning is that, as suggested by research from social psychology, people simply prefer learning from other people, especially those they deem knowledgeable and trustworthy.

One of the key reasons for that is trust, or more specifically dyadic trust, as in a student trusting the teacher, which is a critical antecedent of effective knowledge transfer. Delving a bit deeper into those abstract considerations, group learning situations characterized by high trust and emotionally positive context tend to produce better learning outcomes, as the initially individual cognitions ('I think that…') are slowly transformed into shared cognitions ('We think that…'), ultimately giving rise to shared institutionalized beliefs ('It is true that…'). That mechanism is particularly important to effective transfer of inherently individualized experiential knowledge because that is the very aspect of organizational know-how that tends to be lost when organizational thought and practice leaders are no longer a part of the organization. It is important, however, for experiential knowledge transfer to transcend the typical person-to-person model, as in experienced professionals mentoring junior associates, because that it is unsystematic and thus yields highly variable outcomes (for reasons as basic as personal style similarity or lack thereof, some such partnerships can be highly productive, while others not at all). Instead, the individual thought and practice leaders' experiential knowledge needs to be institutionalized as codified shared beliefs. When that happens, the enhanced know-how becomes an organizational asset, and as such a contributor to the entity's competitive advantage.

The above ideas are not new – numerous organizations used that general line of reasoning as a mean of capturing and popularizing their organizational philosophies; the Toyota Way offers perhaps the best-known example. The Toyota Motor Corporation first captured and detailed the philosophy, values and ideals behind their management and production systems in 2001, calling it 'The Toyota Way 2001'. Those, however, are organization-wide, generally framed ideas, meant to be adapted to different organizational contexts, meaning they frame knowledge transfer as a top-down process, as the knowledge flows from organizations to individuals. Institutionalization of individual-level experience-derived expertise is, on the other hand, a bottom-up process, because the knowledge, or more specifically the sharing of knowledge, flows from individuals to organizations.

But putting too much emphasis on 'the way' can have adverse consequences – the earlier discussed groupthink, conformity-motivated but ultimately dysfunctional pattern of group decision-making, is one such danger. In general, the more of 'our way' type shared organizational beliefs there are, the more ever-present the need for conformity, which can pose a direct threat to cognitive diversity, seen here as active presence of differences in perspective or information processing styles. Like a healthy democracy, critical organizational thinking depends on cognitive diversity because 'our way' type shared beliefs reflect that which worked in the past, and the past is rarely, if ever, a perfect predictor of the future. Especially considering that decision-making-related volatility is the new face of normal, sustaining healthy cognitive diversity is critical to organizations' abilities to effectively respond to frequent and often unexpected environmental changes. Moreover, 'our way' compelled need for conformity can also lead to biased information search, in the form of favouring of information that supports the accepted view.

And yet cognitive diversity can also be problematic. The seemingly ubiquitous cognitive bias can take numerous forms,[2] and greater organizational cognitive diversity tends to give rise to a wider range of cognitive bias manifestations. The same cognitive diversity is also behind pluralistic evaluations of empirical research studies' findings, and even the seemingly objective data analytical results. For instance, probability estimation, which is one of the core elements of statistical inference, can be approached from two fundamentally different methodological perspectives. The first one is Bayesian, which treats probability as a manifestation of the state of knowledge known and available to the decision-maker at decision time; as such, it combines subjective prior beliefs (of the decision-maker and/or decision-influencers) with analyses of past outcomes (i.e. objective data). An alternative probability estimation mechanism, the frequentist approach, limits estimation inputs to just objective data collected by means of experiments or recorded historical occurrences, explicitly barring any subjective (e.g. prior beliefs) inputs. Under many, perhaps even most, circumstances, the two approaches will yield numerically different

---

[2] Daniel Kahneman offers a thorough overview of root causes and manifestations of cognitive bias in his recent book, *Thinking, Fast and Slow* (Farrar, Straus and Giroux, New York, 2011).

estimates, giving rise not only to informational ambiguity, but potentially also to group conflict, both of which adversely impact creativity and flexibility of group thinking.

Overcoming the potential cognitive, behavioural and emotive challenges that may arise in a group setting is tantamount to enabling organizational readiness to learn. As groups mature, which is to say they evolve from a simple collection of individuals into more complex, integrated systems, and develop a stronger sense of identity, they also begin to solidify their behaviours, norms and values. Nurturing the desirable aspects of groups, most notably the diversity of perspectives, all while dissuading forced uniformity is not simple, but it is necessary for organizations to learn in a manner that builds organizational understanding and interpretation of their environment.

3.3.3 Learning Modes

It could be argued that learning is an inescapable part of existence. Though it is tempting to think of learning as episodic and purposeful acquisition of specific know-how, it is in fact more appropriate to view it as a more elementary and persistent process. Upon deeper reflection, it becomes obvious that learning encompasses an almost impossibly broad array of activities ranging from basic behavioural adaptations, as exemplified by learning the painful consequences of touching a hot object, to developing an abstract understanding of concept such as the force of gravity, to fostering a deeper sense of morality or exploring ways of making sense of highly emotional experiences. All told, perhaps 'it is human to err' because learning is such an inseparable part of human existence.

When considered in a more typological sense, embedded in the above reasoning is a hint of distinct learning modes (not to be confused with the four distinct learning modalities discussed in Chap. 1). In a general sense, learning modes can be seen as methods or mechanisms by means of which knowledge is assimilated, which implies the existence of distinct levels or tiers. The currently available neuropsychological evidence suggests two such levels, simply referred to as lower-level learning and higher-level learning. The former entails creating of rudimentary associations of behaviours and outcomes, while the latter encompasses abstracting of cognitive understanding of the underlying rules, mechanisms and norms.[3] When considered from that perspective, organizations, as groups of individuals, constantly engage in lower-level learning associated with repetitive and routine aspects of their functioning, but their higher-level learning efforts, typically taking the form of purposeful

---

[3] This distinction is mirrored in the advancing nature of machine learning, which is the development of computer algorithms that are capable of discerning patterns in data, and improving through experience. The original machine learning algorithms, dating back to the 1970s, were geared towards lower-level learning (those algorithms are still widely used today), while the more recent 'deep learning' applications are geared towards higher-level learning.

efforts to develop cognitive understanding of how internal and external forces and development impact what and how they do, tend to be considerably less persistent. Clearly, enhancing the persistence and efficacy of higher-level learning, especially in the age of hypercompetition and environmental volatility, ought to be viewed as one of the key organizational priorities.

3.4   Organizational Sensemaking

Shaped and influenced by the myriad of institutional and individual-level factors, effective organizational learning mechanisms can be seen as practical adaptations of the conjectures and refutations (C&R) model of scientific discovery. Based on the Bayesian idea of updating prior knowledge with current evidence, while also allowing for introduction of new evidence, the C&R model entails factually validating of prior beliefs followed by subsequent reflection which, in a more procedural sense, can then give rise to generalizable insights. Perhaps the most distinguishing characteristic of this approach is its ability to produce high-information metrics in the form of parsimonious representations of phenomena of interest, which translates into interpretational simplicity, a quality that is particularly important in a group setting. Though undeniably compelling as an idea, when examined in the narrower and often more nuanced applied organizational knowledge creation settings, that otherwise sound approach can be derailed by its conceptual and procedural provisions.

Starting from the premise that subjectivity permeates all aspects of individual-level cognitive functioning, including the seemingly objective rational thinking in addition to highly individualized mental models, all prior beliefs are inescapably individualized, which means that conjectures (formally referred to as hypotheses) are subjectively framed. Moreover, in practical sense more importantly, the degree of subjectivity, or the extent to which the said conjecture framing may be atypical, is difficult to anticipate or even fully appreciate.[4] In fact, that is why results of behavioural studies are notoriously hard to replicate: divergence of perspectives, which encompass how research problems are framed as well as how data are collected, manipulated and analysed, can ultimately render replication efforts non-productive. In the end, overtly similar, even like-sounding, conjectures can ultimately produce materially different knowledge, depending on the 'who' and the 'how' of ensuing analyses. In short, when inescapably individualized conjectures are subjected to seemingly objective refutation, the resultant knowledge can be affected not just by the initial framing of 'what' knowledge is sought, but also by

---

[4]This line of reasoning applies to the sociobehavioural realm of organizational decision-making and other contexts that fall under the broad umbrella of 'social' sciences; it does not extend to natural sciences such as biology or chemistry. Moreover, while the earlier discussed *cognitive diversity quotient* (CDQ) can help with assessing the extent of subjectivity, it does not solve the bigger problem of inherent subjectivity of conjecture framing.

'how' the knowledge creation is conducted, which underscores the importance of better understanding of the underlying mechanics.

3.4.1 Searching for What Matters

The essence of validating of initial conjectures is to correctly differentiate between real and spurious prior beliefs. Abstracting away from numerous underlying methodological considerations, while there are established means of doing so in the form of statistical significance tests, the task of separating informative from non-informative findings is also permeated by subjectivity and by chance. To start, the overtly objective tests of difference (i.e. statistical tools used to differentiate between true and false prior beliefs), namely the F, t, and $\chi2$ statistics, are unduly influenced by sample size. More specifically, the likelihood of finding a 'true', in the statistical sense, prior belief increases as the number of observations used in the statistical testing process increases, everything else remaining the same, which effectively can render even inordinately tenuous knowledge claims 'true' (statistically significant), if the underlying sample size is large enough. And since statistical significance tests are pass/fail, all differences that pass a given threshold, such as the commonly used 0.05 level of significance, are deemed to be equally true. To be clear, that means that a conjecture that is deemed 'false' (not statistically significant) when the underlying sample size is modest, such as for instance 500 records, can become 'true' (statistically significant) when the underlying sample size increases to something like 5000 records, everything else remaining unchanged. What happens when subjectively framed prior beliefs are tested using the enormous datasets that are now commonplace? Even inordinately tenuous conjectures can be validated.

Faced with those challenges, analysts grew accustomed to drawing a line of demarcation between 'statistical' and 'practical' significance, by judging magnitudinally larger effects to be material and smaller one as unimportant. While such reasoning notionally parallels the logic of relying on effect size rather than statistical significance (an idea discussed in more depth in Chaps. 5 and 8), what constitutes 'large' vs. 'small' is, again, subject to individual interpretation. And so while the desire to differentiate between visibly dubious and possibly robust 'truths' is understandable, such subjective second-guessing of objective data analytic outcomes flies in the face of utilizing objective tests of statistical significance in the first place. After all, the goal of statistical testing is to diminish the very subjectivity that is being compounded by *post hoc* subjective overriding of outcomes of statistical tests. A more methodologically sound (not to mention more in keeping with the spirit of the C&R model) approach is to make use of *sensitivity analysis* as a result validation tool. The process is straightforward: The first step is to select the level of statistical significance, such as 95%, which is then followed by recomputing of the test statistic using noticeably different sample sizes, while keeping everything else unchanged – obtaining consistently statistically significant results can then be taken as strong evidence of materiality of effects under consideration.

3.4.2   New Knowledge

Awash in data, organizations desperately want to believe in its transformational value, even though many might be more aptly characterized as being information-rich, but informed decision-poor. All considered, it is relatively easy to compress massive quantities of data into an endless flow of reports that summarize just about anything that can be summarized but answer few, if any nagging questions. Implicit in this reasoning is the progression of data→information→knowledge. In the simplest of terms, data can be seen as facts, information as summarized facts and knowledge as decision-guiding insight. As implied in that progression, moving from 'data' to 'information' is largely contingent on being able to compress detailed data, such as individual product purchases, into aggregates, such as monthly product sales. Given the ubiquity of appropriate tools, under most circumstances transforming data into information poses few, if any, challenges. Moreover, it also lends itself to automation, which in addition to further easing the task at hand also has the added benefit of sidelining the numerous individual-level as well as organization-level informational detractors, such as cognitive bias or groupthink effects. That, however, tends not to be the case with knowledge creation, as evidenced by so many organizations being information-rich but (decision-guiding) knowledge-poor.

In contrast to the definitionally straightforward conception of information, the idea of knowledge is quite a bit more complex (see Fig. 1.3 in Chap. 1). When framed in a more organizationally meaningful terms of 'decision-guiding insights', to be meaningful, any knowledge claim needs to be clearly identified as either tacit or explicit, in addition to also being further described as semantic, procedural or episodic. Implicit in that broad characterization is that 'knowing' is individualized, as illustrated by the following simple scenario: A group of university engineering students followed essentially the same progression through their programme; upon graduating, most accepted employment with established companies, while one chose to pursue the development of technology he conceived, which ultimately spawned a new company. This simple case highlights the highly individualized, elusive nature of knowledge, as evidenced by essentially the same informational inputs, in the form of course content, producing materially different outcomes. On the surface, all hypothetical engineering graduates were equally professionally qualified, yet the innovator among them somehow managed to get more out of common-to-all engineering-related knowledge. Why? Likely because of heightened ability to mentally manipulate the acquired knowledge in a unique way, an elusive cognitive process that could be summarized under the broad umbrella of *thinking*. On the one hand, as a conscious process, thinking is intimately familiar to all, but on the other hand, it is one of the hardest to grasp or objectify aspects of cognitive functioning; in the context of organizational learning, thinking can be characterized as the actuating force that turns potential into realized value. Stated differently, it is that something that transforms generic information into value-creating knowledge; in an aggregate sense, it is the difference-maker that helps to explain why organizations that have access to essentially the same information end up on markedly different economic trajectories. Together with existing informational assets and the ongoing process of learning, innovative organizational thinking frames the organizational know-how, as graphically summarized in Fig. 3.2.

But here comes the critical part: Since the onset of modern industrial age, business organizations entrusted theoretical aspects of their workforce training to educational institutions, while the experiential side of learning was essentially left to chance. By and large, the general college-level liberal arts education offers excellent 'learn-to-think' intellectual training, but aspects of professional training offered by institutions of higher learning, most notably business education, have been criticized as being overly theoretical and divorced from practical problems, jargon-laden and generally numb to the needs of business and other organizations. At the same time, it is rarely a stretch to characterize experiential knowledge prized by practitioners as 'catch as catch can', a portrayal readily supported by the often-considerable professional competence variability across individuals with overtly comparable levels of experience. And just as was the case in the earlier example of engineering students, overtly similar qualifications can produce vastly different effective competency, and thus outcomes. All considered, in the era in which creative use of available information and thoughtful fostering of new knowledge are paramount not just to growth but the very survival of organizations, both commercial and non-commercial organizations now must take a more active interest in how their core constituents – their employees – learn to know and learn to think.

Fig. 3.2 The makeup of organizational know-how

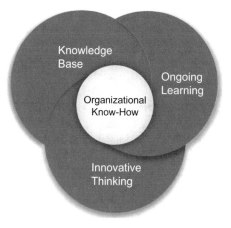

Chapter 4
A Brief History of Learning with Data

4.1 Axial Ages

The quest to understand is one of the most emblematic qualities of mankind. Intuitively, the progressively better ability to unravel the mysteries of the world seems to have evolved in a slow, continuous and nearly monotonic fashion, but closer examination of the history of human sensemaking paints a somewhat different picture. As first discussed in Chap. 1, it might be more appropriate to think of human sensemaking abilities as having advanced by means of distinct leaps, followed by periods of comparative inactivity, a process characterized as *punctuated equilibrium.*[1] Captured by Jasper's notion of the 'axial age', those transformative periods of rapidly advancing cognitive abilities were followed by long periods of comparative lack of development. However, what could be seen as lack of development might be more appropriately portrayed as periods of maturation or germination. Just as seeds need time to grow into plants, those jolts of enlightenment needed time to produce benefits in the form of progress, manifesting itself in gradually deeper contributions to knowledge. After all, it is that cumulative output more than the initial flashes of brilliance that drive scientific and technological advancement.

Figure 4.1 offers a graphical summary of the key axial periods demarking rapid jumps in human sensemaking abilities. The first such major jump took place in a span of about 600 years, starting around 800 BCE and lasting till about 200 BCE, which could be characterized as the period that lit the fire of thinking and cognition.

---

[1] The notion of 'punctuated equilibrium' offers an alternative Darwin's explanation of evolutionary development of species; originally proposed by Stephen J. Gould and Niles Eldredge in 1972, it posits that evolutionary development is marked by isolated episodes of rapid change separated by long periods of little or no change. It is worth noting that the totality of fossil analyses-based evidence appears to lend greater support to punctuated equilibrium than the slow and steady view of evolution posited by Darwinism.

© The Author(s), under exclusive license to Springer Nature
Switzerland AG 2021
A. Banasiewicz, *Organizational Learning in the Age of Data*, EAI/Springer
Innovations in Communication and Computing,
https://doi.org/10.1007/978-3-030-74866-1_4

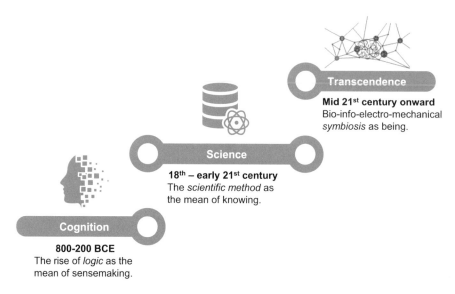

Fig. 4.1 Key cognitive axial periods

It was during that time that the great philosophers of antiquity lived and taught in different parts of the world: Confucius and Lao-tse in China, Buddha in India, and Socrates, Plato and Aristotle in Greece. Their insights can be seen as the first step towards the broader ability to explain and rationalize, rather than fear, the nature. Starting from the premise that human sensemaking is not naturally rational, Confucius, Buddha, Plato and other great early thinkers infused disciplined thinking into collective human sensemaking; their teachings framed the manner in which relevant information in the form of opinions, judgements and facts can be organized and analysed to arrive at a sound conclusion. In other words, while the content of their teachings was certainly important, it was the underlying thought process that provided the much-needed fuel to move mankind forward in terms of cognitive development.

The next 2000 years or so showed comparatively little cognitive sensemaking development; in fact, it could even be argued that the spark of reason began to dim, especially during the period of European history known as the Dark Ages (from the late fifth to the tenth century CE), so named on the account of dogma overtaking reason. And then, for reasons that defy clear explanation, the next axial period began to unfold. The intellectual and creative fog of the Middle Ages (fifth to fourteenth century CE) began to lift giving way to 'rebirth' or Renaissance, a period in European history marked by great social change and the revival of art and literature. The rebirth of interest in arts and literature can be seen as the foundation or a precursor of another broad-based intellectual outburst, this time in the form of the rise of modern science. Known today as the Scientific Revolution, it took hold in Europe in the early modern period, starting in mid-sixteenth century with the publication of Nicolaus Copernicus', a Polish astronomer, *De Revolutionibus*

Orbium Coelestium (*On the Revolutions of the Heavenly Spheres*; as was the norm at that time, the work was originally written in Latin). Effectively sidelining the Hellenic view of the world encapsulated in the belief that human reason is the ultimate source of knowledge, the newly emerging sensemaking modality emphasized objective – meaning, independent of human perception – means of inquiry as the tool of knowledge creation, which ultimately gave rise to science as an autonomous discipline, distinct from philosophy. From Copernicus who famously 'stopped the sun and moved the earth' to Newton who lifted the veil of mystery from the mechanics of the behaviour of inanimate objects, and more recently, from Planck and Bohr who first offered the glimpse of the wondrous, if not outright strange and mindboggling, world of quantum mechanics, to Watson and Crick who shone the light of discovery on the three-dimensional double helix structure of DNA, the Scientific Revolution produced astounding amounts of scientific knowledge and the resultant technological progress.

And now the mankind is once again standing at the precipice of a yet another cognitive axial period, the Transcendence. Though it is too early to do so now, it is likely that the future generations will point to some of our contemporaries as the ones who paved the way of progress in this newly emerging era of technology and data, or more specifically, information technology-enabled cognitive expansion, which will gradually enable mankind to peer into realms that were previously inaccessible to us, and perhaps even answer some of the most existential questions. But in the end, it is not names but ideas that matter, and the unfolding era of Transcendence is about to unleash a completely new ways to learn, made possible by combining of the most human aspects of mankind – the ability to reason, imagine, and feel, in a metaphysical sense – with data-powered technologies that will enable us to transcend our current physical, three-dimensional limitations in search of deeper understanding of the nature of reality, and the essence of our own existence.

Though on the one hand highly reductive (condensing of several thousand years of human civilization to just three incredibly broad periods), the idea of axial age nonetheless sets the stage for a more granular look at the genesis of the contemporary learning with data capabilities. The distinct axial periods of Cognition, Science and Transcendence can be seen as both enablers and consequences of the more tangible socioeconomic progression, one which characterizes both the manner in which knowledge is created as well as the manner in which it is put to use. Manifesting itself in the ongoing evolution of how work is done and lives are lived, that punctuated equilibrium framed progression has profound implications for organizational learning, some of which were implicitly touched on earlier in the context of the multimodal typology of learning with data summarized in Fig. 1.5. But the reasoning encapsulated in the idea of axial periods is highly reductive, in the sense that it only draws attention to the singular and disconnected (as graphically depicted in Fig. 4.1) events. In order to develop a more robust understanding of the currently underway data technology-driven changes, and a deeper appreciation of the gradually emerging bio-info-electro-mechanical symbiosis, it is instructive to examine a more complete progression of the changing nature of work.

4.2   The Changing Nature of Work

Scientific Revolution not only redirected the pursuit of knowledge away from subjective human reason and towards objective inquiry – it also jumpstarted the far more visible Industrial Revolution. Commonly associated with the emergence of steam-powered technologies, Industrial Revolution is often narrowly portrayed as a singular time period demarking the transition from agrarian to industrial societies. However, when the idea of socioeconomic evolution is considered in a wider context of how work is done and lives are lived, the notion of 'punctuated equilibrium' offers an excellent way of capturing the ongoing character of socioeconomic progression. When looked at from that perspective, the latter eighteenth century through early nineteenth century period of rapid industrialization can be seen as the first step in the enduring socioeconomic progression, graphically summarized in Fig. 4.2.

Framed in the broader context of the three distinct axial periods discussed earlier, the more granular depiction of socioeconomic progression can be broken down into three distinct eras: Agrarian, Industrial and Information. When considered from the standpoint of how work is done and lives are lived, Agrarian Era can be seen as a long – stretching as far back as 10,000 years and lasting until the onset of the Industrial Revolution in the late 1700s – largely uniform period organized around producing and maintaining of crops, typically using a combination of hand tools, ploughs and draft animals. When considered from an organizational perspective, agrarian societies were generally structured around relationships that were derived from the holding of land, and religious beliefs were used to justify sharp inequalities, including the unquestioned authority of rulers (e.g. emperors, kings, queens, lords), who were believed to rule by divine power. Not surprisingly, learning was both practical and limited: The ruling classes – most notably, the free citizens of Greece and Rome and later feudal nobility and clergy – tended to acquire knowledge they deemed important to their ability to govern, while working classes acquired what could now be characterized as vocational training, which was quite basic and limited given the simple techniques and tools used during that period.[2] The axial period of Reason discussed earlier was wholly contained in the much broader, time-wise, Agrarian Era.

---

[2] In fact, the present-day notion of 'liberal arts' education emerged in ancient Greek and Roman societies as a manifestation of the different educational needs, and I suppose rights, of the ruling and serving classes. The label itself is derived from the Latin word 'liberalis', or 'appropriate for free men', and it is an expression of the belief that the study of grammar, rhetoric and logic was an essential enabler of free citizens' participation in civic life. That idea was carried forward into the feudal societies that followed in their wake, clergy and nobility (who replaced the free citizens of Greece and Rome as the ruling class) carried forth the core ideals of liberal education, eventually even adding mathematics, music and astronomy to what was considered proper education for the ruling classes. During the same time period, education of first slaves and later commoners was limited to specific, typically servitude related skills, now broadly characterized as vocational training.

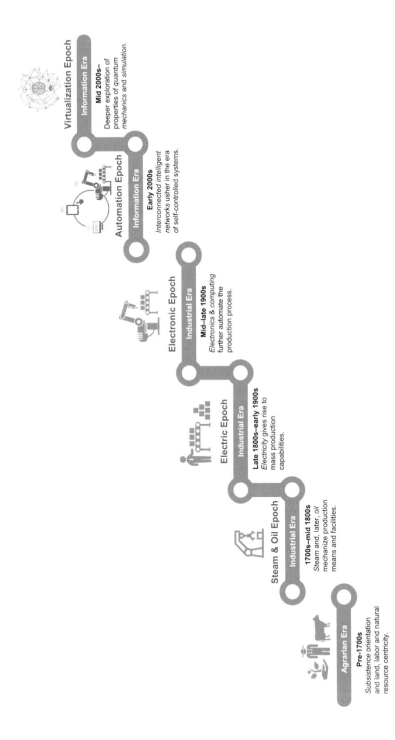

Fig. 4.2 The punctuated equilibrium of socioeconomic progression

The rise of modern science, as encapsulated in the so-named axial period, triggered the emergence and rapid proliferation of initially steam- and later oil-powered mechanization, which effectively brought the Agrarian Era to an end. When considered from the standpoint of how work was done and lives were lived, the post-Agrarian Era period can be more granularly characterized as a comparatively quick succession of three progressively more advanced industrial development phases. Framed in Fig. 4.2 as epochs, each of those three industrial development phases was rooted in a distinct disruptive innovation, and when considered jointly, they comprise the Industrial Era. Focusing on the individual epochs, steam power was quickly replaced by oil, and while it was a technologically significant transition, it nonetheless did not fundamentally alter the core aspects of work and life, a conclusion which suggests a singular Steam and Oil Epoch. Electricity, on the other hand, had a discontinuous impact on how work was done and lives were lived, which implies a distinct, Electric Epoch. And lastly, though rooted in the same source of power (electricity), the Electronic Epoch owes its distinctiveness to another set of transformative, norm changing innovations, most notably transistors and integrated circuits that soon followed, because once again, the resultant rise and proliferation of computing devices and technologies materially changed many key aspects of work and life.

Underpinning the rise of large-scale mechanization, electrification and then electronification has led to an exponential increase in the complexity and sophistication of means of production, which in turn gave rise to the emergence of sprawling industrial organizations relying on trained and skilled classes of workers and professionals. Also born with industrial organizations was the idea of organizational learning, made necessary by the vastly greater (that in the preceding Agrarian Era) operational complexities of those enterprises. As a result of those and other (e.g. political, legal, etc.) changes, the focus, scope and the very essence of learning in general began to change to accommodate the need for a far wider and deeper arrays of required skills and competencies. Perhaps the most visible impact of those transformative changes was on the broadly defined institution of higher learning. Already well-established by then (the oldest still existing university in continuous operation, the University of Bologna, was founded in 1088), institutions of higher learning began to look beyond their liberal arts roots, slowly expanding their educational offerings to include a growing array of specialized, professional education programmes; in a more general sense, the pursuit of knowledge began to take on progressively more occupation-related focus. And the incessant cycle of more knowledge begetting more complex manmade creations which in turn require more knowledge to make and use was started.

One of the defining aspects of the Industrial Era was its focus on hardware. From the early steam-powered engines of the late 1700s to the relatively recent electronic computing systems, the Industrial Era was a period that equated technological progress with physical infrastructure, as a way of surmounting mankind's physical limitations. Steady evolution of manmade technologies', perhaps best exemplified by the emergence of the Internet of Things, began to shift the focus towards the value of intangible information, which gave rise to the currently underway

Information Era, which can be characterized as by mankind's yearning to rise above our cognitive limitations. It began to take shape in the second half of the twentieth century, sparked by the Digital Revolution, which marked a large-scale shift away from analog electronic technologies (i.e. those using continuously variable signal, as in radio waves) to digital electronics (i.e. those using discrete, typically binary signal used in computers). Though tracing its roots to the early commercial computer systems that first began to appear in the 1950s, it was really the arrival of personal computers in the 1980s that transformed electronic computing into everyday utility for just about everyone. The subsequent emergence of the Internet as a new communication modality, soon after enhanced by the utility offered by the World Wide Web interactivity, and further expanded by the rapid maturation of mobile connectivity and the proliferation of interconnected personal and commercial data capturing devices, laid the foundation for the current Information Era. Characterized by intelligent, meaning automated or even autonomous, interconnected networks, the commercial and personal consequences of digital transformation are as monumental as those that characterized the agrarian-to-industrial society transition. Just as steam-, oil- and later electricity-powered machines changed how work was done and lives were lived, self-functioning, interconnected systems – perhaps best illustrated by self-driving vehicles – are now again changing many aspects of work and life, as captured under the Automation Epoch umbrella. But just as steam and oil were but a prelude to Electric and Electronic epochs, the currently underway Automation Epoch is but a preface to the upcoming Virtualization Epoch, the emergence of which can already be seen over the horizon, and which will open brand new avenues of learning (more on that in later chapters).

Perhaps the best way to capture the essentially unbounded and unlimited organizational learning opportunities that are already being spurred by the emerging wave of new digital technologies is by taking a closer look at the domain of augmented and virtual reality. Augmented reality (AR) aims to bridge the gap between the virtual and physical world by adding layers of digital content to physical reality, while virtual reality (VR) is a fully digital experience that can either emulate the physical world or create an altogether artificial experience. It follows that AR is not a fully immersive experience, but an infusion of virtual components into one's physical space, while VR is a fully immersive digital experience.[3] But again, that is just the beginning. Quantum simulators already exist, and though currently those tend to be special purpose devices designed to shed light, no pun intended, on specific problems of physics, the early computers were tools reserved for big science

---

[3] To avoid confusion, this brief overview abstracts away from the three different types of VR, which are fully immersive, semi-immersive and non-immersive. Fully immersive VR is best exemplified by advanced gaming systems, complete with head-mounted displays, headphones, gloves and possible other experience-inducing equipment; semi-immersive VR is best illustrated by professional flight simulators used to train pilots, and perhaps the best example of non-immersive VR is a run-of-the-mill video game. It is worth noting that though some might not think of ubiquitous video games as VR, but a player in a physical space interacting with virtual world certainly falls within the definitional scope of virtual reality.

and big business, though it did not take long for those tools to become commonplace. Moore's law correctly predicted that while the speed and capability of computing devices would double every 2 years, the cost of those devices would be halved during the same timeframe; it seems reasonable to expect that somewhat similar forces will apply to quantum simulations, which will become standard learning tools in not too distant a future.

The possibilities are truly limitless. Recalling the multimodal organizational learning typology outlined in Chap. 1, the two broad meta-categories of reason-based and technology-based learning are each comprised of two more narrowly framed categories of experiential and theoretical (reason-based), and computational and simulational (technology-based) learning. Implicit in that conceptualization is the distinction between physical and virtual realities, but as VR tools and technologies continue to evolve and proliferate, that distinction will likely blur. In fact, there are numerous, well-known scientists and entrepreneurs who promote the so-called 'technological singularity', or simply singularity, hypothesis, which argues that humanity is headed towards the creation of artificial superintelligence in the form of accelerating-self-improvement-capable computers running software-based general artificial intelligence, destined to ultimately far surpass human intelligence. Though at the first glance the idea may seem far-fetched, it is actually quite rational: Since one of the ultimate manifestations of human intelligence is the creation of artificial intelligence, an ultra-intelligent system initially designed by mankind can be expected to continue to improve upon its initial design, ultimately leaving its creator far behind. A perhaps less far-fetched variant of that line of reasoning argues that the seemingly inescapable conclusion to the accelerating technological progress is the emergence of the Internet-based shared cognition. More specifically, within the next couple of decades, humans will be able to connect our neocortex, or the part of brain responsible for thinking, to the Internet; though it might be tempting to also dismiss that idea as a sci-fi fantasy, it might be worth considering that the now ubiquitous addiction to all-things Internet might have seem just as sci-fi-ish a couple of decades ago.

4.2.1 Learning Foci

While the arguments that underpin the idea of punctuated equilibrium of socioeconomic progression are, hopefully, interesting, they are also suggestive of a gradual but persistent shift in the focus of learning and knowledge creation, graphically summarized in Fig. 4.3. When considered from that perspective, in the Age of Reason, the focus of learning and knowing was on governing, because by and large, formal education was limited to the ruling classes. The emergence of industrialized societies, however, democratized access to knowledge because it gave rise to a new professional class of skill and knowledge-based workers, which effectively reoriented the acquisition of knowledge towards industrial production-related endeavours. Now, technological progress is gradually alleviating direct

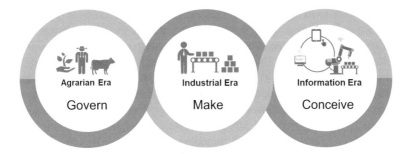

Fig. 4.3   The shifting focus of learning

physical work while at the same time generating vast volumes and rich varieties of data, and the combination of those two related but distinct trends is ushering in the age of *data-enabled creativity*. New knowledge is created, and ideas are tested not just by means of traditional analyses of data, but also through augmented and virtual reality technologies that enable 'experiencing', in virtual sense, reality in a way that transcends physical limits. Hence in contrast to 'traditional' creativity, which is rooted in individual-level reasoning and imagination and thus is constrained by one's experiences and mental resources, data-enabled creativity breaks down those boundaries. It allows one to step outside of one's physical and, when bio-info-electro-mechanical symbiosis is factored in, even cognitive limits, ultimately tapping into quite possibly unlimited realm of virtual experiences.

More specifically, freed from limits imposed by physical experience and organic imagination, human brain can become infinitely more productive because of its enormous computational potential that emanates from its innerworkings that are both analog and digital-like. Neurons, the building blocks of human brain, are analog, meaning they operate using an ever-changing continuum of signals, but their interactions are binary, and thus digital-like, since neurons exchange signals by firing or not. Moreover, unlike digital computers, the brain does not use binary logic or perform binary arithmetic operations; instead, stored information is represented in terms of statistical approximations and estimations rather than exact values, meaning that the brain is ultimately non-deterministic, but instead it is permissive of practically unbounded number of informational possibilities. Stated differently, given its probabilistic, quantum state-like[4] rudimentary information storage and processing paradigm, when provided with richer – i.e. simulated – experiences, human brain's capacity to conceive can grow exponentially. More specifically, making use of augmented, mixed and virtual reality technologies will allow learners to see beyond the boundaries of currently existing reality, which in turn will make it more likely for more learners to conceive novel ideas that may not have emerged in a more traditional learning setting.

---

[4]A key element of quantum theory, which offers a theoretical explanation of the nature and behaviour of matter and energy at subatomic level, quantum state is a mathematical entity described by the underlying probability distribution of outcomes of each possible measurement of a system.

The implications of data-enabled creativity are profound, both for organizational learning as well as learning in general. Even within the comparatively narrow context of learning with data, computational learning, which is now the staple of data science and analytics, will slowly be relegated to the secondary role, while simulational learning will take the centre stage. While some decision-guiding insights will continue to be derived less from direct extrapolation of data patterns and trends, more will come from data and technology-enabled 'what-if' simulations, which will spur unprecedented flows of breakthroughs and tend-resetting ideas. Further compounding the already emerging marketplace dynamics, the only enduring source of competitive advantage will be the ability innovate, but innovation in the Age of Transcendence will combine the power of reason and established science with the power of data-enabled exploration and discovery, which will usher a new chapter in human creative problem-solving.

4.3   From the Age of Reason to the Age of Data

The explosion of interest in data analytics has led to the broad idea of 'learning with data' becoming a household name, but confusion regarding what, exactly, it entails persists. To some, it is the good-ole statistics, while to others, it is artificial intelligence and machine learning; to some, it means simple graphs and charts, while to others, it means complex predictive models. When considered from the perspective of organizational decision-making, learning with data encompasses all those elements, but to understand their respective roles and contributions, it is essential to also understand the underlying developmental journey. Figure 4.4 offers a summary of the gradual, evolutionary maturations of means of making sense of

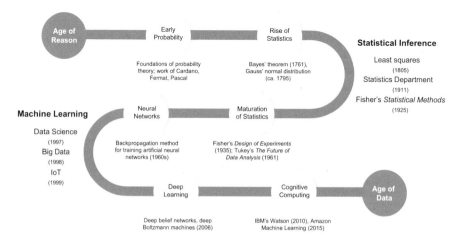

Fig. 4.4 The evolution of learning with data

the ever-expanding varieties and volumes of data, framed in the context of key contributions.

The modern conception of data tends to be tied to electronic storage and computing systems, but in its most rudimentary form, data are simply recordings of states or events. Not too long ago printed telephone books were widely used 'paper databases' containing service subscribers' listings, and filing cabinets were *de facto* databases that kept track of product shipments, medical records and countless other physical (i.e. paper) records. In short, while the modern electronic data infrastructure is a relatively recent development, data as recordings of facts are as old as human record-keeping, the earliest traces of which date back to about 5000 years. All of that is a long way of stating that in a broad sense learning with data can be considered as old as record-keeping, but when considered as a systematic process of transforming raw data into information aimed at reducing uncertainty, another distinct period in socio-cultural-scientific development, the Age of Reason (also known as the Enlightenment),[5] marks the beginning of the modern conception of learning with data, which began with first formal treatments of the notion of probability, as graphically illustrated in Fig. 4.4.

Renewed interest in mathematics, a corollary to the rise of the Age of Science discussed earlier, played a central role in the emergence of the *scientific method* as it is known and practiced today. The 'as known and practiced today' qualifier is quite important, because the roots of the scientific method stretch all the way back to antiquity. When looked from the broad perspective of creating formal knowledge, the emergence of formal method of inquiry began with a famous duo of Hellenic philosophers, Plato and his student Aristotle. Plato viewed knowledge through the prism of things that were visible and those that were intelligible (but not visible), and felt that only the latter, which he termed 'forms', could be ascertained with certainty, using geometry and deductive reasoning; to him, that which could be observed (i.e. physical reality) was imperfect and deceptive. Hence, the Platonic method of learning emphasized reasoning while downplaying the importance of observation, a stance with which Aristotle disagreed. In Aristotle's view, Platonic forms were the fundamental principles governing the natural world, and knowledge of those principles could be gained through systematic inquiry into observable workings of nature. And while Aristotle did not formulate the now-familiar logic of formally posing and testing of knowledge claims, formally referred to as hypotheses or conjectures, his ideas were nonetheless the intellectual spark that, some 2000 years later, led to the rise of the scientific method.

Turning back to the Age of Reason and the emergence of the formal notion of probability, Gerolamo Cardano, one of the lesser-known Italian polymaths of the

---

[5]A period in European history, starting in the late seventeenth century and continuing through the end of the eighteenth century, characterized by rigorous scientific and philosophical discourse, as exemplified by writings of Locke, Kant or Voltaire, and sociopolitical awakening, perhaps best exemplified by the French Revolution.

Renaissance,[6] was the earliest mathematician to formally address the probability of certain outcomes in rolls of dice (primarily because his wide array of interests included gambling). However, since his manuscript, believed to have been written in 1550, was not published until 1663, the credit for framing of foundational ideas of what today is known as probability theory instead goes to a pair of French mathematicians, Blaise Pascal and Pierre de Fermat, who lived roughly a century later. Their 1654 correspondence, which aimed to solve what is now known as the problem of points (the number of turns required to ensure obtaining a six in the roll of two dice), is now widely considered to have laid the foundation of the new concepts of probability and expected value. However, it was Christiaan Huygens, a Dutch physicist, who was the first to publish a formal text on probability theory titled *De Ratiociniis in Ludo Aleae* (On Reasoning in Games and Chance), in 1657, which became the authoritative text on probability for the remainder of the seventeenth century. Still, it was the later work of James (also known as Jacob) Bernoulli, a Swiss mathematician, whose 1713 *Arts Conjectandi* (The Art of Conjecturing) expanded on many of the earlier probability related ideas and, perhaps most importantly, offered what is now known as the Bernoulli theorem, or the law of large numbers.

The next set of seminal developments paving the way for modern statistical learning with data came from Thomas Bayes, an English mathematician and minister, and Carl Friedrich Gauss, a German mathematician and physicist. Starting with Bayes, it is noteworthy that his most famous accomplishment, known today as Bayes' theorem, which describes the probability of an event taking into account prior knowledge of conditions that might be related to the event, was actually never published by him; instead, his notes containing those ideas were edited and published after his death by Richard Price. Still, by providing a mathematically elegant mechanism for probabilistic inference, Bayes' theorem is now an integral part of numerous formal statistical procedures, including Markov chain Monte Carlo (often referred to as simply MCMC) algorithms. In contrast to Bayes, whose contribution was comparatively narrow in scope and, in terms of its impact, has been amplified by time, Gauss is counted among the most influential mathematicians in history. Within the relatively narrow domain of learning with data, he is perhaps best known for what is variously referred to as 'Gaussian' or 'normal' distribution, or simply the 'bell-shaped curve'; a continuous function which approximates the exact binomial (i.e. comprised of two parts) distribution of events. It is the most important probability distribution in statistics – in fact, to many, the study of its properties is almost synonymous with statistics. Symmetric about the mean, the normal distribution shows that values closest to the mean are more frequent in occurrence than those far from the mean; as such, it captures the distributional properties of numerous real-life phenomena such as heights of individuals, blood pressure or IQ scores, and offers a comparative background for description of other, i.e. not normally distributed, phenomena.

---

[6] He authored more than 200 science writings, spanning mathematics, chemistry, biology, astronomy, literature and philosophy; in fact, he is considered to have been one of the most influential mathematicians of the Renaissance.

4.3.1   Statistical Inference

Continuing with the retracing of the development of modern statistical and computational mechanisms of learning with data graphically summarized in Fig. 4.4, the foundational work on probability theory gave rise to an important aspect of the process of creating objectively verifiable knowledge – *statistical inference*. Broadly characterized as learning characteristics of the population from analyses of samples, it is the cornerstone of reason-based theoretical learning (see Fig. 1.5). Implied in that characterization are two somewhat distinct elements of statistical inference: the logic of extrapolation and estimation. The former is encapsulated in the scientific method, or more specifically the hypothetico-deductive embodiment of that general systematic sensemaking approach, which encompasses a set of procedures consisting of systematic measurement of phenomena of interest followed by the formulation, testing and modification of hypotheses. The latter entails the use of analytically derived approximations as unbiased or impartial means of arriving at sound 'most likely' values. The computational mechanics can be somewhat involved largely because of the importance of formally addressing the degree of precision or accuracy, and the corresponding degree of credibility or believability. So framed, formalization of statistical inference had, as depicted in Fig. 4.4, transformative impact on the evolution of learning with data as it contributed the means of 'productizing', in the sense of making practically usable, the ideas of probability theory.

Building of that foundation, the next seminal development was the discovery of the method of least squares, aptly described as the automobile of modern statistical analysis: '…despite its limitations, occasional accidents, and incidental pollution, it and its numerous variations, extensions, and related conveyances carry the bulk of statistical analyses, and are known and valued by nearly all'.[7] First described by Adrien Marie Legendre in 1805 and subsequently expanded on by Carl Friedrich Gauss, it is a mathematical approach to estimating the true value of a quantity through analysis of observed errors, or deviations. Either that very technique or its many derivatives lie at the core of some of the most commonly used statistical techniques, such as correlation and regression (both invented by English Victorian era polymath, Francis Galton); not surprisingly, the discovery of this computationally straightforward but conceptually brilliant estimation method helped to elevate the initially fringe branch of mathematics to a status of a standalone discipline.[8] Further bolstered by work of Karl Pearson (Pearson chi squared test and Pearson correlation), William Sealy Gosset (who published under the pseudonym of 'Student' thus Student's t-test and t-distribution), and Ronald Fisher, who developed the analysis of variance and refined the method of maximum likelihood, but perhaps most

---

[7] Stigler, S. (1981). Gauss and the Invention of Least Squares, *The Annals of Statistics*, 9(3), 465–74.

[8] It is worth noting that the term 'statistics', as used today, was at that time (19th thru early 20th century) used to designate systematic collection of demographic and economic data, whereas the mathematical theories of probability and the logic of statistical inference were usually referred to as 'mathematical statistics'.

importantly systematized all prior developments in his seminal 1925 *Statistical Methods* text, statistics emerged as a standalone domain of knowledge and practice, as evidenced by the emerge of statistics departments at universities throughout the world, starting with the University College London, founded in 1911 by Karl Pearson.

As known and practiced today, statistical learning with data also benefited from Ronald Fisher's work on the design of experiments, which offered mathematical ways of describing and explaining variability under conditions that were hypothesized to reflect that variability, Jerzy Neyman's work on confidence intervals, and John Tukey's work on exploratory data analysis and graphical presentation of data. Shortly thereafter, work on artificial neural networks, which can be characterized as a broadly scoped attempt at emulating the non-sequential functioning of the human brain and nervous system, coupled with rise of modern computing infrastructure opened the doors to alternative approaches to estimation. Rooted in the general idea of serial iteration, as embodied by optimization algorithms (repeated recomputation of a solution until an optimum or a satisfactory solution is found), and further enabled by backpropagation algorithm[9] developed in the 1960s, a new era machine-driven learning with data began to unfold.

4.3.2 Machine Learning

The next key turning point in the developmental journey of learning with data summarized in Fig. 4.4 was the development of optimization-based means of probabilistic estimation, broadly known as *machine learning*. In contrast to the 'traditional' mechanisms of statistical learning outlined above, which can be characterized as tools to be used by human analysts trying to extract insights from data, machine learning algorithms were designed to be capable of improving automatically through experience, i.e. self-guided, systematic analyses of data patterns, and to produce data analytic outcomes with little-to-no human analyst input. Today, machine learning is commonly considered a sub-domain of artificial intelligence (AI), which are technologies that are capable of making sense of, including reacting to, their environment in a manner that resembles human cognition (more on that in Chap. 7). Maturation of machine learning designs and technologies, coupled with the communalization of vast data repositories[10] and the gradual emerge of the Internet of Things (IoT), gave rise to a yet another (statistics being the first

---

[9] In simple terms, backpropagation algorithm is used to train a neural network using a method called chain rule, according to which each forward pass through a network triggers a backward pass, while at the same time adjusting the model parameters' weights and biases.

[10] While the term 'big data' is relatively recent (it was coined in 1998), the idea of large data repositories is comparatively older, as governments, large corporate entities and other select other organizations have been maintaining big data-like repositories long before that term was coined; the last couple of decades saw the proliferation of big data repositories to entities other than governments and big business or NGOs.

one) standalone, learning with data-focused domain of knowledge and practice, data science.

An interesting aspect of the process of maturation of machine learning is illustrated by the celebrated 1997 chess match which pitted the then reining chess world champion Garry Kasparov against IBM's AI system known as Deep Blue. That was not the first time that human champion faced off against an artificial system (all prior bouts were won by human champs), but it was the first a computer prevailed, playing under regular time controls which allow about 3 minutes per move. Though the work on designing chess playing computer systems can be traced all the way back to the 1950s, it was nonetheless an impressive feat of machine learning, considering that Kasparov was considered to be one of the greatest chess players in recorded history. But the truly intriguing part, especially in view of the combinatorial nature of chess, which clearly favours machine-like 'thinking', is the vast informational asymmetry: Deep Blue was capable of processing more than 200 million chess positions per second, which meant it could assess millions of possible move sequences during the allowed time limit, while Kasparov could carefully consider maybe ten sequences during the same 3-minute time limit; mover, IBM designers also had access to hundreds of Kasparov's games, while Kasparov was not given access to the computer's prior games. And, Deep Blue, like all machines, was not hindered by factors such as fatigue or temporary memory lapses. Framed in the context of such staggering asymmetries, that famed 6-game match pitting the best human chess player against the best human-designed chess playing machine learning system produced three draws, one Kasparov win and two Deep Blue victories. One the one hand, many were amazed by the technological marvel in the form of Deep Blue, but on the other hand, it was truly astounding that many years of science and engineering[11] and many millions of dollars of expenditure just barely prevailed at a game that, but its nature, favours machine.

But progress marches on, and if there is one certainty, it is that capabilities of machines to learn will continue to approach human cognition. Characterized earlier as the use of multiple layers of information processing made possible by advanced algorithmic designs, exemplified by deep belief networks (which are directed, or supervised) or deep Boltzmann machines (which are undirected or unsupervised), *deep learning* systems are now capable of extracting features from data in a way that enables creation of broader, generalizable categories. Going a step further, the more recent embodiment of Deep Blue, IBM's Watson, a conversational (i.e. capable of answering questions posed in natural language) AI platform aimed at business applications, exemplifies a yet another evolutionary step in machine learning – a system that completely eliminates any human–raw data interactions. Those heightened capabilities to 'make sense' of patterns embedded in vast quantities of data means that machine learning continues to approach human ability to learn – it also means, however, that outcomes of machine learning are becoming progressively more and more difficult to evaluate in the context of interpretational validity and

---

[11] The Deep Blue project took more than a decade (1985 through 1997) to complete at the cost of over $100 million.

reliability. A case in point is offered by text mining systems now commonly used in legal pre-trial discovery, where the often vast quantities of documents need to be reviewed and summarized. The age-old question of 'how to you know that you know' once again emerges – how does one assure the correctness of summaries produced by a computer system governed by rules that are practically unintelligible? Paradoxically, the manner in which text (and numeric data) mining systems process data and produce data analytic outcomes is simply too complex to be described in a way that would make sense to a human examiner, because by their very design, machine learning algorithms aim to emulate the incredibly complex innerworkings of the human brain. Thus the 'how do you know that you know' enigma: Is it even possible to validate, in the manner that parallels verification of results of statistical analyses conducted by human analysts,[12] the efficacy of data analytic outcomes produced by advanced machine learning/AI systems? And if it is not, what does that mean for organizational learning?

---

[12] Broadly characterized, those can be grouped into two general categories: in-process and in-practice validation. Using predictive modelling as a backdrop, the former is exemplified by technical assessment of goodness-of-fit of models fitted to data, which offers evidence of the extent to which the model captures patterns hidden in data, while the latter commonly takes the form of in-market validation, which offers evidence of the accuracy of model-derived predictions.

Chapter 5
Learning with Data in the Twenty-First Century

5.1 Digital Literacy

The concept of punctuated equilibrium of socioeconomic development discussed in the previous chapter shines the light on the changing nature of skills and competencies required to produce economic value. Given the reliance on mostly manual, simple tools, the Agrarian Era called for very rudimentary skillset, which changed dramatically with the rise of the Industrial Era, and the growing sophistication of means of production that continued to emerge during successive epochs (Steam and Oil → Electric → Electronic) called for progressively more specific and evolved set of skills and competencies. As the next major phase of socioeconomic development, the ongoing maturation of the currently underway Information Era is precipitating widescale 'retooling', which manifests itself in de-emphasis of some long-standing skills, such as handwriting (though still beneficial it is nonetheless far less essential nowadays), and a corresponding dramatic increase of importance of other competencies, most notably data analytic literacy. Will driving, as in operation of motor vehicles, soon follow the suit, given the looming emergence of self-driving vehicles? Within the confines of organizational functioning, perhaps one of the most visible manifestations of that retooling has been the meteoric rise of data science as a new domain of practice and scholarship. Virtually non-existent as recently as two decades ago (the term 'data science' is believed to have been coined in the late 1990s, as noted earlier in Fig. 4.4), it is one widely seen as a core organizational competency.

In a more general sense, the changing nature of how lives are lived and how work is done is continuing to reshuffle organizational skillsets. The impact of the ongoing maturation and proliferation of automated systems is not only profound – it is also multifaceted. On the one hand, more advanced technologies generally offer greater depths and breadths of functionalities, but being able to fully utilize those systems' capabilities calls for specific, often advanced skills. On the other hand, the broadly

© The Author(s), under exclusive license to Springer Nature
Switzerland AG 2021
A. Banasiewicz, *Organizational Learning in the Age of Data*, EAI/Springer
Innovations in Communication and Computing,
https://doi.org/10.1007/978-3-030-74866-1_5

conceived modern networked informational infrastructure is also generating enormous quantities and varieties of data, utilization of which is rapidly becoming the
core element of economic value creation. More specifically, while the expanding
automation is freeing up time previously needed to perform tasks now handled by
artificial systems, the vast treasure troves of raw data generated by the practically
uncountable networked devices hold the promise to fuel the next wave of scientific,
technological and social innovation. However, turning that potential into reality is
contingent of the earlier mentioned retooling: Just as the rise of the Industrial Era
demanded widespread literacy (reading and writing skills) and numeracy, the now
maturing Information Era calls for widespread digital literacy.

An adaptation of a long-standing notion of literacy, defined as the ability to read,
write and use numeracy, the concept of *digital literacy* traces its roots to the 1980s'
emergence and subsequent rapid proliferation of personal computing, further bolstered by birth of the World Wide Web in the early 1990s. While numerous definitions have been put forth, one of the more frequently cited is one developed by the
American Library Association, which frames digital literacy as 'the ability to use
information and communication technologies to find, evaluate, create, and communicate *information* [emphasis added], requiring both cognitive and technical skills[1]'.
Implied in that framing are three distinct building blocks: understanding of data
structures and meaning, familiarity with data analytic methods and tools and the
ability to use information-related technologies. The first two – familiarity with data
(in the sense of how data are organized and their informational content) and data
analytic know-what and how – are frequently lumped together, a practice which can
lead to fuzzing of important distinctions. For instance, it is common to characterize
data literacy as the ability to read, comprehend and communicate data as information, and while it may not be immediately obvious, such broad-scoped framing of
data literacy confounds familiarity with data sources, types and structures (the general data organizing templates used in data storage and retrieval) with the ability to
transform data into information, which entails knowledge of statistical or algorithmic (machine learning) methods and means of analysing data. Particularly in organizational settings, familiarity with data and ability to analyse data tend to be
functionally distinct, and while it is certainly possible to be both data- and analytics-
savvy, that generally requires either very particular type of practical experience or
adequate training in the newly emergent discipline of data science. Barring such
particular experiential or educational backgrounds, knowledge of one (how data are
structured and stored) does not necessarily entail meaningful familiarity with the
other (how to extract meaning out of data). In organizational settings it is common
for 'data keepers', such as database administrators, to command deep knowledge of
data structures and comparative limited data analytic know-how; the opposite tends
to be true of data analytics professionals.[2] All considered, making sense of digital

---

[1] https://literacy.ala.org/digital-literacy/

[2] There are also more structural, data governance and security-related factors that effectively limit
the scope of individual organizational constituents' data familiarity; more specifically, it is com-

literacy calls for explicit differentiation between 'data' and 'data analytic' knowledge.

It is also important to expressly address the distinctiveness of technological competence as the third and final building block of digital literacy. It entails functional understanding of man-made systems designed to facilitate communication- and transaction-oriented interactions, including the manner in which those systems capture, store and synthesize data. In a more operational sense, it entails understanding of planned and unplanned consequences of technology and the interrelationships between technology and individuals and groups. So characterized, technological competence, together with data familiarity and analytic know-how, gives rise to the higher-order notion of digital literacy, graphically summarized in Fig. 5.1.

In view of the considerations raised above, and in the somewhat defined context of organizational learning, digital literacy can be framed as a set of technical, procedural and cognitive skills and competencies that manifest themselves in technological competence, data familiarity and analytic know-how. At a more granular level, technological competence can be seen as the ability to evaluate and use technology as a mean of accessing and interacting with data or information; data familiarity can be conceptualized as the knowledge of sources, organizing structures and types of data; and lastly, analytic know-how can be defined as the ability to extract insights out of raw data using cognitive skills and available data analytic approaches and methods.

There is a yet another aspect of digital literacy that warrants a closer look. As popularly used and understood, the idea of digital literacy tends to confound data- and technology-related competencies with macro-demographic characterizations, in the form of general age group-related propensities. For instance, the label of *digital nativity* recently gained widespread usage, particularly among educators looking to explain, rationalize and possibly leverage natural technology leanings of persons

---

mon, especially among larger organizations, to restrict data access through departmentalization, as a way of diminishing the likelihood of unauthorized, unwarranted or altogether illegal use of data.

Fig. 5.1 Building blocks of digital literacy

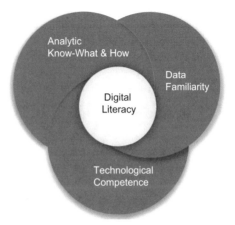

born into the modern technology-intensive society. While well intended and not entirely unfounded, the idea of digital nativity nonetheless confuses technological habits that could even manifest themselves as addiction, with the essence of digital literacy summarized in Fig. 5.1. Yes, all manner of computing technologies and gadgets seem natural to those born into the world that was already saturated with such technologies, but that sense of natural acceptance is distinct and different from the idea of functional understanding of the 'whats' and 'hows' of those tools. Even more importantly, proficiency at using modern communication tools should not be confused with the ability to use those tools to further organizational, meaning work-related, rather than individual, in the sense of social interconnectedness, goals. Examples of that distinction about, for instance, many millions of people can nowadays be considered competent automobile operators – how many of those drivers have a robust understanding of internal combustion, hybrid or electric systems that power automobiles? And by extension, how many could be considered 'mechanically literate'? Some of the same reasoning applies to the idea of digital nativity.

While there is ample evidence supporting the idea that digital natives are more instinctive users of digital communication technologies, there is no clear evidence pointing to a correspondingly deeper immersion with those technologies, in the sense of developing familiarity with data captured by those devices; the evidence is even more scant when it comes to the potential informational value of those data. In fact, there is more evidence supporting the oppositive conclusion, namely, that being a digital native can lead to blissful disregard of the 'how it works' aspects of the networked communication infrastructure, just as being born into the age of automobile has led to widescale usage of those technologies with little-to-no appreciable understanding of the underlying mechanics. In short, feeling comfortable with the modern digital technologies is not necessarily accompanied by a deeper understanding of broader informational utility of those technologies, particularly when it comes to understanding of the nature and the utility of data generated by those tools. The more user-friendly an application the more hidden its operational specifics, which means that, by and large, digital nativity does not entail the kind of immersive knowledge that is needed to be a productive user of data generated by those applications. Given that, what steps should organizations take to transform natural but passive functional users of the modern networked electronic infrastructure into creative, data-enabled organizational problem-solvers? The remainder of this chapter is focused on answering that very question.

5.2 From Digital Nativity to Digital Productivity

To start, it is instructive to look back to the foundational notion of literacy – in an operationally clear sense, what does it mean to be literate? As noted earlier, it is generally taken to mean being able to recognize letters, then words, then sentences, which ultimately enables one to read and write; separately, it also means being able to use numeracy, which in itself is a somewhat expansive idea that encompasses capabilities ranging from relatively tangible skills such as being able to measure or conduct basic calculations (adding, subtracting, etc.), to the considerably more abstractly defined skills such as probabilistic or spatial reasoning. Moreover, literacy is commonly characterized as a binary, meaning 'yes' vs. 'no' attribute, but in developed societies it is more appropriate to view literacy from the perspective of the 'degree of proficiency' continuum. A case in point: Upon completion of elementary education, a person can be considered to be literate, but typically not sufficiently so to be able to, for instance, read and understand technical scientific studies or engage in meaningful data exploration. Given that, it would be inappropriate to lump together those with only elementary school education with those holding advanced university degrees into the one-size-fits-all 'literate' segment, because such generic classification would obscure a number of critical distinctions. Simply put, literacy, both in the original sense considered here and in the more narrow but contemporary context of digital fluency, needs to be looked at as a progressively developing set of skills and competencies, paralleling the earlier discussed idea of analytic literacy (see Fig. 1.6 in Chap. 1).

However, before an attempt is made to understand and map out such a progression, it is essential to first expressly differentiate between what could be considered 'environmental' skill and competency acquisition factors, which encompass common, prevailing surroundings and circumstances, and 'wilful', or individually elected skill and competency acquisition factors. In that context, digital nativity can be seen as an expression of the former, while other core aspects of digital literacy, most notably familiarity with data structures and the ability to extract insights out of data, can be seen as a manifestation of the latter. Being born into an environment permeated by electronic communication and related technologies will, quite predictably and nearly inescapably, result in familiarity with those technologies, but it will take a wilful action to develop comparable levels of familiarity with the other two core dimensions of digital literacy depicted in Fig. 5.1. In short, while it is appropriate to consider digital nativity to be a step towards digital literacy, it is only the first step.

5.2.1   Getting to Know Data

The next step is to systematically develop understanding of data types, sources and structures. Starting with the general types, data can be either *structured*, so called because individual elements follow the familiar fixed, predefined structure where rows equal records and columns equals variables, as illustrated by a set of customer purchase records in Excel spreadsheet, or *unstructured*, where individual data elements do not follow a fixed record-variable format, as exemplified by an extract of

Twitter records. Structured data are easy to describe,[3] store, query and analyse, thus not surprisingly have been the staple of business analytics for decades; in contrast to that, unstructured data are more challenging to store and manage and considerably more challenging to analyse. Consequently, even though the volumes of unstructured data captured today vastly outsize the volumes of structured data (according to IDC, a consultancy, about 90% of 'big data' is unstructured), analytic utilization of unstructured data is comparatively low.[4]

When considered from the informational perspective, structured and unstructured data alike can be characterized as measurements and subsequent recordings of states or events, all of which can be encoded in one of three general formats: numeric, text or symbolic. As suggested by its name, numeric data are encoded using digits, but it is important to note that digital expression does not necessarily imply a quantitative value – in other words, '5' could represent either a quantity (e.g. 5 units of a product) or it could just denote a label (e.g. sales region 5); such potential ambiguity underscores the importance of clear operational definitions (commonly spelled out in data dictionaries noted in footnote #3). The second type of encoding, text, can take the form of words or phrases, but it also encompasses so-called alphanumeric values, which are strings comprised of a mix of letters and digits; in other words, any sequence of letters or letters, digits and even allowable special characters (e.g. $ or &) is considered text data. And lastly, the third and final type of encoding, symbolic, encompasses essentially any encoding that is neither numeric nor text, which typically encompasses a wide array of visual expressions, such as images.

It is considerably more difficult to summarize the vast sources of data available today as far-flung digitization of individual and commercial facets of life manifests itself in already enumerable, yet still continually expanding streams of data. With that in mind, as input into organizational decision-making, data sources can be grouped into several broad but somewhat distinct categories, all possibly of value to commercial and non-commercial organizations. Those include transactional details,

---

[3] Data descriptions tend to take the form of the so-called data dictionaries which offer operational definitions of individual data elements. In a structured data file, variable definitions are the same for all records, which makes the task of preparing a data dictionary as easy for a dataset that contains a large number, such as 100 million records, as one that contains only 100 records; that is obviously not so in unstructured data files, where each row can potentially contain different sets of data values.

[4] While the sheer volume of unstructured data can pose analytic challenges, the comparatively low analytic utilization of unstructured data is primarily rooted in informational sensemaking, as well as validity- and reliability-related considerations. Very briefly, the largely informationally straightforward (due to prevalence of numeric values) and repeating character of structured data lends itself to comparatively easy extraction of patterns and associations, which in turn facilitates fairly straightforward informational sensemaking and objective assessment of validity (truthfulness) and reliability (dependability) of data analytic outcomes; the nuanced and complex informational content of unstructured data (due to prevalence of text), coupled with non-repeating (or at least not predictably repeating) content of successive rows, translates into difficult and speculative patterns and associations extraction task, in addition to which, result validation is often practically not viable.

social networking, machine-to-machine networking (i.e. the Internet of Things interconnectivity), online search and crowdsourced data. Potentially also of interest, at least for some organizations under some circumstances, such as financial services firms looking to expand their credit card holder base or nonprofit organizations looking for new donors, could be special purpose data sources. Those include numerous custom-developed (commonly referred to as 'third party' data) add-on data, as exemplified by the US Census-derived geodemographics, typically used to enhance the informational content of transactional data.

It is worth noting that while the above enumerated general data sources may sound very Digital Age-like, but record-keeping, which is the principal element of data capture, is as old as organized society. Hence at least in principle it is not so much the type of data, but rather the manner in which data are captured that are new, with the notable exception of machine-to-machine networking data. Evidence of bookkeeping, which is recording of financial transactions, was found in the oldest city settlements, circa 2600 BCE Babylon; social networking is also endemic to organized society and thus likely as old, as is the idea and the task of searching and keeping track of those efforts. In fact, even crowdsourcing, which has the appearance of being a new phenomenon, dates back to the onset of Industrial Revolution (more on that later).

Turning back to data sources, when looked at as a general category, *transactional* data is a by-product of modern electronic transaction processing infrastructure, best exemplified by the ubiquitous brick and mortar retail points of sale or online transaction processing systems. Tracing their roots to the emergence of first bar code readers in the mid-1970s, electronic transaction processing systems have steadily spread throughout the realm of business-to-business and business-to-consumer commercial interactions, and those now-generic technologies are the core originator of sales and other transaction-producing exchange data. Considered within the familiar product purchase context, transactional data capture can be characterized either as 'first party', which is when the seller of the product is also its manufacturer, or 'second party[5]', which is the case when the product is produced by one organization (typically its manufacturer) and sold by another (typically, a retailer). Apple's iPhone sales illustrate the difference: When a sale takes place at an Apple Store, which is a part of Apple Inc., the resultant transaction is categorized as first party, but when a sale takes place at an independent retailer, the resultant transactional data is characterized as second party. Typically, first and second party data capture systems capture the same core product characteristics (e.g. model, storage capacity, colour, price, etc.), but second party data are wider in scope, in the sense of encompassing category-wide (i.e. multiple manufacturers or brands) sales.

---

[5]The earlier mentioned 'third-party' data are not included under the transactional data umbrella because those data are not captured but analytically derived from originally captured data. For example, the commonly used geodemographics are block, or roughly 20–30 geographically contiguous households, based averages computed using household-level details originally captured by the US Census Bureau.

In contrast to the linearly steady and uneventful rise and proliferation of electronic transaction processing systems, *social* data had a lot more of a 'made for TV' of genesis. When considered within the confines of the now-ubiquitous social networking, the beginning of the large-scale social interconnectivity and interactions data can be tied to the now-defunct Friendster, launched in 2003. Though known but to a few now, it was considered the premier social media platform in the early 2000s, but was soon eclipsed by rival Myspace, which at one point reached a whopping $12 billion market valuation before it started losing millions of subscribers each month to a rising upstart Facebook, which sent its valuation plummeting to $35 million in a short span of time. And while Facebook was able to establish itself as the premiere social networking platform, the domain of social media grew and proliferated: In addition to networking sites, there are now numerous social review platforms, image sharing and video hosting sites, community blogs and discussion sites and sharing economy networks (vacation rentals, ride share, etc.). The bulk of what has come to be known as 'big data' is a by-product of all manner of personal, special interest, commercial and other social interactions facilitated by the vast array of platforms. It is not an overstatement to suggest that just identifying and categorizing the various 'personal digitization' supporting platforms is a considerable challenge, given the dizzying array of means available to individuals willing to share their behaviours, opinions and attitudes. A single individual may share aspects of his or her personal life on Facebook and professional life on LinkedIn; the same person may search the web using Google, make multiple product purchases on Amazon or eBay and ask probing questions on Quora, special interest sites like Stack Overflow or numerous wikis (hypertext websites that support collaborative editing and managing of content), not to mention occasional 'tweeting' (Tweeter) or 'snapchatting' (Snapchat).

Clearly, what is now generically known as social data but what is really a highly heterogeneous mix of largely unstructured, often difficult to access, and running the gamut of formats – that implausibly varied category stands in stark contrast to the comparatively narrow in scope and easy to manage transactional data. Generated by a wide array of proprietary (Facebook, Google, Amazon, Stack Overflow, Quora, etc.) and open-source (Open Source Social Network, HumHub, Askbot, etc.) platforms and applications, social data contain tremendous amounts of raw informational materials, but those data flows also contain massive amounts of non-informative noise. The ongoing interest in leveraging the content of those vast reservoirs of self-disclosed behaviours, experiences, opinions and attitudes is tempered by numerous operational and analytic challenges associated with extracting valid and reliable insights from pools of highly heterogeneous, in just about every sense imaginable, and noisy informational raw materials.

Somewhat related to social networking is *online search* data, which can be seen as a similar but distinct dimension of digital behaviour. Given the open-ended character of the general notion of 'search', online or physical, the first step in making sense of the notionally unbounded online search is to break that notion down into distinct search types, which are informational, navigational or transactional. Informational search queries are typically focused on finding as many relevant sources relating to a topic of interest; it is the proverbial 'fishing expedition' where the searcher, who might or might not have previous knowledge about the topic, wants to learn more about it. Navigational search queries can take on one of two more specific forms: The first is a known-item search, which is notionally similar to informational search in the sense that while also information-seeking it is narrower in scope as it is focused on a particular source or a set of sources known by the searcher to exist; in other words, the goal of a known-item search is to locate specific resources of interest. The second type of navigational search is a known-page search, which similarly to known-item search is focused on resource location but it is even more specific in the sense that it is focused on a unique URL,[6] as opposed to a resource, such as a document, which can be found in a variety of sources (i.e. the same document can be posted on multiple websites). The third and final type of online search is transactional search, the goal of which is to execute a transaction, such as purchase of a book, which bridges the gap between online search and transactional data summarized earlier.

As noted earlier, *crowdsourced data* and the idea of crowdsourcing in general seem like a newcomer to the Digital Age, but in fact the use of crowdsourcing can be traced all the way back to the onset of the Industrial Revolution. In the early part of the eighteenth century, the British government established the Longitude Prize as a way of enticing the public to submit ideas to develop a reliable method of calculating the longitude of a vessel at sea.[7] Today, crowdsourced data is a manifestation, really a by-product of an attempt to solve a particular problem or to address a specific deficiency. The most commonly seen types of crowdsourcing, and thus crowdsourced data, include macrotasks, as exemplified by web design or application development; microtasks, such as development of specific classification algorithms, funding requests aiming to remove financial barriers; and contests, such as design of a new logo for a team. An interesting aspect of crowdsourced data is that in contrast to other types and sources of data described here, the scope of those data is largely user-shaped, in a way that is somewhat reminiscent of surveys.

The last and perhaps the least understood of all sources of data, the Internet of Things (IoT)-derived *machine-to-machine networking* data is a product of the vast network of physical devices – 'things' – that are embedded with sensors, as well as read-and-write software and technologies, making it possible for those devices to exchange data with other devices over the Internet. Given the transformative impact of the IoT, which, in ways that may not be always clearly visible, is making its way into just about every facet of commercial and personal lives, it should not come as a surprise that machine-to-machine networking data, generated by growing arrays

---

[6] URL (Uniform Resource Locator) is often confused with domain name, which is incorrect as domain name is a part of a URL. In very simple terms, while a domain name could be thought of as particular book, a URL could be considered a specific page in that book.

[7] The prize was won by John Harrison, a clockmaker, who was able to solve the problem using highly accurate watches. In an unrelated but similar crowdsourcing contest, Nicolas Leblanc, a French chemist and surgeon, won a prize established by the French king Louis XVI for developing a process for turning salt into soda ash.

of connected fixed and wearable devices, is in fact a meta-category comprised of several distinct types of data, including status, location and automation data. Status data can be considered to be the most basic type of IoT data, often reflecting basic binary, yes vs. no, states; for instance, is a device turned on or off, is the signal adequately strong or not, etc. Location data is perhaps the most visible and most debated type of IoT data, as it represents tracking of the movement of objects and persons, which gives rise to obvious applications in logistics, ranging from large-scale fleet management to single product/asset tracking, and the often-controversial surveillance. The third and final variant of machine-to-machine networking data, automation, is generally hidden from everyday view as it is used by IoT systems to control devices ranging from household appliances to traffic controls to complex production systems; not surprisingly, it is usually also the most complex facet of that particular source of data.

5.2.2 Making Sense of Data Structures

Setting aside situations where data are used as they are generated, as is the case with self-driving cars and other autonomous systems, in order to be usable, captured data need to organized and stored. The resultant organized collections of structured and unstructured data stored electronically comprise databases, which are then, typically but not necessarily, connected to various retrieval and utilization applications. It is worth noting, however, that even though nowadays the term 'database' conjures up images of electronic storage of data, in a strictly definitional sense, as organized collections of data, a paper phone book is as much a database as an electronic repository of such details.

There are multiple ways of describing databases: by data type, purpose, content, organizational structure, size, hardware-software characteristics, etc. Looking past the numerous technical considerations, when looked at from the standpoint data utilization, the most pertinent aspects of a database are its scope, which is the breadth or diversity of data contained therein; its form, which is the type or encoding of information; and its data model, which is the basic organizational structure of a database. Starting with database scope, the rather generic 'database' label can be used to describe a data warehouse, which as suggested by the name tends to encompass wide arrays of data captured, acquired and maintained by the organization, or it can be used to a data mart, which typically contain considerably narrower, more homogeneous subsets of organizational data that are geared towards more specific usage situations, such as inventory tracking or promotional campaign management. In terms of form, a database can be either structured or unstructured (discussed in the Sect. 5.2.1); numeric, or encoded using primarily digits; text (which may also include some numerically expressed values); multimedia and hypertext, which supports creative linking of diverse types of content, such as text, pictures, video and sound into a single expression; or bibliographic, which is used to organize published sources, such as books, journal or newspaper articles, including basic descriptive information about those items (mostly used in library cataloguing).

The inescapably technical in its meaning data model aspect of databases is an abstract representation of the basic organizational structure of contents of a database, which lays out how data elements are arranged and how they related to one another and to the properties of real-world entities. As could be expected, there are many kinds of data models; some of the more common ones include relational, hierarchical, network, entity relationship or object-oriented. Historically the most commonly used organizational schema, the relational model, divides data into tables, each of which consists of columns (variables) and rows (entities, such as customers, individual transactions, etc.); the model also accounts for the types of relationships between individual data tables, including one-to-one, one-to-many and many-to-many relationships. The hierarchical model organizes data into a tree-like structure, where each record has a single parent, and sibling records are sorted in a particular order which is then used in storing data. The network model builds on the hierarchical model by allowing many-to-many relationships between linked records; thus a single record can be a member or child in multiple sets, which is particularly desirable when conveying complex relationships. A yet another design, the entity-relationship model, is built around distinct types of entities (e.g. purchases, customers, other) and specific relationships that can exist between them; in contrast to relational or hierarchical data models, it tends to be more conceptual in nature and less tied to the physical structure of the database. The last among the most common data models, the object-oriented model, defines a database as a collection of objects, or reusable software elements, with associated features and methods; the most common variants of that model are multimedia (images, videos, etc.) and hypertext, which allows objects to link to other objects, which is now a common online feature.

While the choice of a particular data model is often rooted in non-informational considerations, such as efficiency and the ease of access, it nonetheless has at least an indirect impact on the ultimate informational efficacy of data. Stated differently, data are, at their core, means to an end in the form of insights, when considered within the confines of traditional data analytics, or direct actions, when considered within the confines of data-consuming autonomous systems, such as self-driving cars. It thus follows that to be meaningful, the above abstract descriptions of competing data models need to considered in the context of data utilization.

5.2.2.1   The Data-Information-Knowledge Continuum

Databases exist for two basic reasons: First, they enable an ongoing capture and storage of facts. Second, they serve as platforms for inferential knowledge creation. When both reasons are combined, databases become conduits for transforming raw recordings of states or events broadly referred to as data into information and, possibly, into knowledge.

From the standpoint of users, raw data, as in digitally encoded facts stored in an electronic database, offer very limited user utility because unprocessed data are extremely interpretationally challenging, if not altogether impossible to interpret (in fact, it is hard to think of a person who could visually scan thousands, millions or possibly a lot more of seemingly random values and extract any kind of meaning). Regardless of the underlying format, be it numeric or text, source, such as transactional or social, or organizational model, as exemplified by relational or object-oriented, raw data are rarely informative without thoughtful and purposeful processing, not only because of the often sheer volumes but also due to lack of self-evident differentiation between important and trivial facts.

However, once the raw data are summarized, they are effectively converted into significantly easier to interpret information, which immediately produces sharp increases in user benefits. And while unquestionably more valuable from the sense-making perspective, the effectively generic information is hampered by limited actionability. For example, summing up individual product purchases (raw data) into location- or period-level aggregates (information) certainly increases the benefit and lightens the interpretational challenge, but the resultant information might still be of limited benefit to decision-makers. In the illustrative case of product sales aggregates, the resultant information captures total sales at a location or at a point in time but offers no insights into any cross-location or cross-time variability. In other words, while useful, information such as aggregate sales needs additional refinement before it can become a materially meaningful contributor to organizational decision-making. In the sample case of sales summaries, that might call for adding in the impact of potentially causal factors such as promotional offers, advertising intensity or price changes, to name just a few. Moreover, it is not just a matter of adding in additional pieces of information as standalone elements – enhancing the value of generic information is contingent on being able to look beyond 'what-is' to get a glimpse of 'why it is', which is the essence of the difference between information and knowledge. In a more general sense, information can be conceptualized as a product of efforts geared towards making sense of initially unintelligible raw data, while knowledge can be framed as a product of valid and reliable relating of available information to the decision need at hand.

The data-information-knowledge progression implied value creation process has important usage implications for organizational database utilization. At the most rudimentary level, databases can be used to support ongoing performance reporting through performance dashboards, usually built around a preselected set of metrics of most importance to the organization. Those reports are of particular interest to line managers, who are usually tasked with keeping an eye on operational outcomes, and their format and their content and frequency tend to be shaped by factors such as data availability, industry characteristics and organizational needs. In practice, such basic status or summary reports tend to take the form of standardized score-cards or dashboards produced with highly automated business intelligence software and thus require little-to-no advanced data analytical capabilities, which has been one of the drivers of their widespread usage. As a general endeavour, database reporting tends to be data-type specific (e.g. insurance claim reports are generated from the transactional data, while claim adjustor effectiveness reports are based on a combination of claim development, activity timing and adjuster notes), which means that it is usually difficult to 'cross-pollinate' or combine distinct reporting streams, something that can lead to counterproductive proliferation of standalone reports. All considered, while basic data reporting offers important decision-making inputs, the value of those inputs is largely limited to ongoing operational monitoring and management.

However, to not only prosper but to survive, organizations need to look to the future through the lens of creative strategic planning, which calls for more than operational summaries. Going beyond the mere status quo reporting, a more robust analytical set of processes can help in translating the often-disparate pieces of information into higher-level inferences, framed here as knowledge. Of course, it is not quite as easy as it may sound. As pointed out earlier, the ability to distil large volumes of often source-, type- and structure-heterogeneous details into decision-guiding insights hinges on a combination of a forward-looking informational vision and a robust analytical skillset. Converting raw data into generic information can be handled, for the most part, with the help of highly automated, commercially available database tools, but funnelling the often still voluminous and almost always inconclusive information into unique knowledge cannot be handled by standardized, off-the-shelf database reporting applications, because of reasons ranging from extensive data preparation requirements to the intricacies surrounding multivariate data analyses. Furthermore, the creation of decision-guiding knowledge quite often necessitates amalgamation of multiple, dissimilar data sources into a single yet multidimensional, causal chain, which in turn requires the establishment of cross-factor correlations, cause-effect relationships as well as the more technically obtuse interaction and non-linear effects. It almost always requires a combination of organizational commitment manifesting itself in appropriate resources and a fine-tuned data analytic skillset.

5.2.3   Analytic Know-How

First discussed in Chap. 1, analytic know-how can be seen as a result of combination of application-minded tacit and theory-focused explicit learning, outcome of which manifesting itself in semantic (conceptual understanding), procedural (ability to execute) and episodic (contextual familiarity) dimensions of knowing (see Fig. 1.3). Implied in that characterization is the cumulative nature of data analytic knowledge acquisition, captured (Fig. 1.6) in a three-tiered categorization of data analytic skillset: analytic literacy, competency and proficiency.

Analytic literacy can be seen as a foundational level of tacit and experiential data analytic knowledge, reflecting essential ability to manipulate (i.e. extract, review, correct) data and conduct simple, typically descriptive analyses. *Analytic competency* is attained when greater depth and breadth of explicit knowledge are added to the earlier attained analytic literacy, with specific emphasis on advanced data modelling techniques, most notably multivariate statistical techniques and advanced machine learning algorithms. And finally, *analytic proficiency* is attained by layering meaningful breadth and depth of tacit knowledge in the form of practical experience working with complex data and using advanced data analytic techniques. There are numerous reasons such tiered analytic know-how categorization is important, not the least of which is assuring of the highest possible degrees of validity and reliability of analytically produced knowledge, which is notionally similar to the reasoning used by the actuarial profession.[8] Oftentimes it is difficult, if not outright impossible to discern methodological and computational validity of insights derived from data analytic outcomes because the required technical details might not be included, or the user of information might not have the requisite technical knowledge to make such determinations. And yet, the logic of the informational value chain suggests a need for an objective and fair mechanism of attesting to the efficacy of data analytic inferences, and one possible avenue of pursuing that objective is through formal recognition of professional competencies of those entrusted with generation of informational outcomes to be and used and relied on by others.

The reasoning that underpins the analytic literacy → competency → proficiency progression is rooted in systems thinking, which here manifests itself in an ongoing and cumulative interaction between tacit and explicit knowledge, which in a more operational sense is seen as a product of the interplay among data analytic reasoning, familiarity with data analytic methodologies and proficiency with computational tools. Perhaps the most obtuse of those three core determinants of individual-level data analytic know-how is *analytic reasoning,* framed here as the capacity to use logic as a mean of consciously making sense of a data analytic task at hand. Outcome-wise, well-developed analytic reasoning manifests itself in the ability to develop operationally clear mental maps of what and how is to be done, expressed as a formal analytic plan, the development of which benefits all data analytic tasks but becomes indispensable when tackling more extensive projects, especially those that require multiple collaborators to complete. The reason formal analytic planning becomes indispensable in multi-contributor data analytic undertakings is that when considered in the context of the tacit-explicit knowledge continuum, analytic reasoning, being highly subjective and interpretive in addition to also being shaped by individual perspectives, skews heavily towards the tacit end of that continuum, as graphically depicted in Fig. 5.2. Stated differently, if left unspoken, in the sense of lacking a formal and explicit plan, complex data analytic initiatives can turn into undertakings resembling the mythical Tower of Babel, ultimately failing to reach their goal.

Turning back to the three core determinants of the literacy → competency → proficiency progression, *methodological know-how* can be seen as falling on the

---

[8] The Society of Actuaries, which is a global professional organization for actuaries, uses a series of rigorous exams to bestow two associate-level designations (Associate of Society of Actuaries and Chartered Enterprise Risk Analyst) and its highest designation, Fellow of the Society of Actuaries. That tiered system is widely recognized, and it is used as a formal basis of assuring validity and reliability of actuarial opinions and projections.

Fig. 5.2 Data analytic skills on the continuum of knowledge

opposite (from analytic reasoning) end of that analytic skill continuum. As a distinct skillset, it can be seen as objective and factual and thus as a reflection of the depth and breadth of command of clearly defined and universal body of knowledge of statistical and machine learning concepts and techniques. Falling somewhere between the individually nuanced analytic reasoning and the comparatively universal and objective methodological know-how, *computational know-how* is equal parts explicit and tacit knowledge. Framed, within the confines of data analytics, as familiarity with appropriate programming languages, such as R or Python, and/or applications, such as SAS or SPSS, it combines formal knowledge of programming structures in the form of language- or application-specific logic and syntax (explicit knowledge) and practice-developed application experience (tacit knowledge).

All considered, it is the gradual maturation of analytic reasoning coupled with steady accumulation of methodological and computational know-how that produces the progressively more refined and capable analytic skillset, manifesting itself in the literacy → competency → proficiency progression. It is important to note that the deepening of data analytic skill set is highly individual-dependent, which means that otherwise similarly qualified individuals, i.e. same educational background and comparable levels of experience, could have markedly different data analytic proficiency attainment trajectories. And thus it follows that an organization that is serious about nurturing broad-based data analytic capabilities ought to carefully design its internal (analytic) talent development processes and practices; moreover, in view of the complex nature of that particular skillset, the design of those processes and practices should be shaped by those who command high degree of subject matter expertise. Given the complex nature of such an undertaking, a more in-depth overview of the specifics of the core elements of data analytic proficiency seems warranted.

5.3   Building Data Analytic Proficiency

A point made repeatedly throughout this book is that rapidly expanding automation is reshaping the very essence of economic value creation and, by extension, the makeup of core business competencies. Torrents of data spawned by the ever-expanding electronic transactional and communication infrastructure hide enormous informational potential, but tapping into those insights requires specialized skillset; thus not surprisingly, data analytic proficiency joins communication, teamwork and critical thinking as essential organizational skills. And while, as discussed earlier, familiarity with data should be seen as separate from the ability to analyse data, hidden in the shadows of those two core competencies is the oft-overlooked ability to make the almost always messy and ill-structured raw data analysis-ready. Framed here as data preparation skills, the importance of this critical prerequisite to data utilization cannot be overstated – in fact, it is probably the single biggest contributor to the rise of the now-distinct domain of data science.

5.3.1 Data Preparation

The vast majority of data that are available today, spanning the broad categories of transactional, social networking, machine-to-machine networking, online search and crowdsourcing details, are organized in the manner that reflects how those data were captured, which is often at odds with intended utilization. For example, the ubiquitous UPC scanner-sourced point-of-sales data are captured at the most granular level possible, commonly referred to as SKU (stock-keeping unit), such as 24-ounce box of Kellogg's Frosted Flakes cereal, but analyses of sales trends are usually conducted at the brand level, which requires rolling up of SKU-level data into brand aggregates. This basic example illustrates the importance of a thoughtful and systematic approach to data utilization, the start of which should be anchored in a thorough review of raw data to be used, followed by careful consideration of anticipated usage informed data enhancements and lastly, by clear delineation of specific data analytic steps required to transform raw into analysis-ready data, as graphically summarized in Fig. 5.3.

In an operationally clear sense, data preparation entails three distinct sets of skills: data due diligence, data feature engineering and analytic planning. It is important to note that while those are distinct sets of analytic competencies, in the larger context of data preparation, those skillsets are sequentially related, meaning that the overarching goal of data preparation calls for the data due diligence, data feature engineering and analytic planning progression shown in Fig. 5.3.

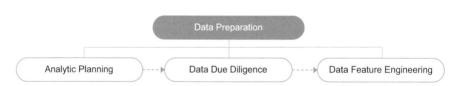

Fig. 5.3 The foundational skillset

5.3.1.1 Analytic Planning

Dwight D. Eisenhower, the Supreme Commander of the Allied Expeditionary Force in Europe during World War II (and, later, the 34th US president), famously stated that '…In preparing for battle, I have always found that plans are useless, but planning is indispensable'. That astute observation very well captures the essence of analytic planning, which is to engage in thoughtful, thorough and systematic review and design of the 'whats' and the 'hows' of contemplated analyses. While overtly the goal of analytic planning is to create an explicit roadmap detailing the overall process as well as the specifics of the individual steps, the true value lies in the reasoning discipline that is typically imposed by the process of planning. To a large degree, that goal is accomplished by explicit spelling out of specifics of data analytic logic along with enumeration of distinct requirements and dependencies; oftentimes, thinking through the details linking data inputs with expected data analytic outcomes can bring to light considerations that might otherwise be overlooked. The earlier example of Kellogg's Frosted Flakes offers a simple illustration: In order to conduct basic analyses of brand (Frosted Flakes) sales trends, the SKU-level (e.g. 24-ounce, 33-ounce and other distinct sizes of Frosted Flakes boxes) data need to first be aggregated to the brand level. Without explicit planning efforts, such basic but nonetheless essential data preparation steps might escape notice, ultimately precipitating avoidable rework and delays.

To yield desired outcomes, analytic planning needs to encompass several core considerations. The first is to link data inputs with stated informational needs and objectives, at the core of which is developing robust appreciation of the informational potential of available data. It entails looking at data at hand not just from the perspective of 'what-is' but also from the perspective of 'what it could be'. For example, a time-stamp (i.e. date and time values that are commonly recorded along with events such as purchase transactions) may be of little informational value in its original form, but when used to compute derived metrics such as 'tenure' or 'duration' can become comparatively more informative. The second key consideration, which also tends to be the most straightforward part of data analytic planning, is review of methodological options, or more specifically, identification of the most appropriate data analytic methodology or methodologies. Here, the focus should be on careful examination of input-related assumptions and output-related interpretational considerations characterizing competing methodological approaches. A closely related next step is clear delineation of expected analytical outcomes, most notably validity and reliability-related metrics such as goodness of fit, classification accuracy and level of statistical significance, all aimed at spelling out specific means and thresholds for establishing non-spuriousness of inferences drawn from data. The final step in the planning process is to expressly link technical results with expected informational inputs, or stated differently, sketching out the logic for translating the oft-esoteric data analytic outcomes into informationally clear decision-aiding insights.

A helpful way of thinking about analytic planning is to consider the resultant plan to be a de facto contract between the analysts and the end user community. Hence a well-crafted analytic plan might also encompass:

1. An explicit delineation of the individual analytic initiatives with a clear linkage to specific informational needs and, ultimately, the organization's strategic objectives
2. An overall completion timeline showing the starting and completion dates for each analytic project
3. An explicit description of how each initiative will improve the decision-making process

Turning back to the initial premise, namely, that it is the planning process rather than the plan that matters most, it is important to not lose sight of that when engaging in analytic planning. It is the disciplined, structured, focused and, ideally, iterative thinking through the entirety of data-to-knowledge process that is the greatest benefit of analytic planning.

5.3.1.2   Data Due Diligence

The first iteration of analytic planning is immediately followed by data due diligence, an open-ended process that entails review of data's *validity* and *reliability*. The former can be broadly characterized as truthfulness, which encompasses factual accuracy as well as logical correctness, while the latter is an expression of dependability or stability of data, or the extent to which individual variables can give rise to stable and consistent insights; taken together, validity and reliability jointly attest to informational legitimacy of data. Interestingly, the interaction between implied importance of validity and reliability can be asymmetrical. More specifically, product purchase data used to predict the probability of an outcome of interest, such as product choice, must be reliable in order to be valid, but does not have to be valid in order to be reliable (e.g. consistent mischaracterization of purchaser characteristics will typically result in invalid causal explanations, but if those misrepresented characteristics, or variables, are stable, the prediction based on those variables will likely be dependable).

In a more operational sense, data due diligence entails review of each individual data element, or variable, from the perspective of implied unit of measurement,[9] the degree of completeness and the reasonableness of values. The *implied measurement* unit is a conjoint of measurement scale, which can be either categorical, which can be further broken down into nominal or ordinal, or continuous, and variable encoding, the most common of which include numeric (digits only), string (text or alphanumeric), date or some form of restricted value, such integers with leading zeros.

---

[9] Since data are defined as measurements of events or states, the term 'measurement', as used here, is meant to communicate that basic characterization of data, rather than implying an active act of quantifying the size, length or amount of something.

The degree of completeness reflects the presence and if so, the extent of missing values, and the reasonableness of values captures the extent to which the range of values for a particular variable can be deemed reasonable.[10] Framed by those three core data considerations, data due diligence aims to assure the validity of data to be used in analyses.

An important, and often overlooked, aspect of data due diligence is *metadata* preparation. Somewhat tautologically defined, metadata is data about data; in more operationally specific terms, it is a summary view of the individual variables contained in the dataset expressed in terms of the key statistical descriptors, such as value ranges and summaries of basic distributional properties such as frequencies, measures of central tendencies and variability as well as coverage and accuracy summarizations. Although historically more familiar to academicians than to practitioners, the concept of metadata is gaining popularity among the latter as the amount and diversity of data contained in organizational repositories continue to grow. That is because, by and large, organizational databases, even those supported by clear data model descriptions and comprehensive data dictionaries, are often just hodgepodges of well- and sparsely populated, as well as consistent, definitionally amended or outright discontinued metrics, ultimately offering limited guidance as to what is analytically usable and what is not – when put in that context, a comprehensive and current metadata can be invaluable.

5.3.1.3   Data Feature Engineering

While data due diligence is focused on 'what-is', the next step in the data preparation process, data feature engineering, is focused on 'what could be'. It is a topic that tends to receive scant attention in texts focused on statistics or machine learning, and it is also largely neglected when teaching the how-to of data analytics, which is an enormous mistake given the direct link between the broadly defined informational efficacy of input data and the robustness of data analytic outcomes. The surprisingly common practice of accepting whatever data are available as ready-to-use input into analyses all too often leads to a predictable outcome: garbage in, garbage out. Extracting meaningful, decision-guiding insights out of data can be likened to assembling of computers or other complex products: The individual components, seen as inputs into the assembly process, are themselves highly refined elements such as motherboards, processors or graphics cards – not raw materials such as silica sand, iron ore or bauxite; no reasonable person would expect to produce a computing by simply mixing the said raw materials, and while less self-evident, the same holds true when it comes to extracting insights out of data.

---

[10] It is important to draw an explicit distinction between the *accuracy* of individual values and *reasonableness* of the range of values for a particular variable – for example, a person's age showing as '56' may ultimately be incorrect (due to data entry or other error), but if the range of age values falls within a typical human lifespan, it would nonetheless be considered reasonable.

   A big part of the reason behind the common neglect of thoughtful data preparation is the highly nuanced and situational nature of data preparation, which is shaped
by a combination of available data and expected data analytic outcomes. Setting
aside, for now, the open-ended expected data analytic outcomes, data can be either
structured or unstructured; can represent captured outcomes of transactions, social
networking, machine-to-machine networking, online search or crowdsourcing; and,
in the sense of individual variables, can be encoded in a wide array of formats
including numeric, string, date, restricted values, free-form text or image. When
coupled with effectively unbound nature of informational needs, that incredibly heterogeneous mix of potential raw informational inputs could give rise to practically
infinite data feature engineering possibilities, which is at the core of the challenge
associated with this critically important but very difficult to frame undertaking. That
said, the essence of data feature engineering can be illustrated by the commonly
encountered problem of restructuring transactional records, graphically depicted by
simple hypothetical example in Fig. 5.4.
   Given that, in a general sense, transactional data can be seen as recording of
individual transactions or sales, in a typical two-dimensional transactional data file,
rows enumerate individual transactions (e.g. items sold), and columns enumerate
transaction describing details (e.g. buyer, price, etc.). As shown on the left side of
Fig. 5.4, such layout mixes purchased 'items' and 'purchasers', which means that
sequential records can either represent multiple items purchased by a single buyer
or purchases made by multiple buyers. While that is an efficient way of capturing
and storing those data, it is analytically problematic because it confounds multiplicity of purchases made by a single buyer with multiplicity of buyers (in a more technical sense, it violates the key assumption of independence of individual rows of
data). Stated differently, in order to be analytically meaningful, individual data
records can either represent different items purchased by the same buyer (i.e. each
is the same buyer) or can represent different buyers (i.e. each record is a different
buyer), but cannot mix both.[11] Consequently, to be analytically usable the mixed
content data file illustrated on the left side of Fig. 5.4 needs to be restructured into a
format illustrated on the right side of that figure. If the goal of ensuing analyses is
to, for instance, examine differences among buyers, the original layout where 'item'
is the basic organizational unit and 'purchaser' is an attribute is transformed into a

Item	Price	Purchaser
8439263692	$ 5.99	John Doe
4339211549	$ 12.50	John Doe
2992201013	$ 34.99	John Doe
1100235234	$ 12.00	Jane Doe
7000923774	$ 40.99	Jane Doe
9923128831	$ 2.95	Jane Doe

Purchaser	Item 1	Item 2	Item 3	Item Price 1	Item Price 2	Item Price 3
John Doe	8439263692	4339211549	2992201013	$ 5.99	$ 12.50	$ 34.99
Jane Doe	1100235234	7000923774	9923128831	$ 12.00	$ 40.99	$ 2.95

Fig. 5.4 Example of data feature engineering: restructuring of transactional data

---

[11] This generalization assumes the so-called cross-sectional data, which are focused on 'who'
(made a purchase, filed a claim, etc.); a common variant is known as longitudinal (also known as
time series) data, which is where individual rows are focused on 'when', as in a point in time.

layout where 'purchaser' is the basic organizational unit and 'item', 'price' and other descriptors are attributed to individual purchases. Needless to say, looking beyond the simplistic example described here, such fundamental restructuring can be computationally and otherwise challenging.

While specifics of preparing data for analyses are inescapably situational, there are nonetheless recurring general data feature engineering considerations. Those can be grouped into data correction, best exemplified by missing value (if there are missing values) imputation and outlier identification and resolution; data enhancements, typically taking the form of recoding of existing into desired values or creation of specific indicators[12]; data aggregation, which entails computing of higher-order summaries, as in rolling up item-level (e.g. SKU) data up to brand level; and *file amalgamation*, or combining all or select contents of two or more separate data files. As could be expected, each of those general data feature engineering activities entails numerous steps which typically require the ability to manipulate data and the ability to correctly interpret coding and meaning of individual elements of data. While those technical details fall outside of the scope of this overview, there are numerous other sources that offer in-depth overview of those considerations; those specifics notwithstanding, it is essential to never lose sight of the fact that time and effort invested in thoughtful and thorough data feature engineering will pay dividends in the form of enhanced validity and reliability of data analytic outcomes, more parsimonious models and explanations, more informative results and more data analytic flexibility.

5.3.2   Derived Effects

Implied in the overview of data due diligence, and particularly in the context of data feature engineering, is the idea of using original data as bases for creating new variables, framed here as (data) derived effects. More formally defined, *effect* is a metric which is created expressly with a specific analytical purpose in mind; as such, raw data-derived effects are inextricably tied to a particular data analytic pursuit. Although in theory there can be an infinite number of such custom variables, they can be grouped into three basic categories of indicators, indices and interaction terms.

Indicators are (typically) dichotomous measures created to denote the presence or absence of an event, trait or a phenomenon of interest. Often referred to in practice as 'flags' or 'dummy-coded' variables, indicators are most frequently expressed as nominal variables, which means their analytic applicability is relatively

---

[12]An example of recoding of existing into new value could be the creation of tiered 'customer value' categories (such as 'high', 'average' and 'developing' groupings) out of a continuous 'customer spending' variable; an example of indicator creation (which commonly take the form of binary, yes vs. no, values) could be derivation of an event indicator, such as 'litigation' from a field containing a date when a lawsuit was filed.

constrained; coding-wise, they can be numeric (e.g. 0–1) or non-numeric (e.g. yes-no), though interpretation-wise the two connote the same meaning. A relatively common indicator example is offered by the use of 'date' field: An automotive insurance company maintains a listing of its policies which includes the key dates, such as the 'date of issuance' and 'claim date'; using the latter of the two dates, a binary indicator can be created so that all policies are effectively categorized as either having incurred a claim (where the presence of a date value would trigger a code of 'yes' or '1') or not (where the absence of a date value would trigger a code of 'no' or '0'). The resultant 'claim-no claim' indicator could then be used as the target variable in claim likelihood modelling.

The second category of derived effect types, indices, is notionally reminiscent of indicators in the general sense of capturing latent properties but is distinct because it is focused on 'rate' rather than 'state'. In more operationally clear terms, indices are mechanisms aimed at resolving the challenge often posed by how outcome tracking data are recorded and stored, which when considered from data utilization perspective confounds cross-sectional and longitudinal structures, as graphically illustrated by a snapshot of official COVID-19 tracking data[13] shown in Fig. 5.5.

The two highlighted rows – date and state – illustrate the confounding of 'where' (state) and 'when' (date) which manifests itself in non-uniqueness of individual rows, which is in violation of the assumption of independence discussed in the context of data feature engineering (in other words, when looked at from cross-sectional perspective there are duplicate 'date' records and when looked at from longitudinal perspective there are duplicate 'state' records). An obvious solution is to reduce the dimensionality of data exemplified in Fig. 5.5, which could be accomplished by eliminating either the 'time' dimension (by collapsing or aggregating date records for each state) or the 'place' dimension (by aggregating state records for each unique

Date	State	Positive	Negative	Hospitalized	Recovered	TestResults	ResultsIncrease
8/11/2020	AL	103851	679838	1506	37923	783689	3122
8/11/2020	MO	188737	848538	1574	26102	1037275	9044
8/11/2020	CA	574411	8611868	6759		9186279	187926
8/11/2020	TX	51039	546143	309	5579	597182	5342
8/10/2020	AL	542792	3506483	6753		4049275	35418
8/10/2020	MO	222588	1672897	2881		1895485	33523
8/10/2020	CA	49329	474591	244	38033	523920	2533
8/10/2020	TX	25100	183305	189	9341	208405	1929
8/9/2020	AL	198248	2950755	1459		3149003	41362
8/9/2020	MO	31730	294939	216	1319	326669	0
8/9/2020	CA	133125	1435439	1335	89083	1568564	20631
8/9/2020	TX	121707	1224637	387	99021	1346344	15068

Fig. 5.5 The simultaneously cross-sectional and longitudinal data layout

[13] The original COVID-19 tracking data sourced from usafacts.org (the source of CDC's data); as shown, data are not valid (values were changed and rearranged to highlight the desired effects) and should only be used for illustrative purposes.

value of date). In that context, the use of index variables aids in minimizing of loss of information by capturing the variability contained in the collapsed dimension.

The rationale behind the third and final derived effect type – interaction – is somewhat more complex. Interactions are combinations of two or more original variables intended to capture the combined effect of those variables that goes beyond their individual impact. The need to create interaction effects emanates from two, largely distinct considerations. The first is comparatively straightforward: To fully explain some phenomena, it might become necessary to account for combined effects of two or more predictors. An obvious example is a thunderstorm, which cannot be fully appreciated without expressly taking into account the combined effect of lightning, thunder and rain. The second interaction effect necessitating factor is a bit more obtuse, as it emanates from the common assumption of predictor independence, the goal of which is to diminish potential explanatory redundancy and by doing so to increase informational parsimony.[14] Inspired by that rational precept, multivariate statistical analyses, such as regression, generally aim to reduce collinearity, or correlations between predictors, which means that if two predictors are highly correlated, only one (typically, whichever one exhibits stronger association with the target variable) will be used in crafting the final data analytic solution. However, unless two predictors are perfectly correlated and thus 100% redundant, the otherwise sound collinearity reduction provision will inescapably lead to loss of some information, but that loss can be minimized when an interaction combining the effect of those two variables is created as a standalone measure.

While the preceding brief overview implies two-way interactions (two-variable combinations), interactions can entail joining together three or more individual variables. Moreover, even if only two-way effects are considered, the number of possible interactions can be quite large because it is exponentially related to the number of variables under consideration. For example, a dataset containing just ten variables can, at least in theory, yield 1013 separate interaction terms.[15]

5.3.2.1 Informational Domain Specification

A distinct but closely related data preparation-related set of considerations is encapsulated in the abstract idea of fully and completely accounting for all potential informational content. It is rooted in the belief that available raw data may not represent all possible data analytic inputs that could be used to produce maximally robust informational outcomes of interest and that additional inputs could be derived from available raw data, as discussed earlier in the context of the three raw data-derived effects. Framed here as *informational domain specification* (IDS), it is an

---

[14] This line of reasoning stems from the principle of parsimony commonly known as Occam's (also spelled Ockham's) razor, which states that entities should not be multiplied without necessity; in other words, if two explanations account for all pertinent facts, the simpler of the two is preferred.

[15] An application of combinatorial computational logic, Number of Interactions = $2^k - k - 1$, where k is the number of variables.

informational mapping process aimed at drafting a conceptual model that accounts for all potential informational output-shaping contributors in the form of the original raw data and any raw data-derived indicators, indices and interaction terms. When considered from the information planning perspective, IDS ties together several competing data analytic considerations, most notably the scope of inputs under consideration, operational comparability of selected variables (e.g. invariant level of aggregation) and informational parsimony, with particular emphasis on reducing the prevalence of non-generalizable factors.

In terms of the outcomes of the IDS process, the informational domain can be just-, over- or under-specified. Ideally, an informational domain is 'just-specified', which is accomplished when the conceptual analytic model encompasses the optimal mix of sufficiently informative and generalizable factors. It is but no means a simple task, primarily because of a general lack of tried-and-true objective benchmarks that could be used to guide the IDS process. Reflecting that difficulty, in practice, just-specified informational domain is one which is neither under- nor over-specified, where under-specified informational domain is one that relies on too few and/or too general inputs which then yield excessively generic informational outcomes that ultimately offer scant decision-guiding value, while over-specified domain is one that makes use of either excessively situational or altogether too many inputs, which ultimately results in unworkably complex or overly situational informational outcomes.

5.3.3   Data Utilization

While important, even critical, data due diligence and data feature engineering are ultimately means to an end, which is extraction of decision-guiding insights out of the vast and diverse data available to virtually all commercial and non-commercial organizations. Characterized here as data utilization, the broadly defined process of transforming initially non-informative transactional, social, online search and other raw data into meaningful informational outcomes is at the core of data-enabled organizational learning. Recalling the earlier (Chap. 1) discussion of adaptive, generative and transformative learning modes (geared towards mostly automatic changes needed to accommodate environmental changes, seeking new decision-guiding insights, and transformative, and shaping new strategic vision, respectively), available data can be used to support a wide range of organizational initiatives from which it follows that manifestly the same data can be analysed in a variety of different ways, all geared towards different types of end user utility. Moreover, the manner in which available organizational data are utilized is also influenced by organizational data analytic maturity, a summary notion that encompasses the presence and, more importantly, the utilization of modern techniques and tools of data management, analysis and communication; Fig. 5.6 offers a high-level categorization of modern data utilization modalities.

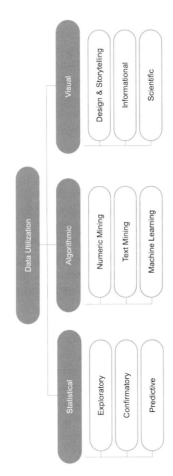

Fig. 5.6 Modes and means of data utilization

The statistical mode of data utilization started out as a series of attempts by renowned seventeenth- and eighteenth-century mathematicians, most notably Pascal, Bernoulli, Bayes and Gauss, to mathematically describe the probability of chance-driven events, and blossomed into its own domain of applied mathematics. Subsequent contributions by Fisher, Pearson, Tukey and numerous other thought leaders expanded that originally narrow focus to also encompass descriptive analyses, hypothesis testing and design of experiments, slowly but persistently building a diverse repository of mathematically sound ways of extracting insights out of data. Given the discipline's long roots, statistical data analyses are the most established, in a sense maturity, of the three broad data utilization avenues, offering a rich array of tried-and-true techniques aimed at discerning, explaining and forecasting.

Although not all,[16] the commonly used statistical methods of analysis of data are subject to assumptions imposed by the underlying parameterized (i.e. expressed as a function of independent quantities called parameters) families of probability distributions. The consequence of that dependence is that in order to yield valid and reliable outcomes, parametric statistical analyses need to assure that input data either fit or can be made to fit (via mathematical transformations) those assumptions, which is not always feasible. An alternative set of data analytic techniques, often broadly referred to as machine learning algorithms and here labelled simply as *algorithmic* methods, does not impose such assumptions on data – rather than being derived from mathematically described (i.e. parameterized) probability distributions, those methods are rooted in mathematical optimization, which aims to select the best element (with regard to a specified criterion) from a set of available alternatives. So, whereas statistical analyses aim to fit the data to a predefined distribution as a way of producing statistical estimates, algorithmic techniques use repeated trials to find the best solution, without imposing any particular assumptions on input data. Implied in that distinction is that efficacy of statistical analyses is highly dependent on data fitting the distributional and other demands of specific techniques, whereas efficacy of algorithmic methods is largely a function of the ability to compute large numbers of successive trials; it is worth noting that the difference between statistical and algorithmic data utilization approaches is not just methodological in nature – the former are human expert involvement-intensive, while the latter lend themselves to considerable amount of automation. Not surprisingly, the rapid proliferation of fast-yet-inexpensive computing power has led to an explosion of interest in machine learning or algorithmic methods of data utilization.

The third and final family of data utilization modes and modalities summarized in Fig. 5.6, *visual* data utilization, is almost entirely a product of technological advancements, specifically data visualization software applications. The difference

---

[16] Nonparametric statistical methods are not constrained by parameterized (i.e. expressed as a function of some independent quantities called parameters) families of probability distributions; some of the commonly used parametric statistics, such as one-way ANOVA or t-test, have nonparametric alternatives (Kruskal-Wallis and Mann-Whitney, respectively). That said, not all parametric methods have nonparametric alternatives, and the latter are not nearly as well-known.

between data visualization and the other two data utilization modes (statistical and algorithmic) discussed earlier is twofold: The first is that data visualization tools are primarily geared towards descriptive summarization, whereas statistical and algorithmic approaches encompass a far broader array of data analyses, as summarized in Fig. 5.6. The second key difference is the ease of use and interpretational clarity – in general, graphical depictions of patterns and relationships tend to be easier to interpret than the more methodologically involved statistical or algorithmic analyses. As such, visual data utilization can be seen as a manifestation of the steadily growing interest in using data as a driver of the ever-expanding arrays of organizational decisions.

The preceding brief characterization of the three core modes and means of data utilization is merely a general introduction, which looks past numerous methodological differences, as well as the main focal areas comprising the distinct data utilization meta-category. A more in-depth examination of each meta-category is offered next.

5.4   Statistical Data Utilization

Tracing its origins to the first mathematical treatments of probability in the seventeenth century, *statistics* is now a mature, well established and also the most methodologically diverse of the three data utilization avenues summarized in Fig. 5.6. With that in mind, as framed here, the statistical mode of data utilization does not encompass the entire domain of statistical knowledge – rather, it is just focused on several distinct sub-domains comprising the core means of statistical dimension of organizational learning with data, defined earlier as systematic and sound extraction of decision-guiding knowledge out of available data. In that still broad context, statistical analyses can advance one of two general objectives: (1) to explore available data for meaningful and material insights, characterized here as 'what-is', and (2) to speculate about the future in probabilistically sound manner, characterized here as 'what is likely to be'. The former can take the form of either descriptive or confirmatory analyses, and the latter falls under the general methodological umbrella of predictive analyses. The underlying factual-speculative continuum, graphically summarized in Fig. 5.7, offers a general framing of the three distinct modalities of statistical learning in terms of their relative degrees of either identifying currently emerging patterns and association (the 'factual' end of the continuum) or making probabilistic projections regarding the future (the 'speculative' end of the continuum).

Fig. 5.7 The factual-probabilistic continuum of statistical learning

It is important to note that the factual-speculative continuum is meant to support 'degree of', rather than 'yes vs. no' type of interpretation. It is meant to communicate that exploratory, confirmatory and predictive analyses all encompass elements of fact-based and probabilistic learning; that said, descriptive analyses skew towards fact-based conclusions, while predictive analyses skew towards probabilistic insights.

5.4.1 Exploratory Analyses

At the most rudimentary level, the goal of data exploration is to summarize 'what-is' using numeric and graphical summaries. Judging by ubiquity of bar graphs, pie charts, histograms, tabular summaries, proportions and averages in newspaper or magazine articles, industry reports and innumerable other communications and the widespread popular usage of the ultimately statistical notion of correlation, descriptive exploratory analyses are the go-to statistical data utilization modality. Discovery-focused analyses can be either univariate or multivariate, where the former are focused on summarizing individual variables, while the latter aim to understand patterns and relationships shared by two or more variables. As such, univariate exploratory analyses are geared towards discerning the informational content of individual variables on standalone basis, whereas multivariate exploratory analyses are focused on capturing cross-variable interdependencies. The earlier mentioned pie charts, histograms, proportions and averages are all examples of univariate analyses, while correlations are perhaps the most commonly seen embodiment of multivariate analyses.

When considered in the confines of the multifaceted idea of digital literacy (Fig. 5.1), the ability to correctly use and interpret descriptive univariate and multivariate statistics can be considered a part of the foundational data analytic skillset. The development of those skills is itself rooted in understanding of basic properties of datasets and the impact of some of those properties, which include manifest qualities such as variable-level measurement scale or implied properties best exemplified by the notion of outliers,[17] on the interpretation of data-encoded informational content. While a given dataset can be either structured or unstructured, and it can contain numeric, text or symbolic data elements, historically, the vast majority of exploratory analyses were focused on structured, numeric data, largely because those data lend themselves to easy to construct and interpret graphical and tabular summarizations. However, the rise and proliferation of unstructured, non-numeric

---

[17]An anomalous value; it is straightforward as an idea, but in practice outlier can be difficult to objectively delineate because there is no clear mathematical definition of what constitutes an outlying value. In principle, two different analysts looking at the same set of data can see different sets of values as outlying; that said, a number of graphical (e.g. normal probability plots) or computational (e.g. Grubb's test for outliers) have been developed in hopes of reducing the degree of subjectivity.

datasets coupled with advances in natural language processing techniques and technologies is continuously stretching the traditionally narrow boundaries of data exploration. And though when compared to highly evolved and well-established structured numeric data exploration approaches and tools, unstructured text data exploration mechanisms are best characterized as emerging, their capabilities, usage and importance are showing sharp upward growth trajectory; still, at present (the early part of 2021), the bulk of data exploration efforts are still focused on structured, numeric and numeric-like (e.g. categorical values that are non-numerically coded but lend themselves to numeric-like analyses) data.

When describing the contents of a structured dataset, it is essential to correctly interpret the implied measurement of individual variables comprising that dataset, which, as noted earlier, is a conjoint of measurement scale, which can be nominal, ordinal (together commonly grouped as categorical) or continuous, and variable encoding, the most common of which include numeric (digits only), string (text or alphanumerics), date and restricted values, such as integers with leading zeros. The variable-level implied measurement is the core determinant of what specific graphical, tabular or numeric form of summarization is most appropriate; that type of foundational due diligence is essential if one is to avoid common data summarization mistakes, such as using histograms to plot categorical data (bar charts should be used for that purpose).

A variable can be either continuous or categorical, and the latter can be further broken down into nominal and ordinal, though that distinction has a more pronounced importance in the context of inferential analyses. Being informationally richer (in the sense of permitting a wider range of mathematical operations, which in turn yields wider array of informational outcomes), continuous variables offer more summarization options – in fact, continuous quantities, such as 'age', can be summarized using numeric, tabular and graphical means. More specifically, summaries of continuous quantities commonly make use of numeric measures of central tendency (mean, median or mode) and dispersion (standard deviation, coefficient of variation), as well as graphical portrayals of aggregate distributions using histograms; it is also worth mentioning that since informationally richer continuous values can be always recoded into informationally poorer categories (the opposite is not possible), originally continuous quantities can also be binned into distinct categories and summarized tabularly or using bar charts. It follows that there are fewer options for describing categorical variables; under most circumstances, bar graphs or pie charts, preferably with embedded numeric raw frequency or percentage values, offer the most effective means of summarizing categorical variables. As noted earlier, it is important to not confuse the similar-looking bar graphs and histograms – the former ought to be used for describing categorical variables, while the latter should be used to summarize continuous values. An easy way to remember the difference is that in a bar graph the individual categories (bars) can be reordered without a corresponding change in informational content (e.g. in a bar graph summarizing, for instance, gender, the ordering of individual categories has no bearing on their frequency), whereas that is not the case with histograms where individual bars represent binned segments of an underlying continuum, which means that reordering of individual bars would have a direct and significant impact on informational content (as it would amount to reshaping of the underlying distribution).

Looking beyond summaries of individual variables, multivariate descriptive analyses are often viewed as a natural next step in summarizing the informational content of a dataset. And once again, the variable-level implied measurement is the core determinant of the underlying how-to, as distinctly different techniques are called for when assessing relationship between two categorical variables, two continuous variables and a continuous and a categorical variable.

Before delving into the how-to of assessing those three general types of relationships, a word of caution: Although methodologically straightforward, analyses of associations can be riddled with hidden traps, as illustrated by Simpson's paradox, which posits that the direction or the strength of a relationship of interest may change when data are aggregated across natural groupings that should be treated separately. It thus follows that care needs to be taken when searching for generalizable relationships; at time, it could be more analytically prudent to settle for narrower but more dependable conclusions. Another hidden trap is the possibility of presence of intervening or moderating factors, which is when an outside variable moderates a particular relationship, something that usually manifests itself in the changing strength or direction of the relationship of interest across values of an outside variable.

5.4.1.1 Relating Categorical Variables

A commonly used approach to assessing the relationship between categorical variables is *crosstabulation* (or crosstabs for short), also known as contingency tables; it relates frequency counts across variables with the goal of discerning underlying relationships. Though the most direct outcome of crosstabulation are concurrence counts, χ^2 (chi square) tests of difference are frequently utilized to assess the level of statistical significance of observed concurrence counts. In a more operational sense, since crosstabs involve construction of matrices, where variables are usually expressed as columns and individual categories as rows, it is usually advantageous to keep the number of categories relatively low.

In a statistical sense, χ^2 is a non-parametric test, which means that it places no distributional requirements on the sample data. That said, however, in order to yield unbiased estimates, the test requires the sample to be random, data to be reported as raw frequencies (not percentages), the individual categories to be mutually exclusive and exhaustive and the observed frequencies to meet the minimum count requirements (the often-cited rule of thumb calls for a minimum cell[18] size of five frequencies; in practical applications, however, a minimum sample size of 50 is more reasonable to assure robust takeaways). The test itself is relatively simple as it

---

[18] A cell is a unique conjoint, or a combination, of a variable and a level; for instance, a 2 × 2 test would be comprised of four individual cells, as there are two variables each with two levels.

compares the difference between the observed and expected frequencies in each cell to determine if significant patterns of similarities (i.e. a relationship) exist between variable pairs; that simplicity, however, comes at a price as the test results are not indicative of the strength of the relationship. With an eye towards remedying that shortcoming, a separate measure, known as Cramer's V, was developed to quantify the strength of χ^2-significant relationships (for crosstabulations larger than 2×2). Cramer's V coefficient value ranges from 0 (no association) to 1 (perfect association); thus its interpretation parallels the well-known Pearson's correlation coefficient, which is discussed next.

5.4.1.2 Relating Continuous Variables

Paralleling categorical variables' crosstabs, *correlation* analysis is a widely used and informationally richer (on the account of continuous variables permitting a wider range of mathematical operations) method of assessing associations between continuous variables. In contrast to crosstabulation which can only detect presence of an association, correlation captures the strength as well as the direction of association between pairs of variables.[19]

Formally defined, correlation is a concurrent change in value of two numeric variables, which represents the degree of relationship between those variables. As it is intuitively obvious, the correlation-expressed relationship can be either positive, when an increase in one is accompanied by an increase in the other, or negative, which is when an increase in one is accompanied by a decrease in the other. A numeric result of correlation analyses is a correlation coefficient, which is a metric expressing the strength of the relationship between the two variables of interest, ranging from +1 (perfectly positive correlation) to −1 (perfectly negative correlation) and centred on 0, which denotes a lack of relationship; Pearson's correlation coefficient is the most frequently used application of the general correlation analysis idea. Lastly, to be considered 'real' or 'material', the correlation coefficient needs to be statistically significant.

It should be noted that in contrast to the comparatively narrow in scope crosstabulation, correlation should be viewed as an umbrella term that encompasses numerous correlation methodologies and techniques. Aside from bivariate correlation summarized above, there are several other expressions of correlation-based associations: Partial correlation measures the association between two variables

---

[19] It is important to not confuse the often-seen multivariable summaries presented by correlation matrices with the fact that all relationships contained in a correlation matrix, regardless of the number of variables comprising such a matrix, are ultimately bivariate or pairwise. This is clearly evidenced by the fact that correlation matrices are two-dimensional (i.e. comprised of rows and columns); thus all correlation coefficients are reflections of associations between a given element in the row and another element in the column; this suggests that trying to capture association among three or more variables would be methodologically complex and practically limiting, primarily because it would necessitate the use of conditional expressions and/or interaction terms discussed earlier.

controlling for the impact of other variables; the similarly named part correlation resembles partial correlation, except that the impact of other variables is only controlled for one of the two correlation measures. Regression analysis, a widely used family of techniques geared towards estimating the relationship between a single dependent variable (which can be either continuous or categorical) and a set of multiple predictors, uses the general logic of correlation analysis to estimate strength of association between the dependent variable's actual values and the best predictions that can be linearly computed from the set of predictor variables (that particular correlation variant is called multiple correlations). Lastly, there is also canonical correlation, the goal of which is to estimate the relationship between sets, rather than individual variables. As such, that particular variant of correlation analysis is primarily of interest to theoretical researchers because correlating sets of variables is ultimately aimed at trying to discern the underlying latent structure, which is largely a theoretical pursuit.

An important consideration that should not be overlooked when assessing the relationship between continuous variables is the nature the relationship, which can be linear or non-linear. A linear relationship is directly proportional, which means that change in one variable will always be accompanied by a proportional change in another variable; when graphed, linear relationship is depicted as a straight line. A non-linear relationship is when change in one variable does not correspond with constant change in another variable; when graphed, non-linear relationship is depicted as a curved line.

5.4.1.3 Relating Categorical and Continuous Variables

The logic of assessing the relationship between categorical and continuous variables is rooted in the core distinction between those two general types of measurements: The former's role is to define specific groupings, while the latter contributes specific measurable quantities that can be used to compare those groupings. For example, in order to determine if meaningful income differences exist for those with different levels of educational attainment, measures of educational attainment and income would first need to be identified, and some basis would also need to be chosen for determining what should be considered 'meaningful' differences. For instance, educational attainment was operationalized using 'level of education' metric, grouping individuals into three non-overlapping categories of high school diploma, 4-year college degree and post-graduate degree, while income was operationalized using a continuous variable of average earnings (the matter of 'meaningful differences' will be addressed shortly). Within that context, the methodological aspect of ensuing analysis could be framed as ascertaining the relationship between the categorical level of education variable and the continuous income variable, while the informational goal could be characterized as trying to determine if material differences exist among average incomes earned by those with only high school education, those holding undergraduate college degrees and those holding graduate degrees. That, in a nutshell, is the essence of relating categorical and continuous variables.

One-way analysis of variance,[20] commonly referred to as one-way ANOVA, is the most commonly used method of conducting the analysis exemplified above. It uses a test of difference known as the F-test to determine if there is statistical evidence suggesting an association between the categorical and continuous variables of interest. More specifically, it compares the means of two or more independent groups, e.g. the mean income for high school only graduates vs. the mean income for college graduates vs. the mean income for holders of post-graduate degrees, to determine if there are statistically significant differences, so as to warrant concluding that there is an association between the categorical and continuous variables of interest. It is important to note that it is a yes vs. no type of a comparison, meaning that a finding of statistically significant difference merely attests to there being some sort of an association, without specifying either the strength or the direction of that relationship (there are follow-ups, known as 'post hoc' tests, that can be used to provide the additional explanatory granularity).

5.4.2   Confirmatory Analyses

A more narrowly framed dimension of exploration minded data learning, confirmatory analyses are focused on testing of knowledge claims stated as formal conjectures or hypotheses, using established tools of statistical inference to assess the validity of knowledge claims. While those knowledge claims can arise out of a wide array of contexts, one particularly common source are exploratory analyses; in that sense, data exploration-generated conjectures are often subjected for formal validation procedures of confirmatory analyses.

Operationally, confirmatory analyses make use of known distribution statistical difference tests, such as F, t or $\chi2$, to compare observed to expected effects with the purpose of distinguishing between spurious and persistent relationships, using a mechanism known as *statistical significance testing*, itself a core element of the scientific method discussed at length in Chap. 1. (It should be noted that in this context, the idea of 'significance' is an expression of probability that what appears to be true is indeed true; it is not an expression of practical importance or materiality.) In a broader data analytical sense, confirmatory analyses can be carried out in a number of methodologically distinct ways, ranging from simple bivariate tests of association outlined in the previous section[21] to more complex multivariate analyses.

---

[20] The 'one-way' designation refers to the fact that the method allows only one categorical, typically referred to as 'factor', variable (as in the example used earlier where 'level of education' is the factor variable); a 'two-way' ANOVA allows two factor or categorical variables.

[21] The same statistical tests or techniques can be used for exploratory or confirmatory purposes as searching for new insights and testing the truthfulness of prior beliefs are informationally, not procedurally, distinct.

The latter encompass a wide array of distinct statistical techniques, which can be grouped under two general umbrellas of dependence and interdependence methods. Dependence techniques aim to test the existence, and, if so, the strength and the direction of *causal* statistical relationships between outcomes of interest (typically referred to as 'dependent' or 'target' variable) and one or more causal factors (commonly labelled as 'independent' or 'predictor' variables). Regression analyses, a broad family of statistical techniques including linear, logistic and ordinal, to name just a few most widely used, offer perhaps the best-known example of dependence confirmatory analyses. Using a dependence method, an analyst would be able to discern, for instance, which of a number of potential influencers of consumer brand choice have statistically material impact on consumer selecting a particular brand and would also be able to discern the direction, as in increasing vs. decreasing the likelihood of choosing the brand of interest, and the strength of impact of individual choice influencing factors. In the way of contrast, interdependence statistical methods aim to test the existence of co-occurrence-based associations, which do not presume cause and effect relationships (in other words, interdependence techniques do not designate variables as predictor and/or target). The widely used correlation analysis techniques[22] are perhaps the best-known application of interdependence statistical methods, used to estimate the direction and the strength of bivariate associations.

Given the core purpose of confirmatory analyses, which is to attest to statistical truthfulness of prior beliefs or conjectures, it is essential to assure the highest possible validity and reliability of data to be used. Moreover, the analysis-specific sample size should be calibrated to detect the minimum effect size of interest (which in practice means statistical power of 0.90 or higher, with 0.95 being the most commonly used threshold); inference criteria, which are conditions for accepting as true or rejecting of individual hypotheses, should be clearly spelled out, which also includes thoughtful selection of data analytic methodology (e.g. correlation to test the association between continuously measured variables or crosstabulation to assess the association between categorical quantities).

5.4.3   Predictive Analyses

The prior two manifestations of data utilization – exploratory and confirmatory analyses – are both focused on derivation and validation of decision-contributing insights, while the third general mode of data utilization – predictive analyses – offers means of projecting the likelihood of future outcomes of interest. For example, making use of its customer database, an athletic footwear company might use

---

[22] Pearson's product moment and Spearman's rank order are the two most commonly used types of correlation analysis; the former calls for both variables to be continuous and normally (i.e. symmetrically) distributed, while the latter is used when both variables are either skewed (i.e. not normally distributed) or ordinal.

exploratory analyses to identify distinct consumer segments and confirmatory analyses to delineate the key demographic and lifestyle attributes of purchasers of high-performance running shoes; the same company could then use predictive analyses to estimate customer-specific likelihood of buying high-performance running shoes. The company could then use the individual-level purchase probabilities as a more economical marketing tool by focusing its promotional efforts on those exhibiting the highest purchase likelihood. The generalized logic of predictive analyses is graphically summarized in Fig. 5.8.

The straight solid line ('random selection') shows that, if selected at random, a given subset of the customer base can be expected to account for a proportional share of purchases, as illustrated by the light-shaded region in the graph shown in Fig. 5.8. However, that does not mean that all individual customers have equal probabilities of purchase – it simply means that when selected at random, a given customer sample's mix of individual-level purchase probabilities will, in aggregate, yield a proportional size of the sample number of purchases; in other words, a randomly selected sample of, for instance, 10% of the customer base will generate about 10% of the overall sales. But given that some individual customers are considerably more likely to purchase a particular product, such as the earlier mentioned high-performance running shoes, if customer selection were to be based on individually estimated purchase probabilities, similarly sized customer sample could be expected to produce materially higher aggregate sales volume. In fact, as posited by Pareto principle,[23] being able to correctly identify customers exhibiting the highest purchase probabilities could allow the athletic footwear company (and many other commercial and non-commercial organizations) to realize, on average, 80% of its

Fig. 5.8 The generalized logic of predictive analyses

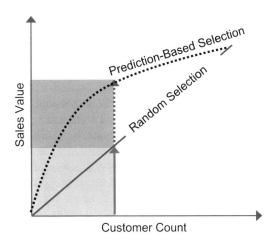

_____

[23] Also known as the 80/20 rule and named after Vilfredo Pareto, a late nineteenth-/early twentieth-century Italian economist, it states that for many outcomes, roughly 80% of consequences come from 20% of the causes.

sales from just 20% of carefully selected customers, as graphically illustrated by the dotted line ('prediction-based selection') and the darker-shaded extension in Fig. 5.8. When considered from the perspective of operational, in particular promotional efficiencies, that is clearly a very compelling idea. It is also the essence of predictive analyses.

5.5 Algorithmic Data Utilization

Walmart, the world's largest retailer,[24] alone processes more than a million of transactions per hour; in total, as of 2020, the aggregate volume of business-to-business and business-to-consumer transactions is estimated to be about 450 billion per day. Many of those transactions also generate pre- and post-purchase online searches, texts and online product reviews, all of which adds potentially informationally rich but hard to manage and analytically process details. And that is but one of numerous sources of data that are generated, captured and stored as a part of the ever-expanding digitization of commercial and non-commercial interactions; as so often discussed under the general heading of 'big data', modern electronic transaction and communication infrastructure produces staggeringly voluminous flows of data. It is easy to see that to be able to productively consume available data, organizations need to look beyond the slow and methodical statistical analyses; however, it is not only because statistical learning is time- and (human) resource-intensive but also because not all data usage situations are methodologically aligned with the logic of the scientific method, which is at the core of most statistical analyses.

Though it may not be always expressly stated, the focal point of the scientific method-centric analyses is derivation of valid and reliable generalizations, a pursuit that can be broadly characterized as theory testing or development. The basic tenets of that broadly defined data analytic approach are rooted in the idea that (1) currently on hand data can be considered a sample[25] and (2) the data analytic goal is to ascertain the likelihood that insights derived from analyses of the available (sample) data can be generalized onto the larger population. And indeed, there are numerous instances where organizational informational needs are aligned with those objectives, with pharmaceutical clinical trials offering perhaps the most visible (particularly during the global coronavirus pandemic) example; that said, there are numerous scenarios where that is simply not the case. For example, an organization seeking to adjust its pricing and promotional strategy in response to new competitor offerings

---

[24] Based on 2020 revenue of $510.33 billion; Amazon.com, with 2020 revenue of 232.88 billion, is a distant second.

[25] While the term 'sample' often conjures up images of a subset of data wilfully selected from a larger population, in a more general sense, any set of data that contains only a subset of larger the universe of entities is effectively a sample. For instance, a customer dataset maintained by a given company typically includes only those buyers of that company's products or services that are known to the company.

needs situationally and contextually specific insights, not abstract generalizations; moreover, to be beneficial, those insights need to be made available as quickly and possible. In that situation, the logic of the scientific method-centric statistical analyses is incongruent with informational needs at hand, and procedural mechanics could also be prohibitively cumbersome. That misalignment becomes even more pronounced in the context of decision automation, discussed in more detail in Chap. 7.

All considered, combination of sheer data volumes, informational-methodological alignment (or lack thereof), decision automation and information timing demands call for an alternative means of transforming raw data into usable insights; one such alternative is offered by the algorithmic dimension of data utilization, commonly referred to as machine learning. Procedurally and methodologically distinct from statistical analyses, it is perhaps best characterized by gradual distancing of processes used to translate unintelligible raw data into useful information from active human cognition. In a more methodologically explicit sense, in contrast to statistical data utilization which is built around mathematically described families of probability distributions, algorithmic data utilization methods can be seen as products of mathematical optimization, where repeated recomputation of a solution is performed until an optimum or a satisfactory solution is found.

Originally rooted in the standard conception of 'algorithm' as a finite sequence of precisely defined, computer-implementable instructions and aimed at performing specific computations, data analytic capabilities of present-day machine learning algorithms have evolved beyond those initial definitional boundaries. More specifically, the simplest algorithms can be characterized as 'static', because they continue to rely on unchanging initial logic to generate essentially the same (type-, not value-wise) computational outcomes, as exemplified by a customer loyalty scoring system that classifies customers into predetermined loyalty segments using fixed logic and classification criteria. Those comparatively simple algorithmic methods have been the staple of applied data analytics for decades, but more recently[26] a more advanced family of data analytic algorithms came into focus, known as 'learning' algorithms. Buoyed by advances in algorithm designs and the combination of rapidly declining cost and equally rapidly increasing power of computing systems, advanced data analytic algorithms are capable of learning from data, in the sense of being able to find progressively better solutions to stated problems. In addition, learning algorithms are also becoming gradually more and more self-reliant, a quality that also stands in a rather sharp contrast to the simpler static algorithms, which can be seen as essentially faster, more efficient means of carrying out repetitive computations. Still, while the now popular moniker of 'machine learning' implicitly refers to adaptive, self-improvement capable means of extracting useful insights out of data, the algorithmic dimension of data utilization summarized earlier in Fig. 5.6 encompasses both learning and static automated algorithms.

---

[26] In commercial availability sense; as shown earlier in Fig. 4.4, the backpropagation method for training artificial neural networks, which is the core design element of present-day machine learning systems, was first developed in the late 1960s.

5.5.1 Data Mining Versus Machine Learning

A casual web search reveals a large array of duelling views and definitions of data mining and machine learning; some see machine learning as a subset of data mining, others see data mining as a sub-domain of machine learning, and still others altogether blur the line between statistics, machine learning and data mining. Given the still evolving character of the domain of data science and analytics, coupled with lack of widely recognized standard-setting bodies[27], such lack of definitional clarity is not particularly surprising, but the resultant ambiguity is nonetheless a problem. With that in mind, in order to first conceptualize and then structure effective means of organizational learning, some degree of meaningful distinctiveness needs to be established separating the different modes of algorithmic data utilization. The goal here is not to dispute any of the existing views, but rather to offer usage-oriented perspective which, at the cost of possibly oversimplifying some aspects of data mining and machine learning, can nonetheless provide easier to follow and implement 'how-to' and 'when to' guidelines.

Figure 5.6 delineates three distinct sub-domains of algorithmic data utilization: numeric data mining, text data mining and machine learning; before delving into each specific area, it is worthwhile to consider the general distinction between data mining and machine learning. While the distinction between the two can get blurry at times due to some shared characteristics, looking beyond those communalities, data mining is framed here as a process of searching for potentially meaningful patterns or relationships using algorithmic, meaning programmed, logic rather than pre-existing human cognitive assumptions or beliefs. Machine learning, on the other hand, is framed here as deployment of advanced computer algorithms that are capable of experience-based self-improvement as a way of producing more informative outcomes. Stated differently, whereas finding patterns in data is the end goal of data mining, machine learning algorithms tend to see data patterns as means to an end, where the end could be either creation of more complex, deeper insights, as exemplified by machine learning-based predictive models, or execution of specific actions, as is the case with self-driven automobiles. That is an important distinction: The goal of data mining is to produce insights meant to inform human sensemaking, whereas the goal of machine learning is to generate decision-directing or decision-actuating information.

Given that general data mining vs. machine learning distinction, open-ended discovery-oriented data mining algorithms tend to be comparatively simpler, in terms of their logical and computational logic, than machine learning systems. As a

---

[27] The emphasis here is on 'widely recognized' standard-setting body; there are numerous organizations that are currently vying to become such norm-bearers, and those range from established (though manifestly not expressly data science focused) entities such as the American Statistical Association and the Institute for Operations Research and the Management Sciences (commonly known as INFORMS) to expressly data science-focused organizations such as Data Science Association, the Data Science Council of America, Association of Data Scientists and the peculiarly named Initiative for Analytics and Data Science Standards.

result, data mining systems tend to be more tailored to general data types, most notably structured numeric or unstructured text data, which is not necessarily the case with more complex machine learning systems; in fact, advanced machine learning application tends to incorporate data mining functionality as a part of their overall design. Keeping in mind the focus of this book, which is an overview of how commercial and non-commercial organizations can systematically and advantageously use available data to make better decisions, the scope of the ensuring overview of data mining and machine learning will be limited to just some key illustrative considerations.

5.5.2 Numeric Data Mining

Structured data, or data that conform to a tabular format with fixed and repeating relationships between rows (records) and columns (variables), and particularly structured numeric data have been a staple of applied data analytics for decades. The repeating character of defined and fixed data layout coupled with analytic friendliness of numerically expressed measures lend themselves naturally to automated data mining (ADM), where static algorithms are used to compress usually large volumes (millions, even billions of records) or raw data into informationally meaningful sets of patterns and relationships. Also known as knowledge discovery in data, or KDD, this particular approach to algorithmic data utilization is, in principle, unconstrained, as its implied goal is to look for any and all relationships.[28] In a more operationally specific sense, the discovery of patterns and relationships in numeric structured data can be focused on one of two general types of outcomes: search for associations, which aims to identify persistent co-occurrences among features (variables, as in data columns), or cluster identification, which is geared towards finding natural groupings of entities (records, as in data rows). A relatively common example of the former is the so-called market-basket analysis, the goal of which is to identify combinations of items that tend to be purchased together, while customer segmentation is a common application of the latter.

In fact, supermarket-basket analysis (search for most commonly purchased together product groupings in supermarket settings) is one of the earliest applications of ADM, dating back to the 1990s. According to the Food Marketing Institute, an average supermarket carries between 15,000 and 60,000 individual SKUs (stock-keeping units, or individual product variants with their own bar code), which means

---

[28] Some of the most commonly used data mining techniques include fuzzy query and analysis, case-based reasoning, and neural networks. Fuzzy, in contrast to crisp, query aims to extract the intent from a relatively vague semantics of a database query, as in finding 'good' or customers vs. identifying all customers whose spending is greater than a specified high-value threshold; case-based reasoning is geared towards identification of records that are similar to specified targets, while neural networks attempt to emulate human brain's learning processes in an effort to extract repeating, meaningful patterns.

that an average 'shopping basket' can contain a myriad of potential SKUs, but some of those product mixes can exhibit significantly higher recurrence. Identifying the most frequently encountered product combination can be beneficial from promotional planning perspective (e.g. discounting individual products carefully selected from non-overlapping groupings in hopes that those 'loss leaders' will generate sales of companion-but-not-discounted items); an analytically simple but computationally demanding task, it offered an almost ideal context for the development of early automated data mining applications. Interestingly, while the task of identifying recurring product mixes is straightforward, the large number of individual products (typically between 15,000 and 60,000) coupled with typically large number of individual transactions can result in thousands of recurring pairings, and there is usually no clear line of demarcation between 'frequently' and 'infrequently' occurring combinations. In that sense, the early application of ADM also drew attention to one of the core challenges of using automated algorithms to sift through large datasets: how-to differential between material and spurious associations. What seemed to be the most obvious solution to that problem, statistical significance testing, was – and still is – poorly suited to the problem at hand.

As discussed earlier, statistical significance tests (F-test and t-test for comparing mean differences between observed and expected values and the $\chi2$ test for comparing observed vs. expected frequencies) offer objective mathematical means of distinguishing between spurious and persistent relationships in the context of sample-to-population generalizability, which is notionally incongruent with the stated goal of many automated data mining undertakings. As illustrated by the supermarket-basket analysis, the 'classic' goal of ADM is to compress vast quantities of low-information details contained in individual data records into high-information summaries in the form of the most frequently co-occurring product purchases; there are no implied sample-to-population generalizations[29]; thus there is no manifest need for statistical significance tests. Moreover, even setting aside the notional incongruity, the efficacy of statistical significance tests is adversely impacted by sample size, more specifically, the larger the sample size, the higher the likelihood of concluding that magnitudinally trivial relationships are significant, in a statistical sense. In practice, the often-large number of cases used in data mining tends to produce a flood of statistically significant associations, many of which are magnitudinally trivial, yet in statistical sense all need to be deemed be equally 'significant'.[30]

---

[29] In a very practical sense, market-basket analysis is intended to capture what is true at a point in time, with the idea that as behavioural patterns change, the analysis should be replicated at future points in time; hence if there was a need to generalize, that need would manifest itself in 'now-to-future' rather than 'sample-to-population' generalizability (statistical significance testing is strictly focused on the latter).

[30] A conclusion that is usually reached based on a test-specific p-value, which is an expression of the probability (hence the 'p' in p-value) that an estimated parameter or an observed difference could have arisen by random chance; in general, the lower the p-value, the greater the statistical significance, or implied truthfulness of the computed estimates. For convenience, three distinct

Implied in record count-related shortcomings of applications of statistical significance, the level of significance should be considered jointly with effect sizes, or magnitudes of derived associations, which is already a common practice in applied analytics, but one that introduces its own biases in the form of subjective analyst interpretation.[31] Moreover, the very idea of implicit hypothesis testing, which is at the root of statistical significance, may be inapplicable in many data mining situations, particularly in situations where data being mined may be more aptly characterized as population than a sample or where analyses do not contemplate making sample-to-population generalizations (see footnote #29). On top it all, considering that the abstractly framed goal of data mining is discovery, the ultimate source of analysis-related uncertainty is not the sample-to-population generalizability-focused statistical significance, but rather it is the possibility of treating as material associations that are ultimately spurious, which can be encapsulated in the idea of *false-positive risk*. Here, there are two well-established, though more computationally involved, criteria that could be used to provide evidence in support of (or against) materiality of data mining-discovered associations: Bayes factor and Akaike information criterion, commonly abbreviated as AIC.

As used here, *Bayes factor* can be expressed as the ratio of the likelihood that the observed association is material and the likelihood that the observed association is spurious. And while the computational details fall outside of the scope of this overview, it should be noted that though notionally straightforward, estimation of the Bayes factor is computationally quite involved, though the Schwartz criterion offers computationally less taxing approximation of the Bayes factor (and some of the open source and other data analytic tools, such as R or Python, now offer easy to use packages). As an estimate, Bayes factor can be any positive number; its interpretation is commonly framed in the context of 'strength of evidence' in relation to false-positive risk, where the higher the estimated Bayes factor, the lower the probability that the observed association is spurious (or stated differently, the higher the probability that the observed association is material). A commonly use set of thresholds is summarized in Table 5.1.[32]

The second of the two p-value alternatives, *Akaike information criterion (AIC)*, may also be an appropriate p-value proxy if competing data mining solutions are under consideration. It assesses the extent to which data mining produced representation, seen as a simplification of reality, in the form of a pattern of observed associations, is a good reflection reality (within the realm of prediction-minded machine learning discussed later, it can be seen as a measure of predictive accuracy). It is rooted in the principle of parsimony, and the essence of its evaluative logic is

---

probability, or p-value, thresholds are commonly used: .01 (1%), .05 (5%) and .1 (10%), with the middle threshold being the most commonly used.

[31] For example, what is considered a 'meaningful' difference between two magnitudes or what is considered to be a 'material' correlation may vary, at times considerably not only among analysts or users of data analytic outcomes but also among usage situations.

[32] An adaptation of a widely cited table in Kass, R.E. and A.E. Raftery (1995), 'Bayes Factors', *Journal of American Statistical Association*, 90(430), p. 791

Table 5.1 Bayes factor tiers

Bayes factor	False-positive risk
3.2 or lower	High
3.2 to 10	Low
10 to 100	Very Low
Greater than 100	Negligible

contained in the trade-off between the goodness of fit (here, the extent to which the mining derived associations 'fit' the data) and the complexity of a particular representation of reality; its goal is to minimize overfitting, which in the context of data mining would manifest itself in identification of non-spurious associations. With that as the goal, the AIC score rewards solutions that are both simple and highly explanatory, and it punishes those that are comparatively more complex; interpretation-wise, a lower AIC score is better.

5.5.3   Text Data Mining

Text data are exceedingly common – by some estimates, many organizations have anywhere between two and ten times more text than numeric data. Understandably, there is strong interest in analyses of text, but that task is fraught with difficulties. Setting aside text-encoded categories such as 'gender' or 'marital status' where text is used to label a predefined set of discrete categories, text data tend to be intrinsically unstructured, largely because of the very manner in which meaning is encoded using text, which requires the use of continuously varying mixes of multiple elements such as subjects, verbs, objects and syntax. The result is a combination of potentially high informational value, which continues to fuel enduring interest in text mining, and high barriers to realizing that potential value. The latter manifests itself in two distinct considerations: (1) identification of associations, which requires patterns of repetition, in the sense of re-occurrence of the same value, and (2) algorithmic discernment of semantics, which calls for the processing algorithm to 'understand' and capture the implied (by unique combinations of terms and their logical connections) meaning contained in individual sentences. And so while the avalanche of online reviews, opinions and attitudes offers a tantalizing source of organizational decision-making-related insights, getting there is a journey that requires an often considerable commitment of effort and resources.

Process-wise, mining of text data entails four distinct steps: (1) retrieval, (2) summarization, (3) structural mining and (4) digitization. Notionally, retrieval can be seen as a combination of sampling, in the sense of identifying records of interest contained within a larger repository, and extraction, or outputting of selected records into a separate file; the end product of the retrieval process usually takes the form of an extract data file. The next text mining process step, summarization, involves condensing of typically large volumes of data; its outcome typically takes the form of an abstract or a synopsis. The next step in the process, structural mining, is probably the broadest as well as most complex aspect of text mining as it involves of (still) textual data into statistically analysable, categorized metadata. Depending on the combination of purpose and the type of data (briefly discussed below), structural mining can take the form of searching for predetermined expressions or attempting to surmise the deeper meaning hidden in the syntactic structure of text. The last general text mining process step, digitization, entails number coding of metadata, or converting non-numeric data values into numeric ones, which is typically required not only for computing basic descriptive statistics but even more importantly for amalgamating of text-derived metrics with structured numeric data. A common example is offered by the task of combining post-purchase product reviews (text data) with purchase details (usually, structured, numeric data) for the purpose of generating the so-called 360° customer view with the help of multisource analytics; a generalized process is graphically summarized in Fig. 5.9.

More general function-wise, text mining can be performed with several different outcomes in mind, such as identification of co-occurrences of themes in a body of text, reducing of documents' content to predefined sets of categories or to emergent (i.e. based on documents' content, rather than being predefined) categories, recasting of textually expressed information into graphics or selecting of subsets of text based on predetermined logic. Method-wise, mining of text data can take the path of either *frequency count and tabulation* (FC&T) or *natural language processing* (NLP). The FC&T approach itself can take the form of tagging of a priori identified expressions, which entails searching a body of data for specific expressions or terms that have been spelled out in advance of the search, or it can take the form of term funnelling, which entails exploratory identification of recurring terms. Tagging requires a substantial amount of knowledge on the part of the analyst, as focal terms have to be delineated ahead of the search; as such, it is deductive in nature. It follows that searching for terms that are believed to be important from the onset is not conducive to uncovering new truths, as the focused mechanics of deductive search make it difficult, if not practically impossible, to notice new recurring expression. On the other hand, since term funnelling requires no prior knowledge, it is inductive in nature and thus overtly focused on uncovering new truths, but it can produce overwhelmingly large quantities of output (tens of thousands of terms and/or expressions is common), which in turn demands substantial amounts of post-extraction processing. As such it is not only time-consuming but also likely to infuse potentially large amounts of rater bias, effectively reducing the objectivity of findings. Those differences notwithstanding, tagging and term funnelling are both only capable finding recurring expressions without considering the context or the way in which those terms are used; in other words, FC&T approach to text mining is generally incapable of capturing semantics-contained meaning, which includes distinguishing between positively and negatively valenced lexical items.

The second broadly defined approach to text mining – natural language processing – aims to remedy the semantic structure processing-related limitations of the FC&T approach by attempting to extract insights from semantic structure of text. NLP is an outgrowth of computational linguistics, a domain of knowledge focused

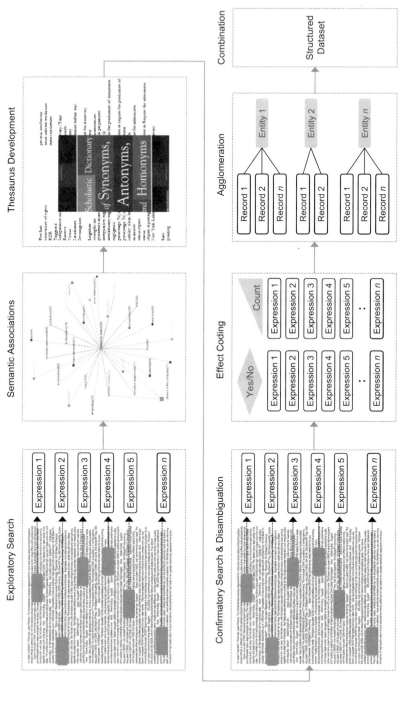

Fig. 5.9 Transforming text data into structured variables

on statistical and/or rule-based modelling of natural language; its goal is to capture the meaning of written communications in the form of tabular metadata amendable to statistical analysis. Inherently inductive, NLP can be particularly adept at uncovering new truths; that said, the task of algorithmically capturing and summarizing meaning of nuanced human communications is inescapably complex, which directly and adversely impact the validity and reliability of results.

Choosing between FC&T and NLP text mining modalities calls for thoughtful consideration of desired informational outcomes and characteristics of input data, as seen from the perspective of the broad text mining approaches. With that in mind, a more in-depth overview of presented next.

5.5.3.1 Frequency Count and Tabulation

As outlined earlier, the goal of the FC&T approach to text mining is to identify, tag and tabulate predetermined terms and/or expressions in a body of text (commonly referred to as 'corpora'), without considering the semantic context of lexical terms. In a more operationally clear sense, it can be either an exploratory tool or a confirmatory tool, the former manifesting itself in the use of term funnelling functionality and the latter manifesting itself in the use of tagging functionality. The main challenge associated with FC&T's exploration-minded functionality is often the sheer volume of identified terms, while the key challenge of confirmation-focused FC&T functionality is the accuracy and completeness of prior knowledge-based search thesauri. At their core, however, both the result volume and lexical completeness challenges stem from a common set of obstacles that stand in the way of algorithmic means of text processing.

The main reason exploratory searches commonly yield overwhelmingly large lexical item volumes and even the most carefully crafted search thesauri are more than likely to deliver incomplete findings is because of ambiguity stemming from wording or phrasing variability and the impact of synonyms and homographs. The word or phrase variability is a syntax (principles and rules used in sentence construction) problem that emanates from the fact that the same term or an idea can oftentimes be expressed with the help of lexically distinct terms. Given the prevalence of that problem, it is a common practice to use the so-called stemming algorithms which take advantage of Latin or Greek roots of English language words to recognize lexical terms' communalities. Dealing with synonyms, which are terms that have a different spelling but the same or similar meaning, and homographs, which are terms that have the same spelling but different meaning, is typically more involved, manual in a sense. The most common approach to addressing those sources of possible confusion is to create external reference categories, which are de facto libraries of terms delineating all known synonyms and homographs for all a priori identified search terms, clearly, a rather substantial undertaking.

Process-wise, the frequency count and tabulation approach to mining textual data makes use of text transformations, which is the process of translating words into digits, where each word is an attribute defined by its frequency of occurrence.

It is built around the notion of 'bag-of-words', which can be seen as data sensemaking-simplifying assumption recasting text to be analysed into unordered collections of words, where grammar and even the word order are disregarded. While disregarding of syntax and syntax-implied meaning typically leads to significant simplification of the mining process, it is important to note that the open-ended nature of unstructured text data nonetheless manifests itself in what could be characterized as unconstrained outcome set. In more practical terms, that means a still formidable disambiguation challenge, primarily in the form of resolving of potential conflicts that tend to arise when a single term or an expression can take on multiple meanings.

Overall, the FC&T approach to text mining entails a considerable amount of analyst input, while at the same time its primary focus is on the identification and classification of identifiable and definable terms and expressions, all while skipping over any deeper meaning that is often contained in the semantic structure of text. As noted earlier, that broadly framed approach offers, in principle, no clear way of contextualizing or otherwise qualifying search-identified terms, beyond merely pinpointing their occurrence and counting their subsequent recurrences. Looked at as a sum of advantages and disadvantages, the FC&T method can be an effective approach to extracting analysable details out of the otherwise inaccessible text data, but it is limited to just handpicking of out-of-context elements. And while in some situations that might be good enough, in other contexts it might be too limited in terms of resultant informational richness, and it could also even lead to an outright misinterpretation of some elements of text or the overall sentiment.

5.5.3.2 Natural Language Processing

The bag-of-words mindset at the core of the frequency count and tabulation approach outlined above almost inescapably leads to lumping together of stem-related but differently valenced terms (i.e. the same term used in the positive vs. negative context); as such, it is the source of the most consequential limitation of the FC&T text mining method. Stripping away of grammar that links together words contained in a sentence, or a sentence-like expression, leads to material loss of information; quite conceivably, it can even lead to an outright misinterpretation of the informational content of a given corpora. All that leads to an obvious conclusion, which is that to more completely capture the informational content of text, in the sense of substantially enriching the depth of text mining insights, it is necessary and to expressly consider the syntactical structure of text-encoded natural language.

Broadly conceived, natural language (which can be spoken, written or signed, though text mining is obviously only concerned with the written aspect of it) is human communication which is distinct and different from constructed languages, best exemplified by computer programming or mathematical logic. One of the key differences between constructed and natural languages is the largely unbounded character of the latter, which is at the core of challenges faced by algorithmic text sensemaking efforts. Formally defined, natural language processing (NLP) is an exploratory process of extracting meaningful insights out of the semantic structure of a body of text; approach-wise, NLP it can take one of two broadly defined forms: supervised automated classification or unsupervised mining. At their core, both types of methodologies expressly consider words, word order and grammar in trying to discern generalizable rules that can be applied to distilling large quantities of text into manageable sets of summarized findings; that said, the two are quite different in terms of their operational mechanics.

The first of the two NLP approaches, supervised automated classification, can be thought of as a pseudo-exploratory methodology as it is a type of supervised learning where a decision rule (an algorithm) is trained using a previously classified text. The training task is essentially that of establishing generalizable rules for assigning topical labels to content, where the efficacy of the resultant algorithm is, to a large degree, dependent on balancing of two important, though somewhat contradictory, notions of accuracy and simplicity. There are two distinct schools of thought that suggest somewhat different strategies for finding the optimal trade-off between those two competing goals: The first of those can be labelled 'knowledge-based', and it relies on codification of expert-derived classification rules; the second one can be labelled 'learning-based', as it relies on experts supplying classified examples rather than classification rules. Under most circumstances, knowledge-based systems are deemed more workable, primarily because the requisite training inputs are more obtainable, as it is often prohibitively difficult to compile adequately representative classification samples that are required by learning-based systems. The core challenge that needs to be addressed by both knowledge- and learning-based NLP systems is that of *overfitting*, which is when the classification algorithm ends up being overly tailored to idiosyncrasies of inputs used to train it. An overfitted classification algorithm will almost always manifest itself in poor generalizability – yet, to exhibit adequate classificatory accuracy, the algorithm needs to be capable of dealing with corpora-level specifics. Striking that difficult balance between the two – once again – contradictory notions of accuracy and generalizability is a challenge that cuts across not just the different modes of algorithmic data utilization, but all modes of computer-aided data utilization.

Unsupervised mining, the second of the two general NLP approaches, can be thought of as a purely exploratory methodology, but even those purely exploratory text mining mechanisms do not represent, in a strict sense of the word, truly independent machine-based processing of human communications. Broadly characterized, unsupervised text mining methods leverage similarity/difference heuristics, such as hierarchical clustering techniques, that are informed by text records' own informational content, rather than relying on information derived from previously classified text. That said, it is important to note that although the mining itself is unsupervised, the general rules within which it is conducted are governed by explicit vocabulary control, which typically takes the form of human expert-constructed thesauri. The individual terms comprising a particular thesaurus are commonly noun phrases (i.e. content words), with the meaning of those terms restricted to that which is deemed most appropriate for the purpose of a particular thesaurus, which in turn is shaped by the context of the search. In terms of lexical inference, or deriving meaning from grammar-connected word combinations, the human expert-provided thesauri also need to expressly define three main types of cross-term relationships: equivalence, which is critical to resolving synonym-related ambiguity, and hierarchy as well as association, both of which are necessary to extracting semantic insights or imbuing meaning to multi-term expressions.

All considered, natural language processing is still a probabilistic science yielding a moderate degree of success; stated differently, given the inherent difficulties of attempting to algorithmically synthesize and summarize nuanced human communications, it is important to have reasonable expectations. In general, the quality of outcomes is highly dependent on the nature of data and the desired utility of findings – the more constrained, focus-wise, the data, the better the quality of outcomes. The reason for that is that narrower text mining context usually results in smaller differences between search inputs, most notably classification rules, examples and thesauri, and the text to be analysed. It thus follows that in contrast to numerically encoded data, which can be analysed with the expectation of highly valid and reliable results, analyses of unconstrained and context-dependent text data cannot be expected to yield comparable (to numeric analyses) levels of efficacy. In fact, the very task of objectively assessing the validity of text mining results is fraught with difficulties, not the least of which is the scarcity of evaluative benchmarks. More specifically, whereas the results of numeric analyses can be cross-referenced in a manner suggesting possible inconsistencies, the use of such evaluative methods is rarely feasible within the confines of text mining.

5.6   Visual Data Learning

While popular wisdom teaches that 'picture is worth a thousand words', everyday experience warns that most pictures are not Van Goghs or Picassos. But given their rapidly evolving sophistication, data visualization tools have the potential to help even the least graphically inclined informational workers –paint like Picasso, at least in a manner of speaking. But there is a catch: Because data visualization tools make expressing data graphically easy, the distinction between 'easy to do' and 'effective' is not always clear. In order for graphically summarized data to be instructive, the creation of graphs, charts and other data visualizations needs to be driven by the story that is to be told, rather than the more mechanical task of summarizing of facts contained in data. But that is not all – design choices and intended messaging of planned visualizations need to be carefully considered, in the context of visual grammar and semantics, as well as the core data considerations.

5.6.1   Design and Storytelling Considerations

While well-crafted visual data representations are easy to interpret, the design of compelling visualizations can be surprisingly complex, because of the numerous, not immediately obvious, design considerations. For instance, what information do frontline workers need to be able to flag potential operational problems is, in all likelihood, substantially different from information needed by the company's executives engaged in strategic planning. Moreover, different 'front lines' – a retail warehouse vs. a hospital vs. a bank – may call for materially different information, just as different manifestations of strategic planning may require different types of informational inputs.

Data visualization design complexities can be simplified with the help of delineation of the key design dimensions. When considered in the qualifying context of organizational usage, visualizations can be characterized in terms of their intended purpose, target audience and content, as graphically summarized in Fig. 5.10.

5.6.1.1   Purpose, Audience and Content

The most rudimentary consideration underpinning design of data visualization is the purpose, or the communication intent. While there might be numerous situational factors that often come into scope, overall, data visualizations can be grouped into two broad categories of expository, which aim to summarize objective information, and declarative, which aim to advance a particular perspective; stated differently, data visualizations can either aim to inform or to persuade. Those broad communication intents are well illustrated by the two widely used data visualization tools: dashboards and scorecards. A dashboard is a templated collection of data visualization elements, or charts, that provide at-a-glance overview of metrics of interest; a common example is a snapshot of business performance expressed in terms of carefully selected mix of key performance indicators (KPIs). As such,

Fig. 5.10 Key data
visualization design
considerations

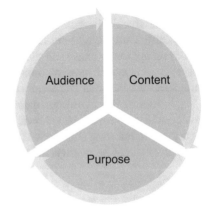

dashboards typically provide whatever information is needed to examine something, without directly asserting any conclusions. A scorecard, on the other hand, is expressly focused on measurement and evaluation; thus rather than merely summarizing 'what-is', scorecards are typically structured to offer a direct assessment of something.

That 'something' points towards the target audience, which tends to be particularly distinct in hierarchical organizations, discussed in more depth in Chap. 2. In those traditional, command-and-control-type management structures, operational tactics-focused line or staff members usually need more narrowly scoped and more granular, in the sense of the level of detail, information than the more broadly and strategically minded senior organizational leaders. That division is not nearly as well pronounced in the more progressive, shared governance organizational models, such as flatarchies or podularities (also discussed in Chap. 2), where the distinction between day-to-day operational and long-term strategic decision-making is considerably more fluid.

In terms of types, in a very general sense, data visualizations can be either factual or conceptual, where the goal of the former is to inform or persuade, while the goal of the latter is to explain or to teach. Notionally related to the earlier discussed purpose, which captures the intent of what is to be communicated, visualization type accounts for the content of what is to be communicated. However, as explored in more detail in the next section, the general factual vs. conceptual distinction hides a rich array of distinct data visualization modalities.

5.6.1.2 Visual Grammar

Adding onto the overview of design of effective visual communications is the choice of what chart to use, which in everyday context is probably the most top-of-mind consideration in data visualization. Broadly characterized here as *visual grammar*, or the choice of appropriate visual elements to communicate the intended meaning, it needs to expressly take into account differences between expository and declarative intents, as well as the differences between factual and conceptual content. Effective expository visualizations tend to emphasize completeness and objectivity, while effective declarative visualizations usually place a premium on sound logic and clarity. Well-thought-out factual visualizations, on the other hand, place a premium on appropriateness and the ease of interpretation of selected charts, in contrast to conceptual presentations which aim to capture the essence of the often-abstract ideas, as graphically illustrated by Fig. 5.11.

Considering that at its core, data visualization is an application of graphical arts to data, the simple examples shown in Fig. 5.14 are just two of many ways of graphically communicating data-derived or data-related information. And while the array of choices might be overwhelming, an organization known as Visual-Literacy.org

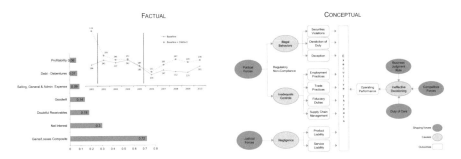

Fig. 5.11   Factual versus conceptual visualization

developed a Periodic Table of Visualization Methods,[33] which, in the manner similar to the familiar periodic table of (chemical) elements, delineates and organizes many common and some not so common forms of visual communication of information. In total, 100 distinct visualization types are grouped into six categories of data, information, concepts, strategy, metaphors and compound visualizations.

Focusing on what could be characterized as a garden variety of organizational data visualization charts, some of the more commonly used graphs include line and bar charts, histograms, scatter plots and pie charts, including its variation known as donut charts. Line charts, which show relationships between variables, are most commonly used to depict trends over time, as exemplified by cross-time sales trends; as of late, it has become customary to use line charts to capture multiple trends, by stacking multiple trend lines. Bar charts are most frequently used to summarize and compare quantities of different categories or sets, as in comparing total sales of different products. Visually similar histograms, which are often confused with bar charts, are best used for visually summarizing distributions of individual variables; the key difference between bar charts and histograms is that the former are meant to be used with unordered categories (e.g. when comparing sales of Brand X to sales of Brand Y only the relative bar height matters, which in the context of the standard two-dimensional Cartesian or x-y coordinates captures the magnitude of the two categories; in that context, ordering of bars on the horizontal or the x-axis is informationally inconsequential), whereas the latter are used with ordered categories, where both the height and order of individual bars combine to communicate the intended information. Scatter plots, also known as X-Y plots, capture the somewhat more obtuse joint variation of two variables; under most circumstances, scatter plots are used to convey a visual summary of either the spread of individual variables or the relationship, as in correlation, between variables. The last of the most commonly used tools of visual grammar are pie charts and their derivative form known as donut charts, which are typically used to show relative proportions of several groups or categories, as exemplified by the distribution of the total company revenue across product categories. It should be noted, however, that in spite of their rather common

---

[33] https://www.visual-literacy.org/periodic_table/periodic_table.html

DigDepth	rateReside	ateCommu	VacantLot	rkedSite_	SarkedSite_	jory_	Excavory_	Wateresponse_	B.fentProbat	ProbabilityTier
3.00	0	0	0	0	0	0	0	0	0.002	Elevated
3.00	0	0	0	0	0	0	0	0	0.002	Elevated
6.00	1	0	0	0	0	1	0	0	0.000	Low
1.00	1	0	0	0	0	0	0	0	0.002	Elevated
1.00	1	0	0	0	0	0	0	0	0.000	Low
0.00	1	0	0	0	0	0	0	0	0.000	Low
1.00	1	0	0	0	0	0	0	0	0.006	Elevated
0.00	1	0	0	0	0	0	0	0	0.000	Low
2.00	0	0	1	0	0	1	0	0	0.174	Elevated
2.00	0	1	0	0	0	0	1	0	0.006	Elevated
0.00	1	0	0	0	0	0	0	0	0.016	Elevated
0.00	0	1	0	0	0	0	0	1	0.014	Elevated
1.00	1	0	0	0	0	0	0	0	0.022	Elevated
3.00	1	0	0	0	0	1	0	0	0.000	Low
3.00	1	0	0	0	0	1	0	0	0.000	Low
1.00	1	0	0	0	0	0	0	0	0.000	Low
1.00	1	0	1	1	0	0	0	0	0.002	Elevated
3.00	1	0	0	0	0	1	0	0	0.000	Low
0.00	0	0	0	0	0	0	0	0	0.010	Elevated
3.00	1	0	0	0	0	1	0	0	0.006	Elevated

Fig. 5.14 Spreadsheet versus GIS-conveyed data analytic outcomes

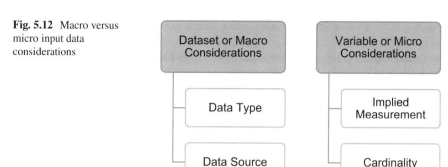

Fig. 5.12 Macro versus micro input data considerations

usage, value of pie charts as a communication tool can be questionable, especially when used with more numerous categories, because it can be difficult to visually surmise magnitudes captured by angled 'slices' comprising those charts.

5.6.1.3   Data Inputs

The preceding discussion of foundational data visualization design consideration in the form of clearly spelled out purpose and thoughtfully selected charting options needs to be followed by equally thoughtful and considerate examination of data to be used as input into contemplated visualizations. Figure 5.12 offers a graphical summary of the key aspects of data evaluation, grouped under a broad umbrella of macro and micro data characteristics – the former draw attention to aggregate properties of data collections (files), while the latter zero in on contents of individual data files. As shown in Fig. 5.12, macro data considerations highlight the importance of robust operational understanding of data type, which can be structured or unstructured, and on data source, which can be transactions, social networking, machine-to-machine networking, online search or crowdsourcing. Micro data considerations, on the other hand, underscore the importance of discerning variable-level implied measurement, which can be continuous or categorical (the latter itself comprised of nominal and ordinal categories), and cardinality, or uniqueness of data values comprising individual variables.[34]

Since building data visualizations most commonly takes place within specialized computer applications, which can range from general-purpose tool, such as Excel, to expressly data visualization-focused applications such as Tableau, developing a clear understanding of data types and sources which is also important from the more pragmatic standpoint of correctly utilizing functionality of those tools. A case in point is offered by the typically simple task of reading in of external data files, where seemingly trivial details can cause major headaches. For instance, structured transactional data commonly use some variant of delimited text; given that just about anything – comma, tab, space, colon, semicolon and slash – can be used to delimit individual values (comma and tab are the most common), it is important to know which specific delimiter is used in the file of interest, and it is also important to verify that the visualization tool of choice is able to read that particular format. A more complex data format example is offered by unstructured web-sourced data, which can be formatted in accordance with many different standards, of which JavaScript Object Notation (JSON) and Extensible Markup Language (XML) are the most commonly used general formatting structures, though it should be noted there are numerous JSON and XML specifications. While there are some visualization applications that accept JSON- and XML-formatted data as input, there are others that do not; in the end, there are times when the only way 'in' (to the data visualization application) is by reformatting of the entire file, which means restructuring it in a way that parses out and organizes individual values in a manner that renders them readable to the application of choice.

Variable-level micro data examination boils down to answering two simple questions: Is a particular data element continuous or categorical (implied measurement), and how variable are the individual values? Just as numerically describing continuous and categorical quantities entails the use of different descriptive statistics,[35] visualizing of continuous and categorical quantities calls for different types of charts. Those choices are particularly important to fully and accurately capture and communicate the essence of the information that is to be communicated.

---

[34] High cardinality indicates high proportion of unique values, as exemplified by 'Customer ID', where each value should be unique; low cardinality indicates high proportion of repeat values, as in 'Gender'.

[35] Continuous values are commonly described using measures of central tendency (mean, median, mode), which describe typical values, and measures of dispersion or spread (variance and its derivative, the standard deviation, as well as range); categorical variables are commonly described using simple frequency counts.

5.6.1.4 Storytelling

The short summary of data visualization would be incomplete without addressing what is, from the perspective of users, perhaps the most important aspect of data visualization: the story. Nowadays, it is not just the volume and variety of raw data that are overwhelming – the same holds true for data-derived information. The proliferation of easy to use, highly automated data reporting tools, coupled with the insatiable desire to summarize and report on just about anything that is countable, creates avalanches of sometime worthwhile but often trivial, if not outright superfluous informational tidbits, which effectively replaces data overabundance with informational overload. Organizations all too often confuse 'reporting' with 'analytics' which typically manifests itself in a rush to funnel anything-data into reports, losing sight of the fact that not everything that is countable should be counted. Whether the goal is just to inform or to persuade, to be effective, data visualizations need to tell a story that matters to someone for a discernible (and hopefully productive) reason. That is the essence of why it is so important to delve into the purpose and the type of planned visualization, to gain proficiency in the basics of visual grammar and to carefully and systematically discern the macro and micro characteristics of data that are to be used: Choosing effective visuals, avoiding clutter and emphasizing key points are commonly accepted hallmarks of a good visual data storytelling, but the harder to templatize logic of what pieces of information should be included and how to weave those elements into an easy to follow story is just as important, especially if the goal is to persuade. The familiar 'more is less' idea is important: Rich varieties of data create a temptation to say a lot, but effective visual data storytellers listen to the voice of informed restraint, always placing quality ahead of quantity.

5.6.2 Informational Visualization

To say that summary graphs and charts are ubiquitous smacks of an understatement – nowadays, it is almost hard to imagine a business report or a presentation that does not include at least some data visualizations. Be it a single histogram, line or pie chart or a multi-item scorecard (illustrated in Fig. 5.13) or dashboard, data visualizations are as ubiquitous as are data themselves. Broadly categorized here as informational visualizations, those graphical insight conveyance tools are used to communicate anything from simple descriptive summaries to results of complex, multivariate predictive models, as illustrated by the sample Securities Litigation Exposure Assessment scorecard in Fig. 5.13.

Aiming to succinctly summarize multiple and diverse sources and types of insights focused on assessing and contextualizing a sample company's exposure to securities litigation (civil lawsuits in which shareholders assert that the company's management violated securities laws), the scorecard depicted in Fig. 5.13 combines the results of company-specific predictive modelling and industry-wide descriptive benchmarking. Content-wise, the former, grouped under the Company-Specific

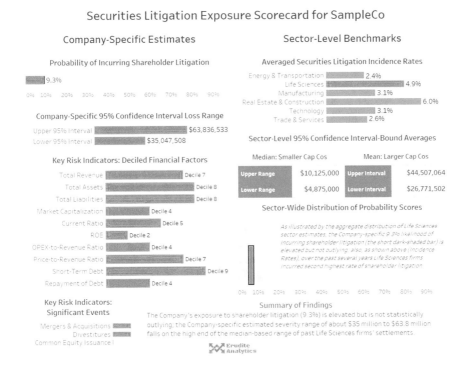

Fig. 5.13 Sample scorecard

Estimates heading, are topline summaries of complex, multivariate statistical models' forecasts, whereas the latter, grouped under the Sector-Level Benchmarks heading, capture comparatively (in the methodological sense) simple, peer-based averages and confidence interval-based ranges. The intended purpose of the scorecard is to offer an easy to grasp, highly conclusive summary of the most pertinent aspects of SampleCo's exposure to the threat posed by securities litigation (which falls under a broad umbrella of executive risk); lastly, the target audience of the scorecard is the organizational executive management and its board of directors.

While the sample scorecard shown in Fig. 5.13 is just that – an example, it nonetheless captures the essence of informational data visualization, which is to use readily recognized data charting and summarization techniques to succinctly and unambiguously communicate specific information. The emphasis is on completeness, clarity and simplicity; the carefully considered use of common visualization tools such as bar charts, histograms and data tabulations (all of which were used in the sample scorecard shown in Fig. 5.13) allows the often methodologically complex and esoteric informational outcomes to be reduced to interpretation-friendly conclusions.

5.6.3   Scientific Visualization

The almost intuitive interpretation of informational data visualizations comes at a cost: a great deal of simplification. In many situations that is acceptable, even desired, as is in the case of securities litigation threat assessment outlined earlier, where the key parts of assessment were derived using methodologically complex multivariate likelihood and severity estimation techniques. In such situations, showing esoteric data manipulation and statistical estimation details would likely do more to confuse than to inform, because such details are only meaningful to those with appropriate technical data analytic knowledge, and though no statistics are readily available regarding the proportion of executive managers and organizational directors with that type of knowledge, anecdotal evidence points towards a clear minority.

However, not all information lends itself to reductionism.[36] In fact, information content of some data analytic outcomes can be materially enhanced through the use of more complex data visualizations, as illustrated by Fig. 5.14.

Shown on the left-hand size of Fig. 5.14 is a partial output of a predictive model built to estimate the likelihood of rapturing buried utility lines in the process of excavation; shown on the right-hand side is the same information rendered using geographic information systems (GIS), a geographically accurate mapping application. As a general application, GIS incorporates geographical feature details with tabular data to map and assess location-related problems. In the example shown in Fig. 5.14, the addition of GIS facilitates display of location-specific (i.e. a physical address) predictive model-generated details, such as the estimated risk level along with supporting details tabularly enumerated in the partial spreadsheet element of Fig. 5.14, on an interactive map, which greatly increases informational friendliness and utility of the otherwise dull and uninviting tabular result presentation.[37]

In a more general sense, scientific data visualization is a mechanism of spatial data display and analysis and thus is appropriate to use when the underlying data contain either explicit or implied geographic features. Those features could take the form of physical address, as is the case in the above example (where each record represents a distinct physical address), distinct terrain features such as bodies of water or mountains, travel routes and other spatial informational details.

---

[36] Rooted in a deeper philosophical reasoning, the practice of reductionism aims to describe complex systems or phenomena in terms of simpler or more fundamental elements; the inescapable loss of granularity is believed to be justifiable in view of simplicity and sufficiency of resultant explanations.

[37] For completeness, being able to project tabular information onto a GIS map requires detailed address (not shown in Fig. 5.14), along with a separate mechanism for translating physical street address, such as 25 Main Street, Bristol, RI 02809, into latitude and longitude coordinates required for mapping.

5.6.4   Graph Data Structures

The overview of data visualization techniques would be incomplete without at least a brief mention of data that are natively graphically rendered. Commonly known as graph data structures (and in aggregate sense, graph databases), those means of capturing and encoding of facts were developed in the early part of the twenty-first century primarily to address the growing complexity, rather than size, of modern data flows. For instance, while the highly disaggregate sensor-generated details are essential to effective control and monitoring of automated electronic systems, their informational utility rarely extends beyond edge computing (computing that is done at or near the source of data, rather than at a remote, centralized location); when aggregated, those data offer greater informational utility to human users, and expressing those aggregates graphically can greatly reduce the complexity of such data.

A graph data element consists of a finite set of nodes, also known as vertices, and a set of edges connecting them; the former represents individual data elements, and the latter encapsulate broadly defined interconnectedness, which, as illustrated in Fig. 5.15, can be either undirected or bidirectional or directed or directional. In a general informational sense, undirected graphs communicate two-way relationships, while directed graphs denote one-way relationships; the exact position, length or orientation of edges typically has no meaning, which is to say that the same graphically depicted associations can be visualized in multiple ways by rearranging the nodes and/or shortening or lengthening the edges.

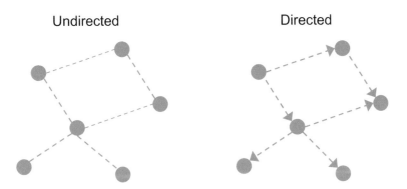

Fig. 5.15 Undirected versus directed data structures

Chapter 6
Managing by the Numbers

6.1 The Evolution of the Art and Science of Managing

The practice of management, defined as a general process of controlling things or people, is inherent to organized living and thus can be considered to be as old as human civilization. The very functioning of early civilizations, the emergence of cities and states, and the building of the great monuments of human communities in the form of pyramids, temples, churches, and palaces required controlled, organized efforts. Not surprisingly, the earliest evidence of somewhat formalized management practices dates back more than six millennia, to around 4500 BCE, and the organized record keeping developed by early Sumerians; however, it was a Babylonian king, Hammurabi, who oversaw the development of more systemic and formally codified control practices some 2500 years later (he reigned from about 1792 to about 1750 BCE). Still, it was the work of a Greek philosopher Xenophon (a student of Socrates), titled the *Oeconomicus* (*Household Management*), that is commonly credited with giving rise to management as a separate domain of study and practice.

Centuries later, the gradual emergence of modern commercial organizations brought with it a need for more formal tracking of organizational resources and commercial transactions, which in turn contributed another key set of foundational elements to the ongoing maturation of the art and science of managing. More than 500 years ago, in 1494 to be exact, a Franciscan monk and a noted mathematician Luca Pacioli[1] published a textbook titled *Summa de Arithmetica, Geometria, Proportioni et Proportionalita* (*Summary of Arithmetic, Geometry, Proportions and Proportionality*). While the work was notable as an elegant summary of the

---

[1] He was also a mathematics teacher of perhaps the most renowned of the polymaths of Italian Renaissance, Leonardo da Vinci, with whom he is believed to have also worked on a book of chess strategy.

© The Author(s), under exclusive license to Springer Nature Switzerland AG 2021
A. Banasiewicz, *Organizational Learning in the Age of Data*, EAI/Springer Innovations in Communication and Computing, https://doi.org/10.1007/978-3-030-74866-1_6

collected knowledge of mathematics of the time, its true claim to fame was the now celebrated 26-page guide for merchants on sound accounting and business practices, built around Pacioli's own codification of the "Venetian" accounting method (born in Tuscany, he lived most of his adult life in Venice, where his book was also published). That guide, separately titled *Particularis de Computis et Scripturis* (*Details of Calculation and Recording*), was the first published detailing of the double-entry bookkeeping system, which rapidly spread throughout Europe and beyond.[2] Today, the idea of separating entity's resources from claims on those resources is known as balance sheet; in a broader sense, Pacioli's work laid the foundation for modern financial statements, which offer a concise way of systematically capturing the key financial aspects of business entities, and by doing so offering objective and unambiguous management targets.

While those early contributions framed the practice of management in terms of general processes and oversight-related considerations, the process orientation of modern management practice can be seen as a result of the far more recent and more applied work of Frederick W. Taylor, an American mechanical engineer. Taylor is widely acknowledged to be the first to approach managing of work through systematic observation and study;[3] in his seminal work published in 1911 and titled *The Principles of Scientific Management*, he summarized his theory of effective job and incentive compensation design as a mean of achieving higher productivity. His ideas were widely adopted and, together with other contributions, most notably those by Frank and Lillian Gilbreth, gave rise to what is now known as scientific management, which can be characterized as controlling by means of analysis and synthesis of workflows. It is worth noting that though groundbreaking, scientific management is inescapably grounded in the Industrial Era's conception of work and management of work, much of which was built around application of direct human effort to industrial production processes; that said, its "numbers" orientation is enduring.

6.1.1 By the Numbers

As evidenced by the brief historical recounting of origins of modern management practice, while the idea of managing by the numbers may seem new, it is not – in fact, the mindset of using objective evidence as the basis for making decisions predates not just the modern informational infrastructure, but even the scientific

---

[2] Pacioli is now celebrated as the father of modern accounting.

[3] Taylor's primary focus was on *time*; thus, his studies came to be known as "time studies"; his contemporaries, Frank and Lillian Gilbreth, who were industrial engineers, were focused on identifying the most efficient *ways* of accomplishing tasks; thus, their research came to be known as "motion studies." What is today known as "time-and-motion" studies, and sometimes erroneously entirely attributed to Taylor, represents a subsequent merging (by others) of the two distinct areas of research.

precursors of that infrastructure.[4] Long before the Internet and electronic data, business organizations began to publish their financials in the form of balance sheet, income statement, and cash flow statements as a way of attracting investors, who used those documents as proof of companies' profit-making abilities. The same organizations also invested considerable time and effort into streamlining workflows as a way of increasing efficiency; in a sense, long before the advent of robotic process automation, commercial enterprises saw their workers as not much more than biological robots...Looking past the numerous and well-known social implications of that mindset, which are acknowledged here but examination of which falls outside of the scope of this analysis, time-and-motion studies of scientific management, together with initially voluntary[5] publishing of financial statements effectively, transformed the management of, at least public, companies into de facto management by the numbers.

But what, exactly, are "the" numbers? Consider Fig. 6.1. The most visible manifestation of the idea of "managing by the numbers" is the use of financial statements, which in the context of the general management setting sketched out in Fig. 6.1 comprise the outcomes of organizational management processes. While there are numerous types of financial statements, the "big three" are (1) balance sheet, which is a summary of the organization's assets, liabilities, and capital, at a point in time; (2) income statement, which captures revenues and expenses during a period of time; and (3) statement of cash flows, which summarizes all cash inflows and cash outflows during a period of time. In that context, "the" numbers are the bottom lines of balance sheet, income statement, and the statement of cash flows; thus, managing by the numbers is tantamount to managing organizations with an eye toward financial statements-captured outcomes.

Fig. 6.1 General management setting

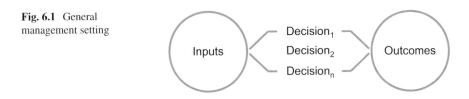

---

[4] While in the sense of philosophy the origins of modern scientific thinking can be traced back to Aristotle, the third of the famous trio of ancient Greek philosophers (Socrates, his student Plato, and then Aristotle, who was Plato's student), the rise of modern science as a larger societal movement is commonly traced back to the Age of Enlightenment (also known as the Age of Reason), which dominated the world of ideas in Europe during the seventeenth and eighteenth centuries.

[5] In the USA, starting in the early 1930s, following the passage of the Securities Act of 1933, which governs issuance of new securities, most notably stocks, to the investing public, and the Securities Exchange Act of 1934, which governs trading of securities in secondary markets (such as the New York Stock Exchange of NASDAQ), all public companies, which are those with securities trading on public exchanges, are required to publish their financial statements each year.

It is considerably less obvious what managing by the numbers entails within the confines of the process dimension of commercial activities. The earlier discussed scientific management ideas are well-suited to repetitive, assembly line-like work geared toward tangible industrial output, but quite a bit less so to other work contexts; in a more general sense, those principles are meaningful primarily to manufacturing organizations, which, as of 2019, accounted for a little more than 11% of the total US economic output, and employed about 8.5% of the total workforce. In other words, scientific management-rooted conception of managing commercial work-related inputs, throughputs, and outputs "by the numbers" is not directly applicable to managing more than 90% of organizational workforce.

An even more fundamental trouble with that Industrial Era-rooted conception of management by the numbers is evident in epistemic and ontological foundations[6] of those ideas, which can be summarized here by simply asking if that overt characterization is in alignment with the implied meaning. In other words, is managing organizations by setting explicit performance targets tantamount to truly managing "by" the numbers, whether those numbers represent production efficiency or aggregate financial results?

In principle, managing "by" the numbers could be expected to notionally parallel piloting a plane in a mode known as "fly-by-wire," which replaces conventional manual flight controls of an aircraft with an electronic interface. In that mode, automated flight control systems use computers to process flight control inputs and send corresponding electrical signals to the flight control surface actuators that control the key functions such as speed, direction, and altitude. Similarly, an organization managed by the numbers could be expected to tie many of its decisions to objective numeric evidence, thus effectively removing the "human factor" from many organizational decisions. In an abstract sense, organizational management could, under those circumstances, be reduced to a simple stimulus-response model,[7] but is that even feasible, and if so, would it be desirable?

In a very general sense, organizational management entails numerous strategic and tactical alternatives; dealing with the resultant uncertainty is the essence of decision-making. Using the fly-by-wire analogy, the rational, in the stimulus-response sense, decision model would entail objectively estimating the relative advantage of each of the available decision alternatives ("inputs" in Fig. 6.1), and then choosing the one with the highest estimated value. Such a mechanically rational approach would effectively do away with nonquantifiable considerations, and it would shield decision-making processes from the ubiquitous cognitive bias, and group-based evaluation-warping effects, such as groupthink. But, and that is a big "but," such a fly-by-wire algorithmic decision-making model is highly dependent on the availability, validity, reliability, and completeness of decision-driving data.

---

[6] Both of distinct branches of philosophy; ontology is concerned with how concepts are grouped into more general categories and which concepts exist on the most fundamental level, whereas epistemology is focused on differences between justified beliefs and subjective opinions.

[7] Expressed through a mathematical function that describes the relationship (f) between the stimulus (x) and the expected (E) outcome (Y), where $E(Y) = f(x)$

While nowadays organizations tend to be awash in data, those data are generally messy and informationally incomplete, meaning are in need of substantial pre-use processing, and even more importantly, only contain a subset of factors that shape or outright determine observed outcomes. And though data preparation is ultimately a manageable problem, informational insufficiency may not be, as not everything that needs to be known (to auto-decide) is knowable – for instance, it is hard to imagine how the highly personal attitudinal determinants of brand choice could be systematically captured along with other, more visible (e.g., price, promotional incentives) choice influencing factors.

6.1.2   To the Numbers

It is hard to escape the conclusion that while "managing by the numbers" is a widely used label, it may be a misnomer or an aspirational goal, at best. Aircraft are routinely flown-by-wire because their onboard computer systems are able to capture the required data that are produced by a carefully designed array of sensors; the ensuring auto-processing of flight details produces electrical signals which in turn cause the flight control actuators to appropriately adjust the key operating elements, such as speed or direction of flight. All that is possible because (1) the required operational decision-making is narrowly defined (getting from point A to B, in generally straight line or via a set of straight lines), (2) the needed data are readily available, and (3) barring a rare malfunction, data are valid and reliable. While such perfect conditions may describe some specific organizational decision-making contexts, the vast majority of organizational choices need to be addressed under less perfect conditions, which renders implementing of managing by the numbers processes far more, even prohibitively, difficult. In fact, what is often characterized as managing "by" the numbers is really managing "to" the numbers.

The ubiquitous financial statements, which offer a concise way of systematically capturing the key financial aspects of business entities, provide natural management targets; similarly, in educational setting, standardized test scores provide natural instructional targets. In both contexts, those are outcomes of organizational decisions (see Fig. 6.1), and decisions that directly impact or even shape those outcomes are commonly made with an eye toward maximizing those outcomes. Business organizations aim to maximize their value, while educational institutions typically aim to maximize test scores-implied quality of their education, now commonly codified in endless rankings. Those, and other, types of organizations do not manage "by" the numbers, but they manage "to" the numbers because their respective numbers are merely performance goal posts, not fly-by-wire-like operational actuators.

Managing to specific revenue or profitability targets is quite common among for-profit business organizations because it offers unambiguous goals and incentives, but it is frequently accompanied by several negative side effects. Focusing on a specific target, such as the attainment of a specific level of profitability, tends to frame success in terms of only meeting that target, often called the "threshold effect," in addition to also diverting attention from other outcomes, such as revenue; even more incredibly, it can also give rise to what is known as the "ratchet effect," which manifests itself in unwillingness to reach the stated target in fear that doing so will lead to the target being raised in the future. Focusing on more relative-minded rankings tends to mitigate the threshold and ratchet effects, but can produce other adverse side effects, perhaps best exemplified by "teaching to the test" by educational institutions that see that as a mean of achieving higher rankings (which are largely based on comparisons of measurable outcomes, such as average test scores).

The roots of the problem of confusing managing *to* the numbers with managing *by* the numbers run deep, as illustrated by the recent case of the Boeing Company, an aerospace-focused manufacturer, and Pacific Gas and Electric Company, or PG&E, an electric utility company. In March of 2019, Boeing 737 MAX, a variant of the company's best-selling Boeing 737 aircraft (itself the best-selling plane in commercial aviation history), was grounded worldwide after 346 people died in two separate crashes. While the underlying fact pattern that led up to those tragedies and the subsequent grounding is complex, some of the more striking aspects of that case include evidence that the company historically renowned for its meticulous designs made seemingly basic software design mistakes that are now believed to have played the key role in those crashes. According to longtime Boeing engineers, what should have been a routine development project was complicated by the company's push to outsource that work to lower-paid contractors, some being paid as low as $9/ hour; moreover, the company's decision to sell as "extra" premium options certain safety features, most notably those known as "angle of attack indicators," may have also played a role. Those Boeing decisions share one thing in common: They are both geared toward shoring up of company's profitability, a key manifestation of managing to the numbers.

A possibly even more egregious example of managing to the numbers gone awry is offered by PG&E choosing profits over safety. As reported by the *The Wall Street Journal*,[8] the company knew for years that hundreds of miles of its electric power lines, some as old as a century, needed to be upgraded to reduce the chance of sparking a fire. Once again, the company's bottom line focus compelled it to not to invest in replacing unsafe transmission lines; the nightmare scenario came true in 2018 when a felled power line sparked what eventually became the deadliest wildfire in California's history, claiming 88 lives, causing more than $30 billion in potential legal claims, ultimately pushing PG&E to file for bankruptcy protection. It is worth noting that while the 2018 wildfire is notable as being the largest, the company's power lines are believed to have caused more than 1500 California wildfires in the span of just 6 years, and time and time again, the pursuit of profit trumped maintenance investments.

While the cases of Boeing and PG&G are just two, fairly extreme examples of management to the numbers gone awry, they are nonetheless illustrative of potential dangers of managing to the numbers, in the form of performance outcomes, rather

---

[8] https://www.wsj.com/articles/pg-e-knew-for-years-its-lines-could-spark-wildfires-and-didnt-fix-them-11562768885

than allowing objective evidence to permeate the full spectrum of organizational decision-making-related considerations, which is the essence of managing by the numbers. While both rely on objective data, the difference between the management philosophies becomes even more pronounced in the context of the comparison of top-down (managing to the numbers) and bottom-up (managing by the numbers) approaches.

6.1.3   To Versus by the Numbers and Top-Down Versus Bottom-Up

Using financial, operational efficiency or other measurable aggregate outcomes as performance targets is one of the hallmarks of the top-down organizational management philosophy, often though not exclusively associated with the centralized organizational structure discussed earlier in Chap. 3. The core feature of the top-down management philosophy is that organizational objectives are first set by organizational leadership, and then are promulgated throughout the organization. Implicit in that philosophy is the "do whatever it takes to make it happen" mindset, the essence of which is not only that ends justify the means, but even more importantly that means do not inform the ends; in fact, means are subsumed under ends. Still, the top-down management approach is widely used, especially among commercial organizations, but even its advocates admit that it leaves an awful lot up to chance; it is a bit like trying to get to climb a steep mountain without trying to understand the challenges ahead. While doing so may make for a more interesting adventure, it also heightens the exposure to known and unanticipated risks.

An intuitively obvious alternative to top-down is bottom-up, which rests on the premise that frontline organizational members are best equipped to set attainable goals. Within the confines of data and information, bottom-up approach uses evidence-driven assessment of what might be feasible as the basis for organizational goal setting, which is commonly framed in the context of range of possible outcomes, typically ranging to "likely" to "stretch." It too has its detractors who argue that it tends to lead to "make a little, sell a little" mindset of underachievement, which certainly cannot be dismissed as a possibility, especially when the in-place organizational culture effectively discourages risk-taking.

Perhaps the best way to illustrate the difference between these competing management philosophies is by taking a closer look at how top-down and bottom-up organizations approach revenue target planning.

Top-Down: A standard bearer for the centralized hierarchical organizational management approach frequently characterizing large, established organizations, the top-down approach entails setting of performance targets by the senior organizational leadership, which are then translated into downstream, progressively more specific targets. For instance, assuming that a business organization wants to grow its revenue by 5% in the coming year, and factoring in the typically inescapable customer attrition (e.g., some current customers might switch to competing products, others may no longer need the product, etc.), the organization estimates that in order to reach the stated net goal of 5% sales growth, it will need to generate about 12% more in sales than it did in the prior year. Making a simplifying assumption that the sales goals are the same across all business units comprising the firm, the individual-level sales quotas would then be raised by 12%. Once finalized, the so-adjusted sales targets are communicated throughout the organization, with the individual business units, and sales professionals within those units, being then entrusted with mapping out their own plans for achieving the stated objectives.

Bottom-Up: More common among decentralized, less hierarchical organizations, this approach is rooted in the idea that those directly engaged in value-producing tasks and activities tend to have more realistic view of what targets might be attainable. To be clear, "attainable" is often expressed as a range – in other words, it is not synonymous with "easy," but may range from "within reach" to "stretch" goals. More important is the process used to derive attainable targets: It starts at the level of individual contributors, and then, using objective evidence, aggregate targets are derived, as illustrated by the so-called sales funnel, which is built on two general sets of inputs: (1) starting effort inputs and (2) past outcomes-derived throughput ratios. For example, a salesperson might feel that she/he can make 60 prospect calls per day, and using the information derived from analyses of past sales efforts, that salesperson can then estimate that, on average, 5% of those calls will likely translate into a "connection" (a manifestation of interest on the part of a prospective customer), about 25% of connections translate into sales meetings, and lastly, about 35% of sales meetings will ultimately result in sales. Using the estimated number of calls and the throughput or conversion ratios, it is then easy to project the most likely number of sales: The initial 60 daily calls will produce about 3 connections (60×0.05), which will lead to 0.75 (3×0.25) scheduled meetings per day, or about 15 meetings per month (0.75×20 monthly workdays), ultimately yielding an estimated 5.2 sales per month (15×0.35), or about 63 sales per year. Using that general logic as the key building block, an aggregate sales forecast can be derived in the bottom-up fashion. The proponents of that approach argue that it naturally lends itself to more realistic planning, because contemplated scenarios are rooted in in the combination of fact-based objective evidence in the form of the throughput ratios and an assessment of what is feasible conducted by those who would be expected to deliver against the forecast. The detractors, on the other hand, tend to question the motives behind self-designated performance targets, pointing to the possibility of gaming the system by those entrusted with rationalizing their own performance goals.

When considered in terms of their respective strengths and weaknesses, top-down and bottom-up philosophies have distinct advantages and deficiencies. Setting revenue or other targets by the top management has the advantage of being informed by larger competitive, economic, and other macro trends, but the managing-from-afar mindset that permeates the top-down philosophy is generally insensitive, even to the point of being dismissive of within-organization, micro considerations, such as staffing levels, skillsets, or resource availability. On the other hand, allowing aggregate revenue or other targets to organically emerge, or "bubble up," has the advantage of being highly sensitive to micro-organizational considerations, but it can be overly inwardly oriented, to the point of being disconnected from larger macroeconomic and competitive factors. When framed in the context of managing "to" or "by" the numbers, the top-down approach skews rather heavily toward the former, while the bottom-up approach is more naturally aligned with the latter. However, in the Age of Data, top-down and bottom-up organizational planning approaches can be more informationally complete, if purposeful steps are taken to leverage the typically readily available data. As depicted in Fig. 6.1, the overall organizational management process can be thought as being comprised of three general elements: inputs, typically in the form of alternatives under consideration, the decision-making process, and outcomes, which manifest themselves not only in the form of financial or efficiency metrics but also through somewhat less obvious expressions, such as customer satisfaction or doing the "right thing," as seen through the prism of social responsibility and social consciousness. The limitations of top-down and bottom-up approaches stem from the fact that the former is largely focused on outcomes of organizational activities, while the latter is primarily concerned with inputs, but given that all three distinct organizational management process elements can be informed by internal and external (to the organization) data, both top-down and bottom-up can be equally rational and informationally complete. Doing so, however, calls for a more imaginative way of looking at available data.

6.2 The Latency Tiers

As many organizations are realizing daily, the promise of "big data" is compelling, but turning that potential into operational reality is fraught with difficulties: As discussed earlier, though informationally rich and ubiquitous, data are also exceedingly voluminous and varied, produced at a dizzying rate, and in need of often significant cleansing and feature engineering. But there is more: To get more out of as-captured data requires looking beyond "what-is," because more thoughtful evaluation of those data can bring into focus data that are implied by as-captured measures. Most notably, those confusingly sounding "data emergent from data" can include features of organizational decisions that precipitated observed outcomes. For example, a decision to discontinue a particular product may not be among the as-captured metrics, but may nonetheless be implied by the sudden revenue drop; in that sense, as-captured data *imply* the existence of additional variables. Still, there could be more potential metrics hidden in the plain sight in the form of potential interaction terms and other *hidden* effects, as graphically summarized in Fig. 6.2.

Within the confines of electronic information processing, or more specifically business intelligence (BI), the notion of data latency captures the amount of time it takes to retrieve data from a data warehouse or other data storage repository; in short, it is a measure of delay. In a more general sense, the idea of latency is meant

Fig. 6.2 Data latency tiers

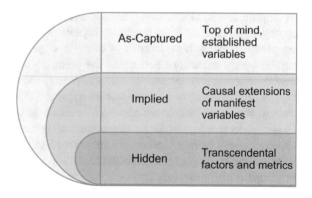

to capture the degree of concealment; in statistics, for instance, latent constructs are variables that are not directly observable but can be inferred from other variables or from patterns in data. As used here, the notion of data latency combines aspects of the general as well as BI-specific conceptualizations; as such, it is meant to capture the potential informational content of data. As briefly discussed above, when considered within the confines of informational content, data can be conceptualized as as-captured, implied, or hidden.

As-captured data are informationally straightforward, encompassing numerous widely recognized and frequently used metrics as exemplified by demographics or purchase metrics. The earlier discussed financial statements encompass an array of revenue, cost, expense, profitability, and related metrics; customer demographics are commonly made up of standard measures such as age, gender, education, etc. As a broad category, as-captured variables usually have straightforward meaning and interpretations, and though there is always an opportunity to look at those familiar data elements in a new, creative manner, many have well-established usage contexts.

Implied data are logical extensions of as-captured variables; in the most direct sense, those metrics contain insights into root causes of observed outcomes. When considered from the standpoint of organizational management, implied data can be suggestive of the efficacy, or lack thereof, of past decisions.

Reaching even deeper into the realm of potential informational content is the third data latency tier: hidden data. It is important to note that it is not data, as such, that are hidden – it is the informational content that is hidden. Social media data offers a good illustration of that idea: Even when narrowed to just communications relating to a particular brand or product, it is still vast hodgepodge of unstructured, variably sourced, and formatted comments, reviews, reports, and the like. Hidden in those details are far more informationally (in the sense of organizational decision-making) meaningful transcendental insights, in the form of product or brand-related customer sentiments.

Given the abstract character of the data latency tiers summarized in Fig. 6.2, the following brief case study built around customer loyalty data usage might contribute to making those ideas more tangible.

6.2.1   Latency Tiers Case Study: Customer Loyalty

For consumer brands, especially those that entail ongoing usage and repurchase, such as packaged goods, brand loyalty is of paramount importance, and thus it is one of the key focal points of those brands' overall marketing strategy. Using the now ubiquitous loyalty cards coupled with other data sources (most notably, third-party data overlays), brands eagerly accumulate customer data, analyses of which support a wide range of retention-minded marketing programs; Table 6.1 shows a sample of some of the commonly captured/available variables. Recalling the data latency tiers summarized in Fig. 6.2, the variables enumerated in Table 6.1 are grouped into "as-captured," "implied," and "hidden" latency tiers; in addition, the implied measurement scale for each variable is described as either "continuous" or "categorical," and each measure also being designated as either "existing" or "derived."

The As-Captured variables are direct reflections of the core demographic and behavioral attributes of the brand's customer base; they are all labeled as "existing" because they either were all directly captured during brand interactions (sales, service, etc.) or were appended shortly after, and their basic analytic properties (i.e.,

Table 6.1 Sample latency tiers

Variable Name	Variable Type	Variable Status	Latency Tier
Gender	Categorical	Existing	
Age	Continuous		
Years of Education	Continuous		
Job Category	Categorical		
Length of Employment	Continuous		As-Captured
Marital Status	Categorical		
Household Size	Categorical		
Household Income	Continuous		
Home Ownership	Categorical		
Brand Tenure	Continuous		
Monthly Brand Spend	Continuous		
CouponRedemption	Categorical		
Active Lifestyle	Categorical		
Movie Rental	Categorical		
Smart Phone	Categorical		Implied
Twitter Account	Categorical		
LinkedIn Account	Categorical		
Facebook Account	Categorical		
Cumulative Brand Spend	Continuous	Derived	
Brand Spend Percentile	Continuous		
High Value Customer	Categorical		Hidden
Future Value	Continuous		
Social Media Active	Categorical		

categorical vs. continuous) reflect their informational content. The Implied measures are also labeled as "existing" because they either were also captured during the brand's interactions or were appended shortly thereafter; though all are shown as "categorical" in the sample depicted in Table 6.1, implied variables could also include continuous measures (for instance, if some of the commonly geodemographic[9] metrics were added into that mix, such as "estimated house value" or "estimated household income"). Implied data features are distinct from As-Captured variables less in terms of their overt characteristics but more in terms of their informational content, as the former augment the informational value of the latter. In other words, the fact that a particular brand customer may have a smartphone or a LinkedIn account may not directly relate to that customer's interest in the brand, but knowing those two pieces of information may offer hints in terms of ongoing communication with that customer.

The Hidden variables are derivatives of either the As-Captured or Implied measures and thus are noticeably different from those two categories. Of the five Hidden variables shown in Table 6.1, four – Cumulative Brand Spend, Brand Spend Percentile, High Value Customer, and Future Value – were all derived from As-Captured variables, and the fifth one was derived from Implied variables, all using relatively common data analytic logic. The four As-Captured grouping-derived metrics are rooted in relative rankings, while the fifth one – Social Media Active – is a product of basic classificatory logic. The specific computational formulas used to derive the five hidden variables are summarized below:

$Cumulative\ Brand\ Spend$ = Monthly Brand Spend × Brand Tenure

$$Brand\ Spend\ Percentile = \left(\frac{CF + 0.5F}{n} \right) \times 100$$

where,

CF is the cumulative frequency below the given score
F is the frequency of the given score
n is the number of scores in the distribution
$High\ Value\ Customer$: IF "Spending Value" $\geq$ "High Value Threshold" $=>$ High Value Customer = 1; ELSE $High\ Value\ Customer = 0$.
$Future\ Value^{10} = \beta_0 + \beta_1 X_1 + \beta_2 X_2 + \dots + \beta_n X_n$
where,

β is a regression weight
X is a regression predictor
$Social\ Media\ Active$: IF Twitter Account = "Yes" AND LinkedIn Account = "Yes" AND Facebook Account = "Yes" $=>$ $Social\ Media\ Active$ = "Yes"; ELSE $Social\ Media\ Active$ = "No"

---

[9] Geodemography is the study of people based on where they live; it attributes the most probable characteristics based on pooled profile of all people living in a small area. In the USA, that small area is usually defined as "blocks," which, according to the US Census Bureau, are "statistical areas bounded by visible features such as roads, streams, and railroad tracks, and by nonvisible boundaries such as property lines, city, township, school district, county limits and short line-of-sight extensions of roads."

It should be noted that the computational logic outlined above illustrates just one of the several possible computational approaches; moreover, the larger goal that inspired sharing of those analytic specifics was to illustrate how existing (i.e., As-Captured and/or Implied) variables can give rise to new variables, using relatively standard computational logic.

Considering the initial informational content of just As-Captured variable set, adding of the initially not present Implied and Hidden variable sets will enable far more actionable loyalty analytics than just relying on the original set of measures. That is because to be cost-effective, loyalty programs need to be built around explicit and factually sound assessment of customer value, as expressed by objective numeric measures such as revenue or profitability, since like all other business investments, loyalty expenditures need to produce positive return on investment (ROI). Embedded in that reasoning is the recognition of, at times considerable, cross-customer revenue and/or profitability variability, which suggests that some customers are a lot more valuable, and thus retention-worthy, than others. Operational implementation of that reasoning, however, is not possible without extraction of properly conceived hidden measures, which in this case capture and summarize key customer-level value manifestations. Customer Brand Spend, which is the first (as listed) of the Hidden variables depicted in Table 6.1, combines Monthly Brand Spend and Brand Tenure (both As-Captured metrics) to summarize cumulative per customer spend, while Brand Spend Percentile further contextualizes that information by casting it in the relative "in relation to others" type context.[11] The High Value Customer variable is a binary "yes" vs. "no" indicator, which is a categorical expression of customer value, in the sense that it groups all customers into those who are valuable enough to warrant special retention efforts, and those who are not;[12] the Future Value measure is altogether different, as it offers a more speculative assessment of expected future customer spending (which is typically a statistical extrapolation of past behaviors). Lastly, the Social Media Active indicator offers an "all-in" summary of the breadth of customers' use of various social media platforms, which is often beneficial from the standpoint of planning customer outreach strategies.

---

[10] There are numerous statistical approaches to estimating future value – shown here is linear regression, which is one of the most basic (and in many regards limited) future value estimation methods.

[11] It is worth noting that cumulative spending may be an appropriate measure of customer value in some situations and, per time period type measure, such as monthly spending, could be a more appropriate expression of customer value. That notwithstanding, being able to see customers from both, i.e., cumulative and per period, perspectives can be expected to give rise to more wholistic assessment.

[12] There are numerous approaches to making that determination, including already in place definitions (e.g., based on past experience, anyone with spending higher than X should be considered "high value"), the use of Pareto principle or the so-called 80/20 rule (according to which, the top 20% of customers often account for 80% of revenue), and others.

6.3   Analytically Enabled Management

One of the greatest minds of antiquity, Plato, argued in his seminal work *The Republic* that kings should become philosophers or that philosophers should become kings.[13] Interestingly, Plato's reasoning is not simply epistemic, meaning it is not limited to the idea of philosophers being more knowledgeable are thus better equipped to make better informed choices – it is also because they would be more inclined to make just choice. His reasoning was rooted in the idea that philosophers far prefer "the life of the mind," or their intellectual pursuits, to the task of governing; thus, if they were to take on the task of governing, it would be more out of desire to seek the attainment of larger goals than just the sheer love of ruling. But if philosophers could not become rulers, then rulers should try to become philosophers, as a way of enhancing their ability to rule justly.

Some 2500 years later, the spirit of Plato's thought-provoking ideas lives on in the modern notion of evidence-based decision-making. And so, in the Age of Data, should business leaders become analysts, or should analysts become business leaders?

It is not a question that lends itself to a simple, direct answer, though the idea of highly analytically literate business leaders is compelling. It is difficult to look past the conclusion that while managing "to the numbers" requires relatively little data analytic proficiency, being able to manage "by the numbers" calls for a considerably higher degree of data analytic prowess. But that is not because sifting through endless arrays of raw data constitutes the best use of senior organizational managers' time; rather, it is because greater understanding of the core aspects of data and data analytics generally translates into heightened propensity to use data analytic outcomes, and greater efficacy of data utilization. That, in turn, can bring to light previously unnoticed opportunities, and it can also be the guiding light in situations where choices are shrouded in ambiguity; robust foundational data analytic knowledge is also essential in identifying situations in which available data are of limited informational value. Recalling the earlier (Chap. 1) discussion of data being comprised of informative "signal" and non-informative "noise," as volumes and varieties of data continue to grow, so do the opportunities to inform and to misinform; hence the importance of analytic literacy now spreads far beyond analysts' offices.

An important aspect of managerial data analytic literacy is a clear sense of the distinction separating "data," "information," and "knowledge." Data are simply facts, which when processed and refined by means of summarization or contextualization become information; the relevant and objective information that offers decision-guiding insights is then elevated to the level of knowledge. In an everyday sense, information offers "what is," which can be conceptualized as the top layer

---

[13]The exact quote, translated into English reads: "Unless, said I, either philosophers become kings in our states or those whom we now call our kings and rulers take to the pursuit of philosophy seriously and adequately, and there is a conjunction of these two things, political power and philosophic intelligence, while the motley horde of the natures who at present pursue either apart from the other are compulsory excluded, there can be no cessation of troubles, dear Glaucon, for our states, nor, I fancy, for the human race either."

of awareness and understanding, while knowledge contributes the more in-depth understanding of "why" and "how." In an organizational setting, information is considerably easier to generate and disseminate than knowledge because it is typically shaped by objective means, as exemplified by monthly sales or donations summaries, which are usually produced using standard data aggregation processes and reported using standard reporting templates. Not surprisingly, modern organizations tend to be awash in information, so much so that the old "data-rich but information-poor" expression is nowadays often paraphrased as "information-rich but knowledge-poor." As implied here, knowledge is elusive because developing robust understanding is person-dependent, in addition to also requiring commitment of effort and robust and relevant inputs. Yet that is the ultimate goal of analytically enabled management: to thoughtfully use available data as means of developing factually sound and deep understanding of phenomena under consideration, and by doing so to contribute toward efficient and effective completion of tasks.

6.3.1   Learning Modalities and Performance of Tasks

The key to avoiding the data- and/or information-rich but knowledge-poor trap is to develop a robust objective knowledge creation infrastructure. The MultiModal Organizational Learning (MMOL) typology first discussed in Chap. 1 (and graphically summarized in Fig. 1.5) can serve as the conceptual foundation, and the four distinct manifestations of organizational learning – experiential, theoretical, computational, and simulational – can be thought of as progressively more sophisticated means of knowledge creation. From the largely observation-based experiential learning to rational cognitive reasoning-based theoretical learning to technology-aided computational learning to technology-enabled simulational learning, the means by which data can be transformed into reasoned organizational actions encompass the human cognition-centric reason-based learning, and the progressively more automated – and even autonomous – technology-based learning. The implied progression can be seen as a product of the combination of greater cognitive acuity, technological progress, and availability of objective data. And although experiential, theoretical, computational, and simulational dimensions can be seen as progressively more sophisticated means of learning, it is important to think of those distinct modalities as complementary pathways to building robust organizational knowledge base.

Within organizational settings, those distinct-and-complementary mechanisms are ultimately geared toward more effective performance of organizational tasks, which could range from open-ended strategic planning and ideation to far more tactical activities such as price setting. There are several competing typologies and taxonomies of organizational tasks,[14] with two of the better-known ones being

---

[14] The key difference between a typology and a taxonomy is that the former is rooted in the idea of a pure type, which is a mental construct that deliberately accentuates certain characteristics, not necessarily found in empirical reality, whereas the latter classifies a set of items on the basis of empirically observable and measurable characteristics.

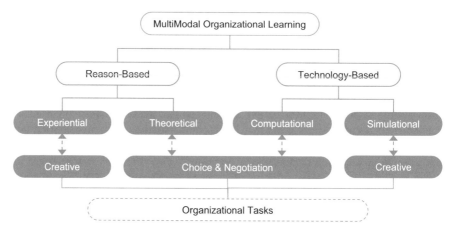

Fig. 6.3 Learning modalities and organizational tasks

Steiner's and McGrath's conceptualizations [1, 2]. Steiner's framework is rooted in the distinction between unitary and divisible tasks, which addresses the key question of whether or not a task can be broken into subtasks; at a more operational level, it expressly differentiates among several, somewhat overlapping, types: disjunctive, which lend themselves to the group adopting a single, winning in the sense, solution; conjunctive, which require utilization of all inputs (e.g., assembling a product); additive, where all inputs are considered but are not necessarily required (e.g., new product brainstorming); and discretionary, where group chooses which inputs to consider. More general in terms of its task groupings, McGrath's typology aims to delineate transcendental, or abstractly framed "pure type," groupings of organizational tasks, which are tasks focused on (1) creation, as in new strategy; (2) choice, as in choosing among competing alternatives; (3) negotiation, which emphasizes cognitive conflict resolution; and (4) execution, or performance of activities. The highly generalizable character of McGrath's typology of organizational tasks aligns well with the comparably general character of the MMOL typology-expressed organizational learning modalities; moreover, fusing learning and task types will further expand the informational utility of the MMOL typology, as graphically summarized in Fig. 6.3.

Within the confines of learning and knowledge-related organizational functioning, the generation of new ideas, in both strategic and tactical realms, can be seen as a "creative" task, while resolving choice-related ambiguity encompasses "choice" and "negotiation" pure task types, the combination of which offers an elegant summarization of the otherwise complex organizational dynamics discussed in Chap. 2. In keeping with the general idea that learning can be reason- or technology-based, creative ideation and problem-solving can be either experiential or simulational, the latter discussed in more detail in the next chapter as data-enabled creativity. The many factors that jointly shape organizational decisions, ranging from the ever-present general sense of uncertainty that surrounds organizational choice-making to availability   and   quality   of   choice-related   information   to   the   myriad   of organizational structure and culture related group dynamics, suggest that choice- and negotiation-related tasks can greatly benefit from the objectivity of theoretical and computational learning modes.

6.3.2 Data Analytic Outcome Utilization

The often hard to clearly express but critical element of fact-based management is the broadly defined ability to utilize data analytic outcomes. The earlier (Chap. 1) discussion of analytic know-how delineated three distinct adeptness tiers: analytic literacy, seen here as a foundational level of data analytic know-how; analytic competency, which is the next level of adeptness reflecting greater breadth and depth of data analytic knowledge; and lastly, analytic proficiency, which represents a pinnacle of analytic adeptness where a robust theoretical foundation is combined with a meaningful amount of practical data analytic experience. Aside from the implied methodological and computational know-how, the journey along the analytic literacy → competency → proficiency continuum also manifests itself in a gradual broadening and deepening of data analytic reasoning, framed here as the capability to clearly rationalize the informational intent, or goal, of distinct data analytic initiatives. That comparatively obtuse aspect of data analytic adeptness is particularly important in the present-day "data-everywhere" reality, which heightens the danger of habitual rather than purposeful analysis and reporting. That tendency commonly manifests itself in willingness to engage in all manner of analyses not because there is a compelling informational need, but simply because the opportunity is there; in short, because it is possible, not because it is desirable.

Though often overlooked, the distinction between routine and purposeful analytics is already quite important, and will continue to grow in importance. As captured in the widely circulated "4 Vs" (volume, variety, velocity, and veracity) characterization of big data, not only is the volume of data staggering but also is the variety; on top of it, both are growing at a dizzying rate. While there is no obvious way of ascertaining how much of the readily available data relates to organizational decision-making, anecdotal evidence suggests that it is usually only a comparatively small portion. In a typical applied organizational scenario, the ratio of informative "signal" to non-informative "noise" is relatively low because the bulk of organizational data represent passively (meaning, not guided by specific informational needs) captured and accumulated recordings of all manner of states, events, and the like – given that, it is important to not confuse richness of raw data with richness of decision-guiding insights. Organizations of all types – commercial, non-commercial, and governmental – instinctively store and catalogue data, and the rapidly declining costs of data storage are making it economically easy to justify. Consequently, many organizations, particularly the older, established ones, sit atop of enormous amounts of data, i.e., are data-rich. But how much of those data riches are or can be linked to some aspect of organizational value creation? Again, anecdotal evidence suggests that, ultimately, a relatively small proportion of organizational overall data holdings are informationally useful due to factors such as recency, accuracy, completeness, applicability, and, of course, explanatory or predictive efficacy. Given that, indiscriminate, routine processing of all available data, inspired more by instinct than by rational choice, can actually impede rather than advance organizational learning. In short, having more data to work with is not automatically advantageous.

The above considerations play a critical role in addressing the general idea of data analytic literacy, as well as its more broadly scoped parent, digital literacy, from the perspective of utilization rather than generation of data analytic outcomes. There is a considerable difference between the depth and the breadth of data analytic know-how needed to conduct meaningful analyses, and the know-how needed to be an informed consumer of outcomes of those analyses. While the earlier discussed data analytic literacy-competency-proficiency continuum is overtly focused on describing the evolution of the former, its general rationale is also applicable to the latter. An analytically literate manager is able to correctly interpret the often-nuanced data analytic outcomes – for instance, when averaging outcomes of interest, when is it more appropriate to express those quantities as an arithmetic mean as opposed a median? It is eye-opening how many seemingly informationally competent organizational decision-makers struggle with that basic manifestation of data analytic literacy. Even those who know that there are three distinct methods (mean, median, and mode) of estimating "average," in the sense of "typical," show an almost instinctive tendency to equate average with the mean, even though in many everyday situations, it is an inappropriate choice.[15] And so while those consuming data analytic results do not need skills that are needed to compute those outcomes, they need a baseline of knowledge that will enable them to draw correct inferences from provided estimates. Consequently, paralleling the tiered levels of data analytic *know-how* (see Fig. 1.6), data analytic *know-what* is also expected to exhibit itself in the form of distinct levels, graphically summarized in Fig. 6.4.

An analytically literate user of data analytic outcomes is anyone exhibiting sufficient understanding of the use of basic, which typically means descriptive, statistical concepts, while a competent user of data analytic outcomes is anyone who also exhibits sufficient interpretational understanding of more advanced statistical concepts, best exemplified by predictive multivariate statistical techniques. And lastly, a proficient user of data analytics is one who also has experiential understanding of the desirability or appropriateness of using specific data analytic tools and techniques – which includes classical statistical techniques as well as machine learning algorithms – to answer specific types of questions under specific sets of circumstances.

---

[15] When the quantity of interest, such as daily sales volume, is symmetrically distributed, i.e., follow the familiar bell-shaped curve known as standard normal distribution, mean is an appropriate choice for validly describing the average (in fact, under those circumstances, all three measures of central tendency is yield approximately the same estimate); however, when the distribution is asymmetrical or skewed, the use of mean as the measure of average would underestimate or overestimate the true value (for negatively and positively, respectively, skewed data).

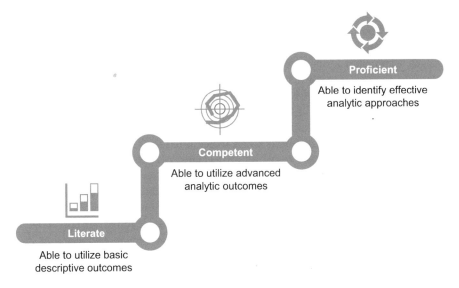

Fig. 6.4 Data analytic outcome utilization adeptness

Recalling the earlier example of the appropriateness of using the mean- vs. median-expressed estimate of the average, the essence of being an erudite user of data analytic insights hinges not just on understanding of whatever estimates are under consideration but also on knowing that those are the appropriate estimates. An organization that is interested in moving beyond just managing "to the numbers," which is quite common among commercial and non-commercial organizations alike, to managing "by the numbers," which is still comparatively rare, should be concerned not just with developing the appropriate informational infrastructure and seeding data-driven decision-making culture – it also needs to assure an appropriate level of data analytic adeptness. And as it is the case with any other facet of learning, an important part of the process is meaningful learning assessment.

6.3.3 Assessment of Data Analytic Adeptness

Broadly defined assessment of learning outcomes is widely acknowledged as important, but the means by which that broad objective is to be accomplished is a perennially disputed topic in educational circles and beyond. Some of the same considerations apply to assessment of aggregate organizational learning, which encapsulates organizations' efficacy at creating, retaining, and disseminating of knowledge aiming at supporting the pursuit of organizational mission. Perhaps the most general, if not generic, approach to measuring organizational learning is to use proxies such as number of patents or aggregate research and development (R&D) expenditures, though there is scant empirical evidence supporting the relationship between the strength of organizations' core competencies and organization-wide learning efficacy. A somewhat more refined, though highly abstract, aggregate organizational learning perspective is captured by the so-called learning and experience curves, which offer a graphical representation of the relationship between the development of organizational know-how and the cumulative effect of repeat performance of know-how-related tasks. Learning and experience curves, however, are incomplete measuring tools because they concentrate exclusively on learning by doing, in addition to which, those generalizations see learning through the prism of measurable results (thus, similarly to the use of proxies like patent counts or R&D expenditures, frame learning in the limiting context of current core competencies). Moreover, reliance on proxies and abstract generalizations ultimately draws attention to outcomes, while more-or-less sidestepping outcome-producing learning processes.

A more on-point approach to assessing organizational learning has been offered by the Kirkpatrick method, named after Donald Kirkpatrick, who first proposed it in 1959 (though its best-known variant, of which there are quite a few, is based on his far more recent 1998 book [3]). The Kirkpatrick method sees the task of measuring organizational learning as being comprised of four distinct levels: (1) reaction, which aims to capture learners' response to learning, typically in the form of satisfaction; (2) learning, which is focused on assessing gains in conceptual understanding and/or experiential skills; (3) behavior, which looks at the extent of learning-inspired behavioral change, typically in the form of utilization of new knowledge and skills; and (4) results, which aims to address organizational impact of learning. An important aspect of the Kirkpatrick model is its focus on examining the effectiveness of individual training initiatives; consequently, those using it to assess the effectiveness of overall organizational learning make a simplifying assumption that the total is a simple sum of its parts, a critical assumption that may not always stand up to scrutiny.

A competing and also more recent (first introduced in 1999) conceptualization is the 4I organizational learning framework [4]. Similar to the Kirkpatrick method, the 4I model also utilizes four distinct assessment foci, but its orientation is more expressly directed toward individual- and group-level cognition. More specifically, the 4I model posits four distinct assessment areas: (1) intuition, which encompasses preconscious recognition of possibilities inherent in individualized understanding of meaning (e.g., understanding the essence of the logic of hypothesis testing); (2) interpretation, which is the ability to explain something through words or actions; (3) integration, which is the development of shared understanding; and (4) institutionalization, which aims to capture the extent to which learning is systematically translated into behavioral change, typically in the form of routines or processes. In contrast to the implicitly static Kirkpatrick's conceptualization, the 4I-embodied learning process is recursive, as it expressly incorporates feed-forward and feedback processes. Implied in this brief characterization is a tacit toggle between the combined effect of individual-level cognition and metacognition, or awareness of one's thinking processes coupled with the ability to actively control those processes, and group-shared understanding manifesting itself through the emergence of

Fig. 6.5 Assessment
acuity of organizational
learning assessment
methods

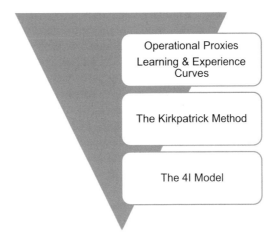

learning-inspired behavioral change. Given its fairly explicit recognition of individual- and group-wide learning processes, the 4I model offers a more nuance of assessing organizational learning, as graphically summarized in Fig. 6.5.

While offering progressively greater degree of assessment acuity, broad organizational proxies, learning, and experience curves, the Kirkpatrick and the 4I models nonetheless all fall somewhat short of the demands of assessing organization-wide data analytic adeptness. That said, the 4I model makes an important contribution to the challenging task of assessing organizational learnings because unlike the Kirkpatrick model and the generic curves and proxies, it expressly addresses *self-regulated learning*, which are individualized and cyclical thoughts, feelings, and actions responsible for the attainment of individual learning goals. Still, more granularity is needed to capture the depth of distinct contributing elements of knowledge that jointly comprise the broadly defined data analytic literacy, and in a more operational sense, also support meaningful differentiation among the three distinct data analytic outcome utilization adeptness tiers summarized earlier in Fig. 6.4. While a good first step, the 4I model does not, either explicitly or implicitly, address factors that trigger the very onset of learning, which the model posits to be intuition, nor does it address the distinction between tacit and explicit knowing.

6.3.3.1 Framing the Degree of Data Analytic Adeptness

The MECE (mutually exclusive and collectively exhaustive) framework is an expression of a general grouping principle that offers systematic means of addressing complex problems. It is a problem-solving tool widely used by management consultants to reduce sensemaking ambiguity, and it can also contribute to the difficult problem of meaningfully and objectively assessing the veracity of organization-wide data analytic adeptness. MECE framework's basic premise is that the best way to approach an analytical (but not necessarily quantitative) problem is by identifying – within the confines of the scope of the analysis – all independent components in such a way as to provide a maximally complete explanation while also avoiding informational redundancies. With that as its general goal, the framework stipulates that each element of knowledge should be informationally nonoverlapping with other ones, while at the same time, the totality of all knowledge elements should yield informationally exhaustive insights. Meeting those rather lofty goals demands unambiguous explanation of the meaning of individual explanatory elements, along with detailing of their measurement properties, all geared toward reducing interpretational ambiguity. In addition, explicit assessment of appropriateness of individual elements of knowledge, as well as sufficiency of coverage of the combined variable set, also needs to be undertaken.

Applying the guiding principles of the MECE framework to the preceding overview of different approaches to assessing the efficacy of organizational learning efforts and mechanisms yields the general organizational analytic adeptness (OAA) framework, which represents an adaptation of the earlier outlined 4I model. The general outline of the OAA framework is graphically summarized in Fig. 6.6.

The recursive intention-interpretation-integration-institutionalization progression, which manifests itself in the corresponding progression of internalization-externalization-socialization-application, and is further framed in the context of "triggers" and "rewards," is the conceptual backbone of the OAA framework. Moreover, the gradual development of organizational data analytic adeptness through individual-, then group-, and then organization-wide assimilation of the requisite knowledge contributes the broader organizational dynamics context, which helps to contextualize the general rationale of the OAA model. Lastly, the model also implicitly accounts for the key tacit vs. explicit knowledge distinction, which is implicit in the evolution of data analytic knowledge from initially deeply subjective, individual-level understanding to group-wide cognition, to finally being expressed as explicit organizational routines and processes. Framed within that broad context, "triggers" are any individually relevant incentives that "spark" the initial interest in developing an understanding of the key elements of data analytic knowledge, while "rewards" are the benefits to the organization that are brought about by what are usually nontrivial organizational changes that are required to embed data analytic-related routines and processes into the larger web of organizational innerworkings.

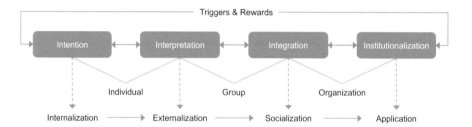

Fig. 6.6 Organizational analytic adeptness (OAA) framework

The OAA framework offers an abstract representation of the general conceptual structure of organizational data analytic adeptness. It is implicitly focused on the creation of organizational data analytic insight utilization infrastructure, capable of enabling systematic and systemic infusion of data analyses-derived insights into a wide array of organizational decision-making processes. Its primary purpose is to inform, in the sense of offering general design and development-related guidance, organizational efforts to formally seed and nurture effective and enduring mechanisms supporting evidence-based decision-making. As such, it should be seen as a blueprint helping organizations reap the benefits of their typically informationally rich, but poorly utilized, data resources.

6.4 Rights and Responsibilities

An important aspect of learning with data is a clear understanding of rights and responsibilities associated with accessing and using data. Given the ever-expanding digitization of personal and commercial interactions, data capture is ubiquitous, and that "in the background" recording of interactions and events is not always evident to those who are digitally tracked. Not surprisingly, just about every country has enacted some sort of data privacy laws to regulate how information is collected, how data subjects are informed, and what control a data subject has over his information once it has been captured. Navigating those laws and regulations can be a daunting task. In the USA, there is no comprehensive federal law that governs data privacy;[16] instead, there is a complex patchwork of sector-, jurisdiction-, and medium-specific laws and regulations, including those that address financial intermediaries, healthcare providers, telecommunications, and marketing. While the Congress has been slow to enact more comprehensive data privacy laws, various federal agencies, such as the Federal Trade Commission which has a fairly broad jurisdiction over commercial entities, have been taking steps to fill that void by issuing privacy-related regulations. Moreover, individual states, territories, and even localities have also been enacting their own laws, the result of which is that the USA currently (as of 2021) has hundreds of data privacy and data security laws that have been enacted by a host of jurisdictions to regulate collection, storage, safeguarding, disposal, and use of personal data collected from their residents.

Needless to say, compliance with applicable data capture, storage (i.e., protection), and usage laws is among the top-of-mind organizational considerations, but forward-thinking commercial and non-commercial organizations are going beyond the so-called black letter laws, or well-established legal rules that are beyond

---

[16] In contrast to that, the European Union enacted (in 2016) what is widely considered the most comprehensive data protection legislation known as the General Data Protection Regulation; it governs the collection, use, transmission, and security of data collected from residents of any of the 28 member countries, and levies fines of up to €20 million (or 4% of revenue) on organizations that fail to comply with its provisions.

dispute, by actively considering ethical data capture and usage implications. It is a manifestation of a slowly changing organizational mindset, one which sees notions such as social responsibility and social consciousness playing a more central role as organizations embrace the idea of "doing the right thing." In the context of data capture and usage, such more socially responsible mindset calls for also examining value-based implications of data usage, which means right vs. wrong judgment. The goal is to go beyond avoidance of noncompliance precipitated punishment by voluntarily instituting positive data privacy practices built around notions of integrity and transparency.

Broadly considered, responsible data capture, stewardship, and usage can be considered in three somewhat distinct prevention contexts: (1) unauthorized data access, commonly referred to as data breach, (2) improper data access, and (3) publication of de-anonymizable data. Unauthorized data access typically, though not always, entails malicious outsiders aiming to steal or damage information, while improper data access entails authorized data users using data in unethical or even illegal manner; publication of identity-discoverable data is comparatively more obtuse a threat as it is a manifestation of publishing – internally or externally – data that were overtly anonymized but that nonetheless contain details that make de-anonymization possible.

6.4.1   Unauthorized Data Access

Confidentiality is a key consideration in implied or explicit consent granted by those whose information is contained in data captured by commercial, nonprofit, and governmental entities. Perhaps the most obvious violation of the promise of confidentiality is an unauthorized data access, commonly known as data breach. The statistics of data breaches are quite staggering: In 2019, more than 2100 individual data breach events have been reported; that number is on pace to more than double in 2020, largely due to a sharp increase in remote operations precipitated by the COVID-19 pandemic. Interestingly, in roughly 50% of individual data breach cases, the exact number of compromised records is unknown; overall, about 62% of breaches are attributed to malicious outsiders, followed by accidental loss (22%), malicious insiders (12%), hacktivists (2%), and state sponsored (2%). The bottom line is that as more and more of commercial and private life is going online, so is crime, and according to experts in the field of data security, some form of organizational data breach is essentially unavoidable. As commercial and non-commercial organizations continue to amass more and more data, and as those holdings become more distributed throughout their organizational ecosystems, organizational exposure to malicious or unintended data breaches will continue to grow.

Securing of organizational data is, in principle, an undertaking comprised of two parts. The first part entails access control in the form of authorization and authentication, while the second part takes the form of obscuring data, which can be achieved via a variety of means, such as data masking, where letters and numbers are replaced with proxy characters; encryption, which uses algorithms (called cipher) to scramble data; or tokenization, where sensitive data are replaced with random, not algorithmically reversible, characters. Aiming to make data unreadable, data obscuring can be considered as a second line of defense against unauthorized access.

An important consideration in data security is access mining, which entails the collection and selling (illegally, typically on the dark web[17]) of access descriptors. Gaining unauthorized access to data is a problem that is so pervasive that there are numerous established (illegal, of course) marketplaces for selling and buying illegally obtained access credentials. For example, one such marketplace, Ultimate Anonymity Services, offers some 35,000 credentials with an average price of $6.75 per credential.

6.4.2   Improper Data Access

In many regards, making sure that those with a legitimate data access rights do not abuse those rights might be the most undefined challenge of responsible data ownership. The reason for that is that illegitimate data access encompasses an impossibly broad array of behaviors ranging from malicious, as in the case of a disgruntled employee's intentional theft, to just plainly stupid, as in the case of a bank employee looking up friends' account balances. Putting in place explicit data usage policies is usually the first, obvious step that an organization can take to diminish that threat. A more formalized approach is to institute explicit access control protocols, which can be vertical, horizontal, or context defined. Vertical access control assigns different access rights to different users, while horizontal access control grants limited access, meaning that specific user profiles are granted access only to specific subsets of data. In a more operational sense, vertical and horizontal protocols can manifest themselves in discretionary access control, where access rights are determined on case-by-case basis; mandatory access control, which groups potential users into explicit categories with differentially defined access rights; or role-based access control, in which functional responsibilities are used as basis for assigning specific data access rights. Altogether different, context-dependent access controls tie access rights to a specific data usage purpose or application, allowing differential access based on the underlying informational purpose.

6.4.3   Publication of De-anonymizable Data

While keeping data secure is generally technologically the most challenging aspect of responsible data capture and usage practices, internal and external sharing of data or data-derived information is likely the most procedurally vexing. Definition-wise,

---

[17] The dark web is the World Wide Web content that is not visible to search engines, and requires specific software, configurations, or authorization to access.

once data leave a secure environment, they can be considered to have been "published," in a sense of having been made available for wider consumption, even if it is just limited to specific organizational stakeholders. Moreover, within the confines of data privacy-related considerations, the definition of publishing is relatively broad as it encompasses not just raw data but also data-derived information, in the form of summaries and other data analytic outcomes. The key threat associated with publishing of data or information where personally identification and other confidential details have been masked or outright deleted is that what might overtly appear to be anonymous, may in fact contain traces of confidential details.

Looking beyond masking of overtly individually identifiable data details, a common step toward further reducing the risk of inadvertent publication of confidential information is to normalize data, which entails reorganizing of individual data elements in a way that reduces redundancies, creates logical dependencies, and groups all related data. Among the core informational benefits of data normalization is reduced variability which gives rise to tighter distributions, thus making individual values seem less unique, or less identifiable. However, data normalization does not eliminate the possibility of there being combinations of additions and subtractions, through a general process of database querying, of individual data records that might cause de facto revealing of private information.

A considerably higher privacy assurance is offered by what is today considered the state-of-the-art approach, known as *differential privacy*. Broadly characterized, it is an approach for publishing of information about a dataset by describing the patterns of groups within the dataset while withholding information about individuals in the dataset. The core appeal of the differential privacy approach is that it offers strong privacy guarantees that are rooted in the fact that it is a semantic privacy notion, meaning it is built around data processing algorithms rather than specific datasets. The basic idea here is that the effect of making a single substitution in the database is small enough to assure privacy in the sense that review of published information is unlikely to reveal if a particular individual's information was included.

Since first being introduced in the early 2000s [5], differential privacy algorithms gained widespread usage not only with traditional numeric and text data but also with newer types, including social networking and facial recognition. Social networking platform operators use a two-step anonymization process: first, computing (using differential privacy methods) parameters of a generative model that accurately captures the original network properties, and second, creating samples of synthetic graphs. A somewhat similar approach is used by providers of facial recognition technologies, which process biometric information using third-party servers. Biometric information delivered to an outside third party can be considered a significant privacy threat as biometrics can be correlated with sensitive data such as healthcare or financial records, but the use of an appropriate variant of semantic differential privacy algorithms can offer strong assurance of privacy protection.

6.5 Learning to Learn

The rapid proliferation of artificial intelligence-powered technologies is slowly but fundamentally reshaping the nature of work, freeing human time-effort while also producing vast quantities of data, which jointly give rise to new human time-effort redeployment opportunities. Those shifts are calling for fundamental rethinking of what it means to "learn," as the traditional acquisition of knowledge-framed conception of learning does not fit the competency-enabled creative exploration type of learning, which is emerging as the dominant driver of organizational value creation. Within the confines of organizational functioning, the idea of competency-enabled creative exploration entails thoughtfully combining the dimension of learning characterized here as "assimilation of enduring truths," best exemplified by acquiring the knowledge of statistical methods used to mine data for new insights, with "discovery of new knowledge" through immersion in raw data. The long-standing instruction- and explanation-focused model of learning offers an effective mechanism for the former, but considerably less so for the latter. Yet within the confines of goal-oriented organizational learning, where creative exploration is bounded, as in Simon's notion of bounded rationality,[18] by time and resource constraints, there is a need to offer a comparable degree of learning guidance, which is the focus of the ensuring chapters.

References

1. Steiner, I. D. (1972). *Group process and productivity*. New York: Academic Press.
2. McGrath, J. E. (1984). *Groups: Interaction and performance*. Englewood Cliffs: Prentice Hall.
3. Kirkpatrick, D. L. (1998). *Evaluating training programs: The four levels* (2nd ed.). San Francisco: Berrett-Koehler Publishers.
4. Crossan, M. M., Lane, H. W., & White, R. E. (1999). An organizational learning framework: From intuition to institution. *Academy of Management Review, 24*(3), 522–537.
5. Dwork, C., Kenthapadi, K., McSherry, F., Mironov, I., & Naor, M. (2006). Our data, ourselves: Privacy via distributed noise generation. In *Advances in cryptology-EUROCRYPT* (pp. 486–503). Berlin: Springer.

---

[18] Proposed by H. Simon as a counterargument to the notion of human rationality, the idea of bounded rationality posits that human decision-making is bounded, or constrained, by cognitive limitation of the human mind, and by the tractability of the problem at hand, and by time constraints.

Chapter 7
Decision Automation

7.1 From Decision Support to Decision Automation

The general idea of 'automation' can be traced back to the automobile industry's pioneering use of automatic devices to control mechanized production lines. The term itself is believed to have been coined by an engineering manager at the Ford Motor Company, D.S. Harder, who used it in an interview with a trade journalist in 1948. Since then, as the use of automated control devices spread throughout industries and spilled into everyday usage situations, the notion of *automation* became a household term. In this chapter, the general idea of data-driven decision-making discussed in Chap. 6 is considered in the context of informational automation, or more specifically, the means and modalities of not only automated insight extraction but also self-contained, autonomous decision-making engines.

Since the 1970s, systems designed to support decision-makers in the decision-making process have been formally recognized as a separate class of technologies known as *decision support systems* (DSS), defined as interactive computer-based systems that support choice evaluation of structured, semi-structured, and unstructured problems. As standalone applications, DSS represent bundles of components including data, statistical or other data analytic models, data visualization tools, and user interface. Those can be either custom-built or developed with the help of the so-called DSS generators, which offer templated platforms geared toward a particular organizational problem, such as promotional planning. While not always the case, it is rather common for DSS to be adaptive to the problem at hand, i.e., capable of providing context-specific decision-making guidance, in addition to also being able to cope with changing decision-making requirements across user types. When integrated with real-time or near-real-time data capture and processing infrastructure, DSS can offer asynchronous, or non-contemporaneous, data inputs, as well as synchronous decision-making support. Additionally, effective decision support systems also include iterative scenario planning methods, most notably what-if

A. Banasiewicz, *Organizational Learning in the Age of Data*, EAI/Springer
Innovations in Communication and Computing,
https://doi.org/10.1007/978-3-030-74866-1_7

analysis tools, to help users fine-tune their choices. And lastly, and often most importantly, DSS need to be built around sound data analytic logic, as illustrated by the buyer loyalty classification example briefly sketched out below.

But when considered in a broader, evolutionary sense, DSS can be seen as a step toward more technologically evolved, meaning self-functioning, systems. After all, many of commercial and non-commercial organizations' decisions are not just routine, but also frequently recurring – tedious and time-consuming manual processing of high-frequency, recurring organizational decisions is not only inefficient but also prone to inconsistencies, even bias. In such situations, there is a compelling reason to look beyond DSS systems that function as advisors to human decision-makers, and instead consider more automated, or even autonomous, means of data-enabled decision-making. However, the more automated a system, the more tailored it becomes, primarily because their informational engines broadly characterized as machine learning (discussed in more detail later in this chapter) require well-defined contexts to deliver the desired functionality. Stated differently, to be meaningful, manifestations of *automated decision-making* (ADM) need to be considered in the context of specific applications. In keeping with that, three distinct examples of progressively more automated ADM applications are discussed next: (1) decision support systems, built around human-supplied rules; (2) decision automation, built around the human-in-the-loop and rules inferred from data model; and (3) autonomous decision-making, built around the human-not-in-the-loop and rules inferred from data model. Framed in the context of mini-cases, the ensuing discussion aims to highlight the core informational aspects of the three ADM variants.

7.1.1   Decision Support Systems Case: Buyer Loyalty Classification

At its most rudimentary level, customer base assessment is primarily concerned with accurate brand buyer categorization and a meaningful segment description. With an eye toward total customer value, which is a function of a sum of purchases accumulated over a period of time (with the further back in time purchases discounted appropriately, such as by using the organization's cost of capital), the overall buyer base is usually divided into *brand loyal* and *brand switching* customer groups. The basis for this categorization is typically the brand-to-category repurchase rate, expressed as a proportion of the brand purchases vis-à-vis total product category purchases over a period of time, or:

$$\text{Brand loyalty} = \left(\frac{b}{c}\right)_{t=i}$$

where b = number of purchases of the brand of interest, c = total number of purchases in the overall product category, and t = the period of time over which the purchases are evaluated, such as 12 or 24 months.

Implicit in this quantification is the equivalence of 'brand loyalty' and 'repurchase exclusivity,' which implies that a truly loyal customer is someone who repurchases only a single brand in the category, as being 100% loyal requires that the number of brand-specific purchases be equal to the number of category-wide ones. While being computationally easy to implement, such simplistic decision heuristic breaks down in situations where consumers can be expected to purchase more than a single brand in the category for reasons other than brand preference. For instance, a consumer with multiple pets may need more than a single brand of pet food, just as purchases made for multi-person households could also likely encompass multiple brands. Correcting for those deficiencies, an enhanced loyalty categorization heuristic has been proposed [1], which expresses brand loyalty as a function of household-level purchases, and expands the brand loyalty operationalization to accommodate the cross-household variability in category purchase requirements (CPR), thus effectively relaxing the repurchase exclusivity assumption. The new metric capturing these effects, CPR, is defined as an estimated number of distinct product types or brands repurchased by a single household in a product category over a period of time; it enhances the heuristic's classificatory validity by differentiating between single and multi-brand loyalty, effectively replacing the original 'loyal' vs. 'switcher' dichotomy with the three-tiered classification of 'single brand loyal,' 'multi-brand loyal,' and 'switcher,' computed as follows:

$$\text{Household brand loyalty} = \sum_{n}^{i=m} h\left(\frac{b}{c}\right)_{t=i}$$

where b = number of purchases of the brand of interest, c = total number of purchases in the overall product category, t = the period of time over which the purchases are evaluated, and h = household purchase pooling.

$$\text{If,} \quad (k=1)_t \Rightarrow \sum_{n}^{i=m} h\left(\frac{b}{c}\right)_{t=i} \geq \text{threshold} = \text{'single}-\text{brand loyal', else if}$$

$$(k>1)_t \Rightarrow \left[\sum_{n}^{i=m} h\left(\frac{b}{c}\right)_{t=i} / k\right] \geq \text{threshold} = \text{'multi}-\text{brand loyal', else if}$$

$$(k=1)_t \Rightarrow \sum_{n}^{i=m} h\left(\frac{b}{c}\right)_{t=i} \quad \text{or} \quad (k>1)_t \Rightarrow \left[\sum_{n}^{i=m} h\left(\frac{b}{c}\right)_{t=i} / k\right] < \text{threshold} = \text{'switcher'}$$

where k = household-level CPR estimate, b = number of purchases of the brand of interest, c = total number of purchases in the overall product category, t = the period of time over which the purchases are evaluated, and h = household-level purchase aggregation.

The customer loyalty categorization case offers a comparatively simple but representative illustration of the basic logic of decision support systems. One of the key features of this and other DSS mechanisms is the centrality of the human-in-the-loop constructive taxonomy, according to which the key system's decisioning characteristics require human involvement (as in delineation of operational definitions of grouping categories used in the loyalty classification case). A step beyond DSS is *decision automation* (DA) tools, which, while still being based on the human-in-the-loop model, leverage decision rules that are inferred from data (as opposed to being human-supplied) with the goal of enhancing the efficacy of routine organizational decision-making. The general idea at the root of DA systems is sometimes characterized as *digital decisioning*, framed as an application of data analytic decision engines to high-volume, repetitive decisions, geared toward elimination of redundancies, enhancing decision-making consistency, and increasing the agility of organizational functioning. A highly visible, and equally controversial, application of the spirit of DA is China's social scoring system, summarized next.

7.1.2   Decision Automation Case: Social Credit Scoring

The Social Credit System (SCS) is a national reputation system developed by the central Chinese government, effectively the Chinese Communist Party, initially trialed in 43 'demonstration cities' and districts starting in 2009, before launching as a notional pilot in 2014. The system takes the familiar credit scoring idea and extends it to all aspects of life with the goal of judging citizens' overall behavior and trustworthiness. In accordance with such goals, the SCS provides rewards or punishments as feedback to individuals and companies, based not just on the lawfulness, but on the morality of their actions, covering economic, social, and political conduct. The Big Brother or Black Mirror dystopian parallels aside, as an application of data analytics, the SCS resembles a straightforward, preprogrammed rule-based system, such as the earlier outlined brand loyalty classification; however, each of the 43 'model jurisdictions' implements the program differently. For example, under the Rongcheng City model, everyone is assigned a base score of 1000 points on the credit management system which is linked to four governmental departments, and points are then added or deducted on the system by human government officials for specific behavior, such as late payment of fines or traffic penalties. There are in total 150 categories of positive conduct leading to additional points on the system, and astounding 570 categories of negative behaviors leading to point deductions for individuals (which by itself is clearly suggestive of the predominantly punitive orientation of that system).

Moreover – and that is a key point – the analytic logic of the SCS system is not at all transparent, which when coupled with the lack of power checks and balances that typifies all totalitarian communist governments has profound and far-reaching implications. For instance, those with low social credit rating scores may not be eligible for loans and certain jobs or may be denied the opportunity to travel, while those with high scores enjoy benefits such as cheaper public transport, free gym facilities, and priority treatment and shorter waiting times in hospitals. In that sense, the SCS system can be seen as the most all-encompassing and technologically sophisticated manifestation of the idea of social engineering, geared toward solidifying and assuring the Communist Chinese Party's dominance over Chinese society. The system is still work-in-progress, and currently the Chinese government is in the process of forming partnerships with private companies with sophisticated data analytics capacities to further evolve this social control metastructure. Among the best known of those collaborations is the government's work on the Sesame Credit system with (Chinese) tech giant Alibaba; one of the key aspects of that collaboration is an automated assessment of potential borrowers' social network contacts in calculating credit scores. Sesame Credit (officially known as Ant Financial Zhima Credit) combines information from Alibaba's database with other personal information, such as online browsing history, social networking, tax information, and traffic infringement history to automatically determine the trustworthiness of individuals. In other words, it is not just one's own behavior, but it is the mere fact of being even somewhat remotely socially connected to those with low scores that negatively impacts on one's own score, and thus one's economic and social well-being. When considered from the perspective of personal freedom, it is frighteningly dystopian application of the idea of decision automation.

It is not surprising that the ultimate goal of the SCS initiative is to create a single, countrywide unified social credit code, one that is less of a score and more of a permanent record, linked to each citizen's unique identity. While, as noted earlier, having high social credit score will entitle one to perks such as cheaper public transport, free gym facilities, and priority medical treatment, the case of Liu Hu, a Chinese journalist who writes about censorship and government corruption, offers a vivid illustration of consequences of having low social credit score: Liu has been arrested, fined, and ultimately blacklisted as he found that he was named on a List of Dishonest Persons Subject to Enforcement by the Supreme People's Court, rendering him 'not qualified' to buy a plane ticket or even to travel on some train lines, in addition to also being unable to buy property or take out a loan.

Clearly, the SCS system raises a number of highly worrisome questions, so much so that it is difficult to imagine such system being contemplated by open, democratic societies – still, it is a highly visible application of data analytics-based decision automation. And it also illustrates an important point, often raised by ethicists, that just because something can be done does not necessarily mean that it should be done…

While largely driven by automated computational algorithms, the SCS system can nonetheless still be characterized as an application of the general human-in-the-loop model because it combines system-generated and human inputs. It is conceivable, really, likely, that advances in fully automated natural language processing and related aspects of data analytic automation will slowly lessen, and eventually obviate the need for human input, at which point the SCS will become an application of human-not-in-the-loop model of decision automation. That said, there already are systems that fall into that broad *autonomous decision systems* (ADS) category, best illustrated by the emerging technology that continues to feed attention and fuel imagination: self-driving cars.

7.1.3   Autonomous Decision Systems Case: Self-Driving Cars

At the turn of the twentieth century, a new technological marvel – the automobile – was beginning to replace the horse-drawn carriage as the dominant mode of transportation. Though initially met with skepticism, the new invention was aided by the subsequent introduction of traffic rules and smooth new asphalt street surfaces, all of which ultimately paved the way, no pun intended, for its mass acceptance. And while automobiles' qualities such as speed, versatility, and reliability continued to evolve ever since the technology's late nineteenth-century inception, the technology did not fully deliver on the promise embedded in its name. Needing a human operator to function as intended, the automobile has not been truly 'auto,' as in 'self' – until the recent emergence of the still prototypical but finally auto- or self-driving cars.

While it is difficult to pinpoint the exact singular event that sparked the race to develop fully self-driving cars (if there even is such as singular flash-point), the DARPA[1] Grand Challenge, first held in 2004, is likely one of those events. The initial (there were several since 2004) Grand Challenge targeted the development of fully autonomous ground vehicles capable of completing a substantial off-road course within a limited time, later expanded to also include autonomous operation in a mock urban environment. The core challenges faced by developers parallel the key accident causes of human-operated vehicles: recognition and decision errors, which can be seen as manifestations of two key situational awareness problems of failure to detect and failure to understand, respectively. Development of systems that are capable of surmounting those difficulties encompasses the use of complex machine learning algorithms training on huge datasets that include millions of road images and related details, further aided by lightning-fast processors, and rapidly evolving computer vision leveraging massively parallel processing units.

In a sense, however, the problem is unlike a typical machine learning application where some degree of error is deemed acceptable. Even the most advanced commercial deep learning applications, such as Google's Assistant, Apple's Siri, or Amazon's Alexa voice recognition systems, are highly but not fully accurate, which is simply not good enough for self-driving cars. The challenge posed by the complexity and breadth of possible driving conditions and the demand of perfectly correct recognition and equally correct response stretch beyond typical machine learning situations, where learning is based on what is effectively a subset of all possible informational inputs and results are probabilistic, thus prone to error. The currently in-use commercial deep learning systems mentioned earlier are generally recognized as being high-functioning embodiments of informational autonomous but nonetheless limited-purpose applications, trained on and meant to be used within confined informational contexts. In fact, the key challenge in machine learning, including deep learning, is development of human-like abilities to abstract generalizable understanding from incomplete details, in a way that reaches far beyond the immediate meaning of those details. Further complicating the task at hand is the need to blend into the hustle and bustle of human society, which calls for situational awareness, where the key challenge (as it relates to autonomous vehicle development) is failure to detect and/or failure to understand.

---

[1] The Defense Advanced Research Projects Agency, widely considered as the most prominent research organization of the US Department of Defense (many modern technologies, including the Internet, were born there)

The numerous self-driving car development projects currently underway all aim to endow their systems with the capability of ostensibly performing the same functions as a human driver, which in terms of the system controlling algorithms include the following capabilities: self-localization and self-navigation using 3D mapping technologies, identification of static and moving obstacles, classification of obstacles and other information using machine vision, generation of current knowledge-based road condition predictions, evaluation of those predictions against observed details, and, finally, actuation of automated actions such as turning, braking, or accelerating, as appropriate. In a more machine learning sense, those specific steps entail performance of image and object recognition, classification, and grouping; in other words, the vehicle-controlling algorithms have to first understand how many objects exist in their view and then classify them into action-evoking categories such as pedestrians, cars, traffic lights, etc. In the information processing sense, that calls for boosting of informative (signal) parts of images over non-informative (noise) parts of those images in order to, for example, be able to distinguish between a pedestrian which is standing and moving in the direction of the vehicle's path, or a pedestrian and a feature such as a post or a traffic light at a crossing. Moreover, the underlying algorithmic logic also has to be able to do all that in varying weather, light, and situational conditions, which calls for additional capabilities in the form of adaptive, unsupervised segmentation using data from different (e.g., cameras, lasers, etc.) sensors. And yet, that is not all: Recognition and classification of human and objects in the vehicle's proximity must go beyond simple processing of visual input through the vehicle's cameras and simply applying the resultant labels to objects using a database stored onboard – it requires autonomous interactions with electronic systems and databases outside the vehicle, such as external GPS-based guidance.

Still, there is more. Looking beyond image processing, the vehicle-controlling algorithms also have to solve a range of dynamically arising problems, such as avoiding potential collisions by predicting the vehicle's own trajectory as well as trajectories of nearby vehicles. All of those computations have to then be swiftly and accurately actuated thru generation of appropriate steering, acceleration, deceleration, and breaking commands, all while also relating past driving conditions to the current destination to account for traffic congestion and preemptively look for alternative routes, or adjust driving speed.

In the nod to those and other challenges, the Society of Automotive Engineers (SAE) identified five levels of automation in computer-assisted driving:

Level 0: No automation; it includes functionality that assists the driver to regain lost control of the vehicle, such as automatic breaking systems.

Level 1: Function-specific automation, where a system fully controls one function.

Level 2: Combined function automation, where a system controls at least two functions.

Level 3: Limited self-driving automation, where the driver cedes full control under specific conditions.

Level 4: Full self-driving automation, where the driver is not expected to become involved throughout the duration of the trip.

At the time of this brief overview (the latter part of 2020), numerous Level 3 platforms have already been commercially deployed, but no Level 4 platform is yet commercially available, suggesting that, at least in the commercial realm, fully autonomous systems are still under development.

7.1.4   ADM Adoption Hurdles

The promise of evidence-based, and thus better informed and more objective, organizational decision-making is universally compelling, but in practice many decision automation projects fail to turn that promise into operational reality. While reasons can often be highly situational and nuanced, there are several recurring themes: disruptive character of digital innovation, the nature of human-machine interactions, trust and regulatory considerations, and accuracy-related misconceptions.

Startups and born-digital organizations are natural early adopters of informational innovation because they are typically not being burdened by legacy systems, workflows, and structures, thus having everything to gain and next to nothing to lose by embracing new decision-making approaches and technologies. That is not the case for large, established organizations with entrenched systems and cultures, but it is not just a matter of organizational inertia – those organizations need to find ways to innovate that are not disruptive to their operations or customers, which typically entails slower, incremental change process (i.e., one, though certainly not the only, reason wider embrace of renewable energy is slower than many of us would like it to be).

The second key of automated decision-making, or ADM, adoption hurdle takes the form of human-machine interactions. Automated decision systems are often seen as 'job eliminators,' even though truly effective deployment of those systems entails augmenting, not replacing, human decision-making. For example, insurance claim managers tend to be charged with handling large numbers, typically a lot more than 100, of cases at any given time, which obviously severely limits the amount of time and attention they can allocate to a single claim. In that context, thoughtful deployment of ADM can benefit the overall claim processing efforts by providing auto-processing of routine claims, making it possible for human adjustors to invest more time and care into nonroutine, more complex claims.

The third key ADM adoption hurdle centers on issues of trust and regulatory compliance, which is particularly profound for organizations that operate in highly regulated segments, such as healthcare or financial services. Intuitively, informational automation should be appealing to organizations that need to provide, typically on ongoing basis, evidence of compliance, but many organizations are gripped by fear that such systems will not produce adequately robust outcomes. It could be called the human bias – it is more comforting to trust a human expert than it is to trust an expert system, even if there are numerous and glaring examples of the former's recurring failures to deliver. And so even though well-designed expert systems might, ultimately, be less error-prone, human decision-makers tend to be more forgiving of human fallibility.

The final key hurdle, accuracy-related misconception, comes into play across a wide range of analytics-derived insights. In general, detractors of relying on data analytics point to imperfect accuracy as a reason for disuse, which not only glazes over the many limitations and biases of human sensemaking but even more importantly, it factually misstates the actual efficacy of data analytics-derived insights. To use the earlier example of an insurance claim manager, those who resist adoption of ADM tend to point to instances of incorrect claim classification as the reason for not trusting such systems. Again, setting aside the overwhelming amounts of evidence that clearly point to relatively common claim evaluation errors made by human adjusters (which typically take the form of labeling claim as 'standard' only to see it develop adversely, in the sense of becoming very costly), an automated claim processing system that has, for instance, a modest 60% accuracy rate is still likely considerably more accurate than a human examiner for two key reasons: First, it can simultaneously, objectively, and tirelessly take into account as many factors as are available, while its human counterpart is not only bounded by time, fatigue, and subjective perception – that human evaluator's cognitive abilities are also inescapably reductive due to the limitations of the brain's channel capacity,[2] which

---

[2] Human channel capacity is the idea originally proposed in 1956 by G A Miller, a psychologist, who placed a numeric limit of 7 ± 2 of discrete pieces of information an average human can hold in his/her working memory. According to the proponents of that notion, the reason the number of informational 'chunks' a person can simultaneously process in attention is finite, and fairly small, is due to limitations of short-term and working memory, which can be likened to physical storage restrictions of a computer system, or a physical container for that matter.

effectively limits the number of distinct claim attributes that a human evaluator can actively consider while trying to make an evaluation. Second, the commonly used logic of evaluating the stated classificatory accuracy of ADM systems, such as the aforementioned 60%, is rooted in mathematically flawed logic that confuses disaggregate chance with aggregate probability. Yes, a human adjuster choosing at random might have 50-50 odds of correctly labeling individual claims as 'low' vs. 'high' risk, but no, that does not mean that the said adjustor's aggregate correct classification rate can be assumed to be 50%. Without getting wrapped up in mathematical details, the 50% chance could be considered valid only if the claim handler were to make just a single evaluation – handling multiple claims calls for multiple evaluations which in turn compounds the overall classification error. Using a simple case of just three, 50-50 (meaning choosing at random) claim evaluations will yield 12.5% aggregate probability of being correct ($0.5 \times 0.5 \times 0.5$), not 50%. And while some evaluations might have higher than the random 50-50 odds of correct assignment, the compounding of individual errors has a powerfully dampening effect on aggregate accuracy. All considered, a seemingly diminutive 60% accuracy of the earlier mentioned predictive algorithm might look quite compelling when contrasted against a properly framed base rate.

7.1.4.1 The Next Step

Overcoming ADM adoption hurdles does not necessarily assure success. In order to be considered a true organizational asset, ADM systems need to deliver desired, or at least noticeable, decision-making-related capability gains, in an accountable and communicative manner, as graphically depicted in Fig. 7.1.

Informationally capable ADM system provides the 'right' information at the 'right' time in a way that supports maximally correct inferences or actions. In other words, it does exactly what it is expected to do. The next ADM system requirement – accountability – is more complex and nuanced. In the everyday human

Fig. 7.1 The forces shaping ADM's efficacy

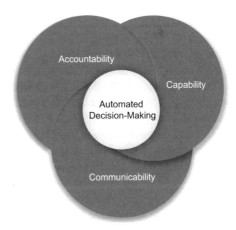

Accountability

Capability

Automated
Decision-Making

Communicability

context, accountability is a reflection of responsibility; as such, it can be seen as a social obligation emanating from the idea of free will – to be accountable is to be responsible for one's choices, which implies consciousness or being self-aware. Hence, within the confines of artificial systems governed by mathematical logic rather than free will, to be 'accountable' is to be 'traceable,' which is necessary to assuring validity and reliability of algorithmic outcomes. In other words, to be trusted, the computational logic of ADM systems' results need to be discernable to human examiners.

The final of the three core determinants of ADM adoption, communicability, is perhaps the most fluid. In very general terms, it refers to the degree to which the system can support effective and efficient human-computer interactions – that requirement is fluid because it encompasses not only being user-friendly but also 'behaving' in an ethically and morally responsible and responsive manner. Here, the idea of free will surfaces again because behaving implies choice, but as noted earlier, artificial decision-making systems are governed by rules of mathematical logic, and those do not encompass moral or ethical considerations. Stated differently, ADM systems are simply not capable of moral judgment. Moreover, it flies in the face of reason that an artificial system that has no intrinsic sense of worth can be expected to 'understand' the sense of worth that humans attached not just to physical objects, but even more importantly, to intangible ideals or beliefs. Still, those are tremendously important, in fact critical, considerations that need to be thoughtfully and soberly examined *before*, not during or after, autonomous systems are entrusted with decision-making powers. Those considerations, now commonly summarized under the heading of 'human-machine interactions,' are both complex and important, and thus will be addressed in more detail later in this chapter, after first taking a closer look at the core aspects of those systems' innerworkings.

7.2   The Heart of ADM: Machine Learning

An intuitively obvious definition of learning frames that important notion simply as the acquisition of knowledge. As discussed in Chap. 1, the idea of knowledge can be understood in the context of sources, which can be either explicit or tacit, and dimensions, of which there are three: semantic, procedural, and episodic (see Fig. 1.3). And although not explicitly called out, cognition is also implicitly embedded in that human-centric, understanding- and remembering-based framing of learning, which begs an obvious question: can the same, or at least closely related, framing of the general idea of 'learning' apply to artificial systems?

Loosely defined, *machine learning* is the process by which computer algorithms can translate patterns and relationship in data into outcomes in the form of inferences or actions, without the help of explicit instructions. Notionally, machine learning exhibits the core aspects of (human) explicit and tacit human learning: Explicit 'knowledge' is embedded in individual algorithms' programming architecture, while tacit knowledge is acquired through exposure and experience, which in the case of computer algorithms is typically characterized as training. Moreover, similar to human tacit knowledge, machine learning is interpretive as well as constantly accruing through exposure to different sets of data. In fact, the most popular type of machine learning algorithms are artificial neural networks (ANN), a family of methods so named because the design of their computational units aims to mimic intelligent functions performed by networks of biological neurons, most notably the human brain. ANN algorithms are trained using now-common large datasets to identify suitable actions (typically within relatively narrow and well-defined context), in a manner generally resembling human decision-making. Those and other types of machine learning algorithms, such as decision trees, naïve Bayes, nearest neighbor, or support vector machines, can be seen as 'engines' powering larger automated decision-making systems discussed earlier, geared toward acting without the need for direct human intervention.

Broadly conceived, meaning including deep learning, machine learning also exhibits the three distinct dimensions – procedural, semantic, and episodic – of human learning. Core elements of procedural knowledge are almost always a part of an algorithm's design, while the key elements of incremental self-improvements (procedural learning) are added through ongoing ingestion of data-derived patterns and associations (training), along with concept generalizations and predictive inferencing (semantic learning), as well as instance recognition (episodic learning). Of course, that is not to say that machine can 'know' in the same conscious sense as a human, but is that distinction important in the context of organizational sensemaking? More on point, is that distinction important when it comes to making routine decisions, where efficiency, informational thoroughness, and consistency might be of more immediate concern?

In addition, human vs. machine learning comparisons might also be obscured, at least in part, by some persistent definitional fuzziness. Buoyed by what has been nothing short of phenomenally explosive growth of all manner of information technologies, many technical terms related to data science and analytics have been thrust into popular vernacular, the consequence of which has been excessively variable usage of terms such as 'machine learning,' 'deep learning,' and 'artificial intelligence' (now commonly abbreviated as AI), so much so that nontechnical audiences can find themselves somewhat confused by how any one of those tools differs from others. While this book is only concerned with organizational learning-related implications of ADM technologies, it is nonetheless beneficial to anchor that discussion in a somewhat uniform understanding of the general typological structure, which is summarized in Fig. 7.2.

The very general lines of demarcation outlined in Fig. 7.2 are merely meant to offer some high-level definitional clarity, as the more in-depth discussion of artificial intelligence, which stretches far beyond automated decision-making systems, falls largely outside of the scope of this book. That said, the two key ADM-related sub-domains – machine learning and deep learning – are of key interest to organizational learning, and thus are discussed next.

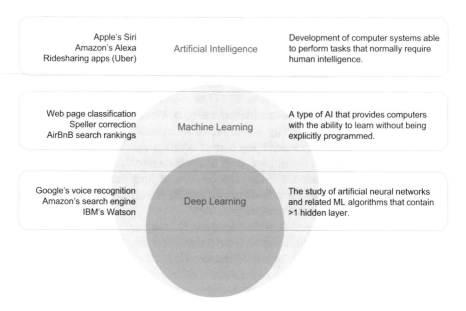

Apple's Siri
Amazon's Alexa
Ridesharing apps (Uber)

Artificial Intelligence

Development of computer systems able
to perform tasks that normally require
human intelligence.

Web page classification
Speller correction
AirBnB search rankings

Machine Learning

A type of AI that provides computers
with the ability to learn without being
explicitly programmed.

Google's voice recognition
Amazon's search engine
IBM's Watson

Deep Learning

The study of artificial neural networks
and related ML algorithms that contain
>1 hidden layer.

Fig. 7.2 Artificial intelligence vs. machine learning vs. deep learning

7.2.1   The Key Modalities

When considered as an informational tool, the domain of machine learning can be
seen as the coming together of modern computational power and vast amounts of
diverse and readily available data. As implied by the term 'machine learning,' ANN
and other algorithms have built-in (i.e., programmed) capabilities to learn, which in
this context means being able to identify and utilize data patterns using appropriately selected training datasets. Technically, training can make use of one of the
following three modalities: supervised, unsupervised, or reinforcement learning. It
should be noted that those three algorithm training approaches are somewhat analo-
gous to distinct modes of (human) statistical learning, more specifically exploratory
vs. confirmatory, rather than being suggestive of the degree of human involvement
in the training process (i.e., as outlined below, there is no human 'supervisor' in
supervised learning).

Supervised learning occurs when an algorithm is trained using labeled data,
which are data that include expressly identified and labeled, hence the name, correct
outcomes, or conditions of interest. For instance, when an artificial neural network
or a decision tree or a support vector machine algorithm is being trained to identify
'high value' customers in a given customer dataset, in addition to customer-
describing predictor variables (which might include demographic, lifestyle, and
past purchase details), the training dataset also needs to include a variable that
clearly labels each customer as being 'high value' or not. In other words, the 'supervised' aspect of learning refers to the fact that human analyst-created examples, in

the form of the said 'high value' designations, provide the machine learning algorithm with unambiguous learning targets. In that context, training of the customer value-predicting algorithm, and by extension supervised learning process in general, boils down to identifying and fine-tuning of parameters that enable maximally accurate classification of categorical outcomes or estimation of most likely values of continuous quantities.

In terms of many, perhaps even most, procedural aspects, the general *unsupervised learning* process is not materially different from that of supervised learning – in fact, the key difference between those two machine learning modalities is that, as graphically summarized in Fig. 7.3, unsupervised learning does not have the benefit of labeled data. Of course, while in terms of the high-level process schematic depicted in Fig. 7.3 that appears to be a relatively minor difference, in terms of how learning happens, it is not.

More specifically, when an unsupervised learning algorithm is fed with the same customer-describing details – except for human analyst-created labels, such as 'customer value' – the algorithm has to discover those differences based on available details, such as some combination of behavioral, demographic, and lifestyle variables. In a more difficult scenario, the algorithm being trained might even have to decide how many classes or categories there might be in the data, which usually entails grouping of records in a conceptual space defined by some set of clustering variables, a challenging undertaking that carries high risk of over- or under-identification. Overall, the 'unsupervised' designation is meant to signal the ability to learn from data without the help of human analyst-supplied examples. Implied in this general characterization is that being able to perform more speculative data analytic tasks, such as exploratory customer classification, calls for a differently designed algorithm, since regardless of whether it is considered supervised or unsupervised, an algorithm is ultimately just a finite sequence of specific, computer-implementable instructions. Hence, while process schematic-wise supervised and unsupervised learning appear quite similar, that overt similarity masks considerable algorithm design differences.

The third general variant of machine learning, known as *reinforced learning*, is governed by slightly more complicated design logic, which decouples 'actions'

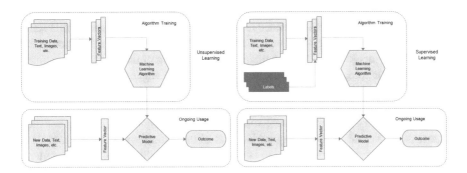

Fig. 7.3 Supervised vs. unsupervised learning

from 'rewards' with the goal of maximizing rewards rather than taking the right action (something that, in the context of the earlier discussed customer value example, might take the form of correctly classifying customers into 'high value' vs. not categories). In many applied contexts, that is merely a technical distinction that renders the description of the relevant algorithms slightly more complicated – for example, a particular reinforced learning algorithm might have to take several sequential actions to maximize the outcome of interest. Of perhaps greater interest is that the ability of a given machine learning algorithm to conduct the decoupling that is the central feature of reinforced learning, and which entails making sequential decisions that are related to each other, is suggestive of the ability to 'think' ahead in time. It is illustrative of a higher form of embedded data analytic logic, which is a feature that typifies game playing machine learning algorithms. For example, a chess playing reinforced learning algorithm needs to be able to balance the breadth of moves it considers with the depth of move-specific analysis, given that official chess matches tend to be timed, typically allowing 3 minutes per move. Furthermore, that balance might change throughout the match as the number of plausible moves diminishes; thus, contrary to the other two machine learning approaches, the emphasis in reinforced learning is on combining multiple decisions in a way that produces the greatest overall benefit. It follows that the key aspect of that mode of artificial learning, and especially its reward maximization functionality, is a complicated function of multiple, interdependent choices that emerges in the process of learning from the combination of the initial algorithm architecture and patterns derived from data. In a very general sense, it is the key design and functioning difference that separates 'traditional' machine learning and the more advanced deep learning.

Older machine learning approaches were, and still are, implemented in narrower contexts (as exemplified by the narrowly defined customer value classification task discussed earlier), where the role of different computational elements that together gave rise to artificial learning was at least somewhat transparent. The newer, more algorithmically advanced *deep learning* systems tend to be comparatively opaque in terms of the exact number of data analytic layers and the complexity of each layer, as measured in terms of the number of distinct computational units. While in principle all machine learning approaches aim to, to a greater or lesser degree, emulate the general functioning of human brain, the design of deep learning algorithms delves a lot deeper (no pun intended) into the biochemical processes of that functioning. Consequently, neuroscience, which studies the structure and function of the nervous system and brain, plays a major role in crafting of deep network architectures. For example, neuroscientific research evidence suggests that certain brain areas situated away from sensory regions nonetheless respond to different stimuli, such as scenery or human faces, which suggests 'division of labor' when it comes to making sense of different visual stimuli, and ultimately distinguishing between categories of stimuli. Importantly, while areas of the brain responsible for the initial processing of visual stimuli react to any input placed in the visual field, they tend not to be able to classify the observed stimuli until they receive the appropriate input from higher brain regions, which are able to distinguish between different categories of visual stimuli. In other words, in order to make sense of visual stimuli, the brain combines signals from several areas that interact in a larger network. Levering such neuroscientific evidence, designers of deep learning algorithms can develop hierarchical data analytic and sensemaking architecture capable of carrying out computationally challenging tasks such as image classification.

While offering highly utilitarian capabilities, perhaps best exemplified by the now commonly encountered voice recognition systems, deep learning technologies are already able to replace humans with intelligent agents that can perform tasks and make the right decisions. However, lacking human-like cognitive abilities, those artificial learning algorithms are not capable of understanding or interpreting inputs in the way a human does, which opens the Pandora's box of moral, ethical, and legal issues. While exploring the full extent of those considerations is worthy of its own book, or even a series of books, a relatively simple example might shed some light on the many complex questions that will continue to arise as the march of automation continues. When an artificial system executes an action that, unintentionally, results in harm to a human or damage to property, who is responsible? For instance, if a self-driving vehicle strikes a pedestrian through no fault on the part of that pedestrian, who should be held responsible? In today's context where vehicles are operated by human drivers, the operators are responsible for any injuries or damages, but when the system (i.e., the self-driving vehicle) operates autonomously, there could be many accident-contributing factors, including the system's own design (which itself might be quite complex, as multiple parts of the overall system could come from different developers and suppliers), behavior of others, nearby systems, availability of proper and adequate traffic controls, etc. Moreover, what is the scope for the user's involvement in the decision process, if the vehicle's occupant was found 'tinkering' with the system settings?

7.3   The Soul of ADM: Human-Computer Interaction

Unlike most other tools devised by mankind, electronic computers offer a diverse array of usages; moreover, the manner in which humans interact with computers can be characterized as an open-ended, evolving dialog between human users and our electronic companions. In a sense, computers are unlike any other tool, the vast majority of which, even those as sophisticated as automobiles or airplanes, continue to offer essentially the same functionality. At the risk of stating the obvious, modern cars and planes are considerably faster, safer, and more dependable now than those built a century or so ago; still, their core functionality has not materially changed. That, however, is not at all the case with computers: Just focusing on one of the many computer usage contexts, analyses of data, it was not long ago that computers' utility was limited to helping analysts interpret large amounts of data through creation of summaries of key outcomes, or creation of forward-looking forecasts. Today, advanced computer-based data analytic systems are slowly obfuscating the need for direct human input altogether; moreover, those systems are not just processing vast data streams, but they are progressively more and more able to make sense of those data in human-like manner. A direct consequence of computer-enabled automation is that not only how work is done that is fundamentally changing – how lives are lived is also being slowly reshaped by the ever-expanding techno-informational infrastructure. Those are among the reasons pointing toward the importance of expressly addressing the nature of human-computer interactions (HCI) is one of the key aspects of the evolving nature of organizational learning.

HCI is a complex, multifaceted bundle of considerations, encompassing highly visible physical design choices and the often-hidden data processing, analysis, and usage logic, perhaps best illustrated by China's Social Credit Scoring system discussed earlier in this chapter. In recent years, the physical design aspect of HCI has been at the forefront of computer systems' design; as computing systems became more and more interactive, the need to carefully study the design of human-computer interfaces became progressively more pronounced. At the core of that aspect of HCI is the *dilemma of wicked problems*, which are, typically, social or cultural conundrums that are extremely challenging because of incomplete or contradictory knowledge, large numbers of diverse opinions, and the interconnected nature of these problems with other problems. As such, wicked problems tend to have no precise definition, are highly context-dependent, rarely offer opportunities for trial and error, and commonly have no optimal or provably correct solutions. Even collection of design-informing information can be problematic, as there are often multiple data capture and analysis approaches, which can ultimately yield incompatible insights.

It is important to note, however, that while the design of physical computer interfaces[3] is clearly important, within the confines of organizational learning, it is the utility and capabilities of automated data analytics that are of primary concern. Given the nontransparent, i.e., 'black box,' operational characteristics of machine/deep learning systems, it is typically more difficult to assess the face validity, as well as the explanatory or predictive validity, of ADM systems-produced data analytic insights. In a way of a contrast, the use of statistical analysis tools, such as regression models, lends itself to a relatively easy assessment of face validity because of the high degree of analytic transparency of those methods. To continue with the regression example, standard regression outcome[4] spells out all assessment-related details, such as statistically significant predictors, their individual weights, and the like – in contrast to that, artificial neural networks and other advanced machine learning algorithms simply do not do the same, primarily because the manner in which they function (nonlinear, multilayer processes govern by probabilistic logic) does not lend itself to such level of data analytic transparency. In fact, the

---

[3] A key, public-facing aspect of physical computer interface design is focused on application programming interface, or API design, which, firstly and foremost is manifestations of a digital platform's (such as Expedia, eBay, or Salesforce) architecture and governance choices encompassing core design specifics such as partitioning, systems integration, decision rights, control, and pricing; secondly, API design is also an expression of the means by which digital platform providers create third-party developer and end customer ecosystems.

inability to examine the innerworkings, in the sense of an underlying explanatory or predictive model, of automated analysis-derived outcomes has been one of the major obstacles to wider acceptance and usage of ADM systems, which has important HCI design implications: It is not just 'how' or 'how much' information is communicated through the interface, but 'what' information is communicated that needs to be tailored to different user types. There is often a considerable difference between information needs of technical, such as data scientists, and non-technical, such as business managers, audience groups.

While obviously front and center, tangible design choices are not the only considerations that matter in the context of human-computer interactions – there are numerous profoundly philosophical questions that linger in the background that also need to be addressed, albeit in a far more tacit manner. Just how human-like are artificial intelligence (AI)-based ADM systems? Can they 'think' in a way that parallels, or at least approximates, human thinking? The human parallel questions are particularly vexing because being human-like is not inherently beneficial, especially when it comes to objective and consistent decision-making; in fact, one of the core benefits of ADM is avoidance of human decision-making shortcomings. More specifically, there is little doubt that it would certainly not be desirable to endow ADM systems with cognitive biases, limit their abilities to recall all pertinent information, or impose channel capacity-like limitations on their information processing capabilities; at the same time, it would be desirable to endow those systems with the human ability to generalize and reason abstractly.

In the narrower context of a system's ability to interact in a human-like manner, that question was famously tackled by Alan Turing, a renowned (most notably for the design of a machine that helped to break the German Enigma Code during the World War II) British mathematician who, in 1950, proposed what has come to be known as the Turing test, designed to test a machine's ability to exhibit intelligent behavior.[5] By and large, the human-like aspect of HCI design is an open question that does not lend itself to a single answer; it is also worth noting that as of the latter part of 2020, no AI system has been able to pass the Turing test in a way that is generally accepted. All considered, there are compelling reasons to believe that, when looked at from the perspective of diffusion of innovation, ADM systems are still in the early stages of dissemination.

When examined in the context of the generalized diffusion of innovation model graphically summarized in Fig. 7.4 (and originally developed in the early 1960s by

---

[4]This is a broad generalization as there are multiple distinct regression methodologies, including linear, logistic, and ordinal, all of which are mathematically distinct, and thus ultimately produce different result evaluation assessment-related outcomes; those differences notwithstanding, all those methods are analytically transparent.

[5]The test was called the 'imitation game' by Turing; it is conducted in a controlled setting with test subjects, a person and a computer program, being hidden from view of the judge whose job is to distinguish between answers given by the two subjects (it should be noted that the test does not check the ability to give correct answers – it is only concerned with how closely answers resemble those a human would give). If the judge cannot tell the difference between answers provided by the computer and the person, the computer has succeeded in demonstrating human intelligence.

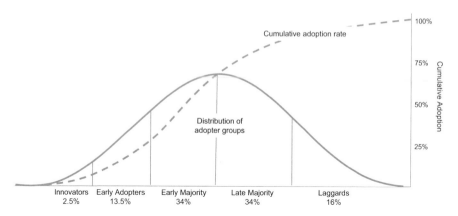

Fig. 7.4 Generalized diffusion of innovation process

E. Rogers), the currently available evidence places ADM systems' adoption in the 'innovators-early adopters' range, which account for only about 16% of the cumulative adoption rate. Why so low? Innovation adoption research points to organizational barriers, such as motivation and ability, the compatibility of the new system with already-in-place systems, and the ongoing assessment of the use and applicability of the new system. At the individual user level, a different conceptual framework known as technology acceptance model points toward perceived usefulness, or the degree to which a person believes that the new system will benefit his or her job performance, and perceived ease of use of the new system.

There are, however, two distinct adoption-related aspects of advanced data analytics-based systems that are not accounted for by either the diffusion of innovation or the technology acceptance model – those are trust and challenged conveyance of meaning. The former is a manifestation of decisional control, which is the ability to determine outcomes, while the latter is a more nuanced notion which centers on the degree to which the manner in which ADMs make, and in case of autonomous systems, actuate decisions. Descriptions of the functioning of machine learning systems tend to make extensive use of anthropomorphisms, or attributions of human traits to nonhuman entities not only to liken the functioning of those technologies to human sensemaking, but also because English, and quite likely most other languages, simply lacks the appropriate words to discuss advanced data analytics from the standpoint of artificial systems that, on the one hand, were designed to be human-like in terms of their sensemaking, while on the other hand, lack human cognitive capabilities. Doing so, however, can have an adverse side effect of implicitly pitting human decision-makers against ADM systems, which can trigger defense mechanisms in the form of attempts at discrediting the efficacy of those systems. In fact, recent research[6] addressing acceptance of AI applications suggests that adoption resistance of human users increases as the target of acceptance moves from the

---

[6] Most notably, Klumpp [2]

idea of accepting the competence of a particular system to accepting the efficacy of decisions produced by that system, to ultimately accepting the autonomy of AI systems. Stated differently, as the degree of what is perceived as personal intrusion grows (acceptance of a given system's autonomy is generally deemed to be more personally intrusive than acceptance of that system's general competence), the resistance increases.

7.3.1   *The Question of Accountability*

Compared to rule-based DSS algorithms, automated decision-making systems are opaque, meaning their decision logic lacks transparency, which has been a major adaption barrier. If an automated system generates what might be shown to be a clearly incorrect conclusion, such as granting a preferential terms loan to a high-risk applicant, who or what is at fault? Is it the system's original architect, the analyst responsible for its training, or its user? The idea of personhood being applied to nonhumans dates back to the early nineteenth century in the USA when the US Supreme Court described corporations (in the 1819 Dartmouth College vs. Woodward case) as '…an artificial being, invisible, intangible, and existing only in contemplation of the law…[as a] mere creature of law.' Does that mean that the same general legal principle extends onto ADM systems, which, after all, fit the definition of 'artificial beings' even better than corporations do? That is a complex question, especially when the impact of the human element is to also be taken into account, be it in the context of the human-in-the-loop design considerations (which are focused on whether or not there is human involvement beyond the initial algorithm creation) or in a more general sense of ADMs being ultimately manmade. While autonomous systems might be overtly self-controlled, the machine learning algorithms that comprise the 'brain' of those systems are not. Supervised machine learning requires a fairly direct human involvement as it relies on human-provided outcome labeling, and while unsupervised learning does not require direct human involvement, analysts are nonetheless commonly 'in-the-loop' to steer data and variable selection; reinforced learning is perhaps least dependent on ongoing human involvement, but training of those algorithms also calls for periodic human input.

The question of ADM systems' accountability is illustrated by Microsoft's 2016 experience related to the company's release of an artificial application, a chatbot called Tay.ai designed to interact with Twitter users and to learn from these interactions, into the online social sphere. Within just 24 hours, Microsoft had to deactivate Tay's Twitter account due to a flood of retweets of racism comments on Tay's feed, many of which included additional offensive content added by the chatbot. While, regrettably, such commentaries are not uncommon, what is noteworthy here is the ability of advanced forms of AI, such as Tay, to mimic human behavior, which shines the light on a yet another intricacy of human-computer interactions: How to orient the design of self-functioning artificial systems to the redeeming side of humanity while avoiding the many dark aspects of human nature? It is one thing to train ADM systems to replicate, and even improve upon human behaviors geared toward performance of tasks rooted in objectively framed skills and competencies, such as driving, it is an altogether different, and far more difficult task altogether to those systems to behave in morally sound and ethical manner.

Such broad questions rarely have direct answers, but when considered within the confines of organizational learning, at least some of the answer may be found in the notion of mental models. First discussed in Chap. 3 in the general context of learning and remembering that ethereal aspect of human cognition also has important HCI implications.

7.3.2   *Human Versus Machine Mental Models*

Broadly characterized as abstract representations of the real word that aid the sensemaking aspects of cognitive functioning, particularly reasoning about problems or situations not directly encountered, *mental models* have profound implications for organizational learning. Inspiring ideas and biased perspectives alike emanate from those highly individualized sensemaking aids, which are neither objectively discernible nor reviewable, but which have direct-yet-elusive impact on individual-level cognitive functioning. As currently understood, mental models are a product of ongoing learning; in a sense, those abstract constructs can be seen as higher-order generalizations of ongoing ingestion of knowledge, perhaps best characterized as sensemaking templates.

When considered from the perspective of organizational decision-making, or more specifically from the perspective of cognitive diversity, mental models-rooted cognitive differences can lead not only to distinct creative problem-solving outcomes but also to materially different interpretation of the same set of objective facts. Simply put, mental model variability can contribute positively, through diversity of ideas, or negatively, through biased, irrational judgment, to organizational sensemaking. (The much talked about cognitive biases, or instances of sensemaking conclusions that deviate from rational judgment, can be explained through the idea of mental models, which in that context can be seen as subconscious, rationally imperceptible 'instincts' that can compel otherwise well-intended individuals to reach ill-founded conclusions.) All considered, mental models capture much of what it means 'to be human,' in the cognitive functioning sense, which underscores the importance of expressly accounting for that unique aspect of human sensemaking when addressing human-computer interactions. With that in mind, the cognitive diversity quotient (CDQ), first introduced in Chap. 3 and shown again below, can be of value when trying to advance the art and science of training of machine learning, especially deep learning, algorithms. More specifically, given the importance of developing sound abstract representations of reality to effective functioning of deep learning algorithms, CDQ can offer an objective and regularized assessment of cognitive diversity embedded into individual datasets.

$$CDQ = \sqrt{\frac{\Sigma_{i=1}\left(Perspective_i - baseline\right)^2}{Number\ of\ perspectives}}$$

The pooled individual vs. baseline perspective variability captured by CDQ can be interpreted as the aggregate expression of perspective divergence. While in strictly human cognition context CDQ can infuse elements of analytic logic into often ill-structured mechanics of organizational consensus building, within the confines of training of machine learning algorithms, it can contribute to more thoughtful feature engineering of training data. In other words, to avoid the same mental traps that can warp human sensemaking, the implicit reality contained in machine learning input data needs to be carefully crafted, which runs counter to the commonly seen impulse of indiscriminately feeding data into those systems, and letting them 'figure it out,' which is akin to teaching (humans) by means of bringing learners to libraries and turning them loose to learn what they need to learn on their own. If human education is a carefully crafted process, where thoughtfully selected and structured information is gradually brought to bear throughout the learning process, why would training of artificial learning systems be well served by treating artificial learning systems as receptacles of any and all data and fed to them without the benefit of learning strategy? Just because machines can process vast quantities of data at an incredibly rapid pace does not mean that their sensemaking processes do not require the same thoughtfully designed learning strategy, which starts with carefully crafted data feature engineering.

7.3.2.1   Data Feature Engineering

In the Age of Data, the long-standing GIGO, or garbage-in, garbage-out, principle holds true more than ever, because data play a bigger role now than ever before. The idea that is consistently overlooked is that if competent and experienced human analysts struggle to make sense of dirty and noisy data, human-created systems will struggle even more. Moreover, the advent of 'big data' seems to have shifted the focus of attention from the 'what' to 'how much'; intoxicated by volumes of data, nowadays routinely described in hard to relate to magnitudes (e.g., terabytes, petabytes, exabytes, etc.), data users are often inclined to look past the importance of seeing data as a collection of individual measures, known as variables, where each such measure captures some small aspect of observed reality, be it a sales transaction or a posted opinion. If the quality of those individual measures is poor, the size will not make up for that, and machine (as well as human) learning will yield outcomes of low informational quality. Moreover, the quality of data is not synonymous with correctness or reasonableness of individual values – for instance, a fully populated (i.e., no missing values) dataset where all values appear to fall within reasonable ranges can still be of low quality, if the individual variables are not effect-coded. The idea of effect-coding is one of the key manifestations of the divide that separates efficient data capture and storage, and analytic data utilization; it

Occupational Category	Professional	Sales	Labor	Service	Agriculture	Crafts
Professional	1	0	0	0	0	0
Sales	0	1	0	0	0	0
Labor	0	0	1	0	0	0
Service	0	0	0	1	0	0
Agriculture	0	0	0	0	1	0
Crafts	0	0	0	0	0	1

Fig. 7.5 Disaggregation of compound variables

represents disaggregating of compound variables into more informationally revealing indicators, as graphically illustrated in Fig. 7.5.

The essence of variable transformation illustrated in Fig. 7.5 is best described
using a simple example: One of the variables contained in a dataset of interest,
'Occupational Category,' is comprised of several distinct categories: sales, professional, labor, service, agriculture, and crafts; when used as input into analysis aiming to predict brand choice, such multicategory variable needs to be broken out into
multiple, category-level indicators to be analytically meaningful.[7] Some machine
learning algorithms will perform such recoding automatically, while others will not.
Moreover, as a standalone variable, the Occupation Category measure does not take
into account any qualifying considerations such as gender, age, or income – in other
words, the impact of different occupations on brand choice might be moderated by
additional qualifying considerations, which means that a 'professional-young-
female' effect might have very different predictive impact on brand choice than
'professional-older-male' effect, but in order for those effects to be captured in
(machine) learning processes, the initially singular Occupational Category variable
needs to be disaggregated, or broken up, into as many standalone indicator-coded
(the commonly used 1 vs. 0 numeric indicators that correspond to yes vs. no) measures as there are levels of that variable. As shown in Fig. 7.5, the Occupational
Category measure, initially comprised of six distinct categories (Professional, Sales,
etc.), is decomposed into six standalone, occupation category-specific measures
which will allow creation of the aforementioned interaction terms (e.g., professional-
young-female); it should be noted that the six effect-coded variables replace the
original Occupational Category measure.

Not many machine learning algorithms automatically contemplate all possible
interaction terms implied above, primarily because in a larger dataset, the number
of potential cross-variable interactions can quickly become unmanageable,[8] and
many of the resulting random combinations would not make sense. And that is

---

[7] In more operationally clear terms, that requires recoding of the original multicategory variable
into as many dummy-coded (yes-no or 1-0) binary indicator variables as there are categories comprising the original variable.

[8] The general computational formula to be used here is $n!/(r!(n - r)!)$, where n is the total number
of possibilities and r is the number of selections; assuming just ten possibilities for the three-way
interaction (occupation-gender-age category) used in the example would produce 1000 possible
interactions.

precisely why thoughtful and careful data feature engineering is so critical to effective training of machine learning algorithms – in order for ADM systems to develop deep and thorough enough understanding of the dynamics of, in this case, brand choice, the input data used to those algorithms should include the addition of carefully crafted (by a knowledgeable human) interaction terms.

7.3.3   Right Tool for the Task

Although the importance of correctly aligning ADM tools with decision-making tasks is intuitively obvious, misuse of decision-aiding tools is not uncommon, and it has been attributed to factors ranging from misalignment between task demands and features of the technology to specific characteristics of ADM systems. That challenge arises, at least in part, from the problem of easily stoked, but often unfounded, exuberance.

The IBM Watson, the company's flagship AI system that first burst into public view with its impressive 2011 Jeopardy! quiz show victory, offers an instructive illustration here. Soon after its game show-winning coming out party, IBM announced that its system, capable of understanding natural language, would become an AI doctor, painting a picture of its breakthrough technology reducing medical diagnosis errors, optimizing treatments, and even alleviating physician shortages. According to the company, the first commercial applications were to be market-ready in about 18–24 months, counting from the 2011 announcement. It did not happen. Not in 2013 or 2014; in fact, it still has not happened in the early part of 2021. Thus far, what was to be a brilliant AI doctor can, at best, be characterized as an AI assistant capable of performing only certain routine tasks. Reasons behind that underperformance are numerous but tend to center on machine learning systems' (in general, not just IBM Watson) struggles to make sense of complex medical information, about 80% of which is unstructured text, full of jargon, shorthand, and subjective statements. Consequently, a decade after IBM's boastful promise, only a small handful of machine learning tools have been approved for use, and those are limited purpose image processing applications focused on narrowly defined tasks, such as analysis of x-rays and retina scans.

Implied in the obstacles encountered by IBM AI developers is the difficulty of machine-encoding of human capabilities grouped under a broad umbrella of critical and creative thinking. Expertise, be it in medicine or any other domain of knowledge, cannot be expressed as a simple sum of known facts because, as illustrated in Fig. 1.3, knowledge manifests itself in a multifaceted manner that encompasses distinct sources (explicit vs. tacit) and dimensions (semantic, procedural, and episodic). Advanced ADM systems excel at mastering some of those dimensions, most notably explicit, semantic, and procedural knowledge, but struggle with the nuanced, immersive nature of tacit and episodic knowledge; hence, the brief overview of the key decision automation-related ideas would be incomplete without expressly addressing the creative of ADM.

7.4   Data-Enabled Creativity

The MultiModal Organizational Learning (MMOL) typology, first discussed in Chap. 1, expressly differentiates between reason-based and technology-based learning. The former manifests itself as either direct observation-based experiential or scientific investigation-rooted theoretical learning, while the latter can be either computational, which entails direct analyses of available data, or simulational learning, which takes the form of computer-rendered imitations (see Fig. 1.5). When considered from an organizational perspective, technology-based learning dimension draws attention to the fact that the learner should not be assumed to be a human – as discussed earlier in this chapter, artificial systems' abilities to generate and utilize information are quickly approaching, and in some contexts even surpassing, human sensemaking capabilities.

To be operationally more meaningful, the MMOL framework needs to be placed in the qualifying context of distinct types of organizational learning tasks, because organizational pursuit of knowledge is ultimately geared toward performance of specific activities. As discussed in Chap. 6, there are numerous task grouping schemas; the ones offered by Steiner and, separately, by McGrath exhibit the closest taxonomical relationship to the spirit of MMOL typology, ultimately giving rise to the operationally expanded organizational learning framework summarized earlier in Fig. 6.3, and expressly differentiating between open-ended 'creative' and evaluation and/or confirmation minded 'choice and negotiation' task meta-categories.

Data-driven learning and creativity have been traditionally seen as distinct and different, but the essence of the ideas presented here is that information technology-driven changing nature of work compels a fundamental rethinking of not only the notion of 'learning' but also the concept of 'creativity.' The long-standing conception of creativity equates that vaguely defined idea with the use of imagination, which in turn is (usually implicitly) framed as forming of new ideas, not already present to or in the senses. Such characterization of creativity begs a fundamental question: Is creativity conjuring up something completely otherworldly, in the sense of not having been experienced, or is it creating a new something – be it an abstract idea or a work of art – by combining what are ultimately known building blocks? Evidence from the field of child psychology, which focuses on cognitive and behavioral human development, points to the latter as being a more plausible framing. Given that, it seems quite reasonable that immersion in computer-supported and/or rendered 'reality' could spark ideas that might not otherwise arise, thus effectively adding a new dimension to the traditional, conjured by the mind, conception of creativity. Consequently, as graphically illustrated in Fig. 7.6, creativity can be conceptualized as a phenomenon rooted in making sense of either observed reality (experiential learning) or computer-rendered simulations (simulational learning).

The experiential-simulational duality of the expanded conception of creativity has direct implications for organizational value creation. While experiential, reason-based learning offers a natural avenue for the pursuit of creative problem-solving, not all phenomena can be physically experienced; moreover, ideas that grow out of

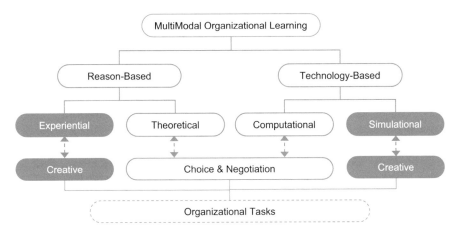

Fig. 7.6 Expanded MMOL typology of organizational learning

careful observation of what-is are far more likely to fuel continuous rather than discontinuous innovation. To be sure, being able to think creatively within the confines of experiential, reason-based learning modality is essential to organizational learning, but at the same time, it might be insufficient for organizations more unique sources of competitive advantage. This is where technology-based simulational learning comes in, contributing a platform for establishing and sustaining previously inaccessible dimension of creative ideation. Data-driven, immersive computer-rendered simulations offer virtually limitless ability to go beyond that which either currently exists or can otherwise be experienced by the senses, by allowing deep learning systems to conjure up all manner of associations, contexts, scenarios, and the like. In fact, there are ample reasons to believe that the anticipated emergence and proliferation of virtualization and simulation platforms will trigger the next great explosion of human creativity, one that at some point in the future will be looked upon as another cognitive axial period.

Discussed in Chap. 4, the cognitive axial periods were the grand human civilization development trend resetting eras that were triggered by the seemingly sudden and comparatively concentrated emergence of unique factors. The evolutionary journey that started with the great thinkers of antiquity (the Age of Reason), centuries later followed by the explosive rise of scientific inquiry and methods (the Age of Science), and currently manifesting itself through the rapidly unfolding and vast techno-informational infrastructure (the Age of Transcendance), will ultimately see mankind rise above our limitations, not just the biological ones but also the earlier discussed experiential ones. Newly emerging technologies such as data-enabled quantum simulations will enable us to peer into worlds and realms that are physically inaccessible to us, which will challenge and stimulate out minds to, once again, see further and go further. The Age of Transcendence is about to unleash completely new ways to learn, made possible by combining the most human aspects of mankind – the ability to reason, imagine, and feel, in a metaphysical sense – with technologically rendered contexts that will enable us to transcend our physical, time, and space limitations in search for deeper understanding of the nature of reality.

7.4.1 Reason-Algorithm Collaboration

Turning back to the expanded MMOL typology depicted in Fig. 7.6, and more specifically the prospect of computer simulation-enabled creative problem-solving, it is easy to get excited about the possibility of using such capabilities to shed light on some of the most existential questions of mankind, but should business and other nonscientific organizations be as excited, given the comparatively mundane nature of their informational needs? The answer is a resounding 'yes!'

Commercial organizations that seek to grow and prosper, and social organizations that aim to further different social and societal goals, pursue their goals in competitive environments; thus, their ability to advance their agendas is tied directly to the efficacy of their functioning, which includes the use of already in hand or potentially available informational resources. Business companies can win more customers, and nonprofit organizations can secure the necessary resources by thoughtfully structuring organizational tasks in a way that combines the core advantages of informational automation – speed, scalability, and consistency – with core human competencies, most notably adaptively creative problem-solving. As illustrated by the earlier discussed DSS (loyalty classification) and ADM (self-driving cars) cases, modern data analytics-driven systems can augment, or even replace, direct human effort, which has the obvious benefit of freeing human time to be redirected toward addressing other problems. But all that is old news – modern, well-functioning organizations are already well acquainted with those ideas, as decision automation is now commonplace in high frequency of interaction contexts, such as consumer lending or insurance claim processing. That said, there is a yet another, often overlooked, dimension of human-computer interaction (HCI), one where humans and computers are engaged in in a two-way interaction, termed here *human-computer collaboration (HCC)*.

As discussed earlier in this chapter, HCI sees the manner in which humans interact with machines as one-way phenomenon, primarily concerned with making computers and other devices as easily usable as possible. Clearly, that is a critical consideration when designing human-facing interfaces of computing systems, but it tacitly dismisses the idea of collaboration-like, two-way human-computer interactions. To put it in terms of organizational ecosystems, HCI frames computers as 'suppliers' of specific informational inputs, rather than 'partners' in the knowledge creation process. And while that framing is reasonable when computers serve as de facto high-end electronic calculators (which, to be sure, is still the core and tremendously valuable functionality of computing systems), it is too restrictive in situations in which advanced computing systems are capable of self-controlled intelligent behavior. Of course, as noted earlier, not being conscious, artificial computing systems are not (yet?) capable of truly independent 'thought,' meaning that their informational output is ultimately constrained by the combination of their programming and data from which they learn. Still, those systems can process vast quantities of data tirelessly, accurately, and at a dizzying speed, all of which raises a tantalizing possibility: What if human informational workers and artificial computing systems could engage in teamwork-like collaborations? Could the current conception of 'machine learning' be expanded to also encompass adaptation of ANN and other machine learning algorithms to the nuances of individual analysts' areas of analytic focus? In other words, could machine learning entail not just algorithms extracting meaningful patterns out of data, but also learning about the informational needs of their human partners? All considered, what if the current conception of data scientists' and business analysts' skillset was to be expanded to include not only the traditional computing competencies but also HCC-framed teamwork? Would that offer organizations the long sought-after ability to monetize their data assets?

The idea of monetizing organizational data is compelling to all, but to date, only a handful of organizations found effective ways of doing so, and most of those organizations, such as Google or Facebook, are themselves products of the Information Age. Most companies' business models are not built atop of data flows, which means those organizations are not innately positioned to extract direct economic benefits out of their data holdings. For those organizations, the path to extracting economic value out of their data assets is through broad organizational embrace of the idea of data-enabled creativity. Operationally, that means going beyond the tried-and-true modes of data utilization, which nowadays encompass anything from outcome tracking and reporting mechanisms to predictive analytic capabilities. In the past, brilliant breakthroughs were often painted as random flashes of light in the dark of night, but going forward, breakthroughs could become far more predictable outcomes of systematic pursuit of discontinuous innovation, rooted in meaningful human-computer collaborations. There is nothing outlandish about such an assertion as it has long been known that innovation-supportive organizational practices and culture produce steady flows of cutting-edge innovation (perhaps best illustrated by the 3M Company discussed in Chap. 2); HCC is merely the next step in that evolutionary development.

7.4.2   The New Frontier of Imagination

The preceding reasoning is rooted in an implicit assumption that once realized, a breakthrough idea will recognize as such, but there are ample reasons to be skeptical. Why, for instance, are some works of art celebrated while others are looked at with disinterest, or barely noticed? Or why are the same works go unnoticed at first, only to become artistic superstars later? As is well-known, Vincent van Gogh, the famed post-impressionist painter, created some of the most expensive paintings to have ever sold,[9] he even has a major museum devoted entirely to his works (the Van

---

[9] *Portrait of Dr. Gachet* sold in 1990 for $82.5 million, which in 2020 dollars would be about double that, or $161.4 million

Gogh Museum in Amsterdam), yet he did not enjoy any of that success during his lifetime, in spite of completing more than 800 paintings (more than 2100 individual artworks in total) in a span of just a decade. What makes his *Starry Night* or his *Irises* so 'creative?' Along similar lines of reasoning though in far less artistic sense, the category of modern sport utility vehicles, or SUVs, is commonly believed to have been sparked into existence by the introduction of 1990 Ford Explorer, a norm-setting design. Yet the 1984 Jeep Cherokee was remarkably similar (both were four-door, comparably sized and shaped vehicles with essentially the same or comparable attributes); why then does the Explorer get the creative glory?

Is some form of success required to earn the 'creative' badge? It is a bit like the familiar sounding 'if a tree falls in a forest and no one is around to hear it, does it make a sound' quandary. Questions like that do not lend themselves to straightforward, objective answers, but there is value in considering them, and that value is the development of a meaningful perspective on which the idea of data-enabled creativity can be operationally developed.

Whether or not some form or success or wider acceptance are deemed necessary, creativity tends to be assessed through the prism of 'stuff,' such as the above van Gogh's paintings or category-defining automobiles. In an organizational setting, that translates into output, be it in the form discrete products or services, or decision choices. That perspective, however, overlooks an increasingly important, latent aspect of creativity: *situational sensemaking*. The current highly volatile, hyper-competitive socio-politico-economic environment rewards organizations, seen here as human collectives working together in pursuit of shared commercial or societal goals, that exhibit superior abilities to problem-solve and to adapt to the ever-changing environmental realities. If volatility and change are ongoing and largely unpredictable, organizational ability to respond is ultimately contingent on individual organizational constituents' situational sensemaking, or considered more broadly, creative problem-solving capabilities. In that particular context, creativity is construed to be the ability to identify emerging opportunities and threats, and adaptively respond to observed or perceived changes in one's environment. While that framing certainly does not preclude innovative product or service ideation, it is nonetheless primarily focused on widening and deepening of individual-level cognitive thinking and emotive feeling capabilities, seen here as founding enablements of situational sensemaking. And lastly, it is a crucial element of the earlier discussed human-computer collaboration.

The above reasoning is rooted in the belief that, at its core, human creativity entails finding new and meaningful arrangements of already known elements (rather than conjuring up otherworldly ideas); in a sense, it could be likened to finding new patterns in old data. An essential enabler of being able to do that is imagination, generally understood to be a process of forming new ideas or representations not present to the senses. While the difference between creativity and imagination is not always clear, the former is generally seen as a purposeful process, which means it tends to be confined by some pragmatic bounds, while the latter is usually neither purposeful nor confined by any type of bounds. Hence, it is that unbounded nature of imagination that grants it the freedom to conjure up new concepts, perspectives, or associations that are unlikely to be produced without it, and imagination is also the aspect of creative problem-solving that can be stimulated by means of computer-rendered simulations.

It is important to note that what is described here represents a fundamentally different conception of computer simulation-based learning than the one used to train pilots, nurses, or business managers, all of which are based on the idea of artificial rendering of accurately depicted physical reality, with the goal of emulating the conditions those professionals are expected to encounter on-the-job. The stated goal of data-enabled simulations discussed here is to do the opposite – to create scenarios, associations, or conditions that may not have yet been encountered, with the goal of stimulating imagination, and ultimately, creative problem-solving. Operationalizing of that idea, however, is contingent on closer alignment of human and machine learning processes, which in turn touches on the idea of 'semantic triangle,' first suggested by Ogden and Richards [3] nearly a century ago. It offers an abstract representation of how linguistic symbols are related to the objects they represent, and when framed in the contemporary context of human and machine learning, its core premise is graphically illustrated in Fig. 7.7.

As graphically depicted in Fig. 7.7, the primary sensemaking mechanism in human learning relies on associating symbols, such as words, with typically contextual meaning; on the other hand, the primary machine learning avenue relies on grouping of specific examples, or points of reference, into larger categories, which is seen as the ability to generalize. Stated differently, being self-aware, humans develop conscious understanding, while self-awareness-lacking artificial learning systems are not capable of developing human-like understanding, framed here as cognitive sensemaking. Simply put, the association between, for instance, 'task involvement' and 'result quality' can be expected to evoke a particular meaning in a human, which can be seen as a product of an intuitively obvious, implied abstract relationship, easily grasped by most. To a machine, however, the same association is not 'understood' in terms of such elusive, transcendental qualities – instead, it merely belongs to a particular class of associations, which means that artificial intelligence manifests itself in the ability to create semantically and practically

Fig. 7.7 Semantic triangle

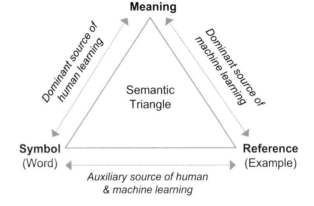

meaningful groupings, rather than human-like cognitive understanding. With that in mind, the idea of simulation-based learning is predicated upon the assumption that given the many potential classes of associations that machine learning systems can produce, at least some can be expected to fall completely outside of the evoked or consideration set of human meanings. In other words, by creating new but algorithmically plausible (i.e., consistent with embedded computational logic) classes, machine learning modality can suggest potential new ideas that reach beyond what a given human learner could be expected to conjure up on his or her own. In that sense, artificial systems-based simulations can help humans 'imagine' the otherwise unimaginable.

To summarize the preceding discussion, the expanded, data-enabled conception of creativity is captured in Fig. 7.8.

People, and thus organizations, have been learning from printed sources for the past 500 or so years, from mass audio sources such as radio for about 100 years, from audio-visual television for roughly 70 years, and from computer-rendered online sources for about 25 years. Simulation-based learning is merely the next, logical step in the evolution of human cognitive development. One of the most celebrated scientists of the modern era, Albert Einstein, famously noted in a 1924 interview that '…imagination is more important than knowledge…knowledge is limited, whereas imagination embraces the entire world, stimulating progress, giving birth to evolution.' Using technology to further stimulate the very human imagination that, unaided, has already produced so much will undoubtedly propel mankind to creative heights that, at the moment, may seem, well, unimaginable.

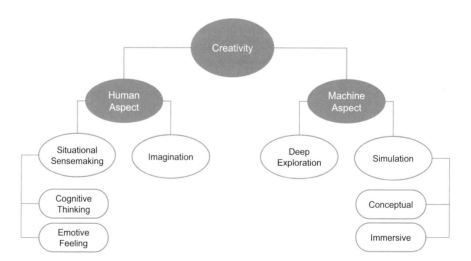

Fig. 7.8 Expanded conception of creativity

References

1. Banasiewicz, A. (2013). *Marketing database analytics: Transforming data for competitive advantage*. New York: Routledge.
2. Klumpp, M. (2017). Automation and artificial intelligence in business logistics systems: Human reaction and collaborations requirements. *Journal of Logistics Research and Applications, 21*(3), 224–242.
3. Ogden, C. K., & Richards, I. A. (1927). *Meaning of meaning*. New York: Harcourt, Brace & Company.

Chapter 8
In Machine We Trust

8.1 Algocracy Anyone?

In the early part of the twentieth century, the progressive movement in the United States was promoting the view that what it saw as democratic but partisan and corrupt system of government should be replaced with a more just bureaucracy staffed with appropriately trained and impartial cadre of professional administrators. It was a variant of the general idea of epistocracy, or the rule of the knowledgeable, which has long been a topic of debate among political philosophers and which can be traced to Plato's idea of 'philosopher kings' (see footnote #83). As evidenced by endless streams of surveys, such as the American National Election Studies, the average voter is woefully ignorant of economic and social science basics needed to evaluate political candidates' policy proposals, which leaves them easy prey for closed-minded ideologues and unscrupulous politicians. Not surprisingly, the current democratic systems where political power, in the sense of voting rights, is equally distributed among all – the well-informed as well as the ignorant – citizens prompt the fairly obvious question: Would the society be better served by a system that conditions access to political power on knowledge that is deemed necessary to make an informed choice[1]? And if rule by the knowledgeable would indeed be better, could it mean that rule by truly impartial and far more capable of considering all pertinent information ADM algorithms would be even better?

---

[1] Interestingly, a variant of that idea actually existed in the United Kingdom until about 1950, where select universities, such as Oxford, had their own constituencies; those with a degree from one of those universities could vote both in their university constituency and their place of residence (in other words, elite had two votes). In the United States, numerous southern states as well as some northern states, such as New York and Connecticut, enacted various forms of voter literacy tests until all such tests were permanently outlawed by the US Congress in 1975.

© The Author(s), under exclusive license to Springer Nature
Switzerland AG 2021
A. Banasiewicz, *Organizational Learning in the Age of Data*, EAI/Springer
Innovations in Communication and Computing,
https://doi.org/10.1007/978-3-030-74866-1_8

Nick Harkaway, a British author, tackles that very question in his 2017 science fiction novel, *Gnomon*. His epic, multithreaded story is set in near-future London which is now governed by a pervasive yet overtly benign total surveillance system called the Witness. The system is enabled by over 500 million cameras, microphones and other sensors that watch and listen to all citizens all the time and feed all data into impartial, self-teaching algorithms that constantly classify all inputs, but do nothing unless a need arises to act. There is no pretence of non-interference in one's personal affairs – everyone is fully and equally seen by the Witness; the citizens understand and accept the value of the system, trading full transparency of their behaviours (and thoughts, as the Witness is also capable of reading citizens' minds) for unparalleled security.

Embedded in the theme of *Gnomon* is the idea of algocracy, or the rule algorithms, which can be seen as a digital embodiment of the long-standing notion of epistocracy. It is tempting to view algocracy as a seemingly inescapable coming together of the idea of informed decision-making and rapidly evolving informational automation, offering unbiased, efficient and consistent means of making wide arrays of decisions. And in fact, algorithmic decision-making is already commonplace in many aspects of life such as insurance underwriting, credit scoring, marketing and social service eligibility determination, to name just a few. However, those are relatively narrow contexts that lend themselves to high degrees of standardization, where a particular set of characteristics can be used to relatively accurately, consistently and efficiently produce conclusive outcomes, such as loan or social services eligibility. However, when the idea of using ADM algorithms is considered in broader contexts, numerous and substantial difficulties arise, as readily illustrated by personal freedom-suppressing Chinese Social Scoring System or IBM Watson system's inability to deliver on its once boastful promise of becoming the first AI doctor.

Those challenges, however, are not just limited to the inability of artificial neural network-based systems to reason, in the human sense, which typically entails drawing inferences from data that go beyond associative grouping. Using automated decision algorithms in broader decision-making contexts also entails addressing critical questions of algorithmic bias, decision logic transparency, accountability and the concentration of power in the hands of algorithm operators.

8.1.1 Algorithmic Bias

An argument made throughout the preceding chapters, at times explicitly and at other times implicitly, was that algorithmic decision-making is unbiased. That indeed is the case insofar as the execution of computational logic – a given algorithm can be expected to execute the same data processing and analysis steps over and over again exactly the same way, which means that each data record, such as a loan application, will be evaluated in exactly the same manner. It is in that narrowly defined context that an algorithm can be considered unbiased. However, when a more expansive view of machine learning is taken, the possibility of bias arises, at two distinct points in what could be considered the algorithm life cycle: the initial programming and the ensuing algorithm training.

When designing a given algorithm's logical and computational structure, designers make choices that reflect their cultural and other assumptions and, often implicit beliefs, which are then embedded within the logical algorithm structure as unstated opinions. The Chinese Social Credit System (SCS), discussed in previous chapter, offers perhaps the most jarring example here: The system takes the familiar credit scoring idea and extends it to all aspects of life with the goal of judging citizens' overall behaviour and trustworthiness. High social credit score entitles one to perks such as cheaper public transport, free gym facilities and priority for shorter waiting times in hospitals, while low social credit score can result in a person being unable to buy property or take out a loan and even being barred from travelling by plane or train. Such person-specific details are relatively easy to access because, under the guise of forging a 'public opinion environment where keeping trust is glorious', the Chinese government is making each citizen's score publicly accessible. More than just making those scores accessible, it is common for electronic billboards to publicly shame those with low social credit scores while glorifying high scorers.

When compared to the dystopian SCS system, the 'nudging' approach employed by numerous digital commercial enterprises in the United States, such as Amazon, Apple, Twitter or Facebook, may seem quite tame. Still, those and other electronic platform organizations employ what can be considered user surveillance to exercise softly coercive power, aimed at steering their users in the direction of desired behaviours. Their automated algorithms decide what constitutes more relevant online search results, make online product purchase recommendations, suggest social and even romantic matches, generate music playlists, advise on best routes to reach destinations and the list goes on. Logic embedded in those algorithms already has a profound impact on everyday choices, yet it is almost completely hidden from view, and it can be unapologetically biased.

An even less apparent example of a potential, unintended algorithmic bias is offered by the credit scoring algorithms utilized by independent credit bureaus such as the US-based TransUnion, Equifax or Experian, all of which use credit worthiness proxies, such as education, employment status, type of residence, etc. While the choice of those proxies may be empirically warranted in developed countries like the United States, the same evaluation logic is sometimes used in developing countries, where it can result in enduring lack of access to credit by large segments of society. Here, applications that were designed with fairness in mind may in fact lead to manifestly unfair consequences.

The second of the two key contributors to algorithmic bias is the process of algorithm training, typically taking the form of the earlier discussed machine learning, which is when the computational logic of a particular algorithm 'meets' the data. As discussed earlier, automated algorithms learn by means of identifying and generalizing patterns derived from data fed into those algorithms, which means that data selection and any data pre-processing, most notably data feature engineering, will have a direct and quite possibly significant impact on subsequent algorithm functioning. It is easy to imagine well-meaning analysts preparing machine learning input data making choice that, in a way that is ultimately unbeknownst to them, are distorted by various cognitive impulses such as mental filtering, where either positive or negative details are amplified, or by a phenomenon known as global labelling, where specific attributes or events are categorized as universally good or bad.

It thus follows that algorithms can be intentionally or consciously tilted towards a particular ideology or perspective; even more importantly (because it is less obvious), algorithmic bias can also arise unintentionally, even in situations in which fairness and equality are overtly stated design goals. A case in point: The stated objective of various credit scoring systems is to provide qualified borrowers with fair and unabated access to capital markets, which on the surface sounds fair and equitable except that in some situations it may inadvertently produce unintended and undesirable consequences. All considered, it is possible that the idea of a truly objective algorithm could be practically unattainable, given that the very design and training of decision-guiding or decision-making algorithms tend to be perspective- and value-laden.

Still, there is more. Even looking beyond potential data selection and preparation-related distortions, machine learning algorithms are inescapably retrospective, as vividly illustrated by online recommendation engines. While those systems are now capable of generalizing from specific products onto broader categories of products, the underlying recommendation logic is still not capable of 'thinking' in a manner that goes beyond identification of communalities hidden in recent behavioural and other patterns. Product recommendation algorithms' interpretation of 'what comes next' is still a simple extrapolation of 'what happened last', which is a continuation of a long-standing, basic business analytic logic (i.e. the most recent behaviour tends to be the best indicator of next behaviour). While within the confines of brand choice the most recent choice pattern might indeed be the best predictor of the next choice, it would be myopic to assume that such coarse logic applies to all choices.

8.1.2   Decision Logic Transparency

Machine learning algorithms are inescapably opaque. In fact, there is an inverse relationship between their accuracy and understandability of their analytic logic; the most evolved, deep learning algorithms are almost completely unintelligible to anyone trying to discern the exact 'how' behind their informational outcomes. In a sense, the computational logic of deep learning systems can be likened to a person knowing why something makes intuitive sense, but being unable to clearly elucidate the underlying reasons. With that in mind and considering that artificial neural networks (ANNs), which are the most commonly used category of machine learning algorithms, are expressly designed to mimic the human brain's parallel, non-linear processes, that lack of analytic transparency should not come as a surprise. Interestingly, even if those underlying analytic causal chains could be, somehow, described, many of those explanation would be implausibly complex, which becomes apparent upon a closer consideration of the general ANNs' architecture, graphically summarized in Fig. 8.1.

It should be noted that the schematic depicted in Fig. 8.1 contains a very small number of 'nodes', which are computational units or functions with input and output connections (nodes correspond to neurons in the human brain and are depicted by the light- and dark-shaded ovals in the above graphic). It is common for the number of nodes to number in hundreds or thousands, and some of the more advanced algorithms are comprised of millions of nodes organized into many layers and linked by a dizzying number of connections – as pointed out earlier, even if one were able to detail all of the individual associations that jointly produced an outcome of interest, the resultant map would be so busy and complex that it would be of little informational value. It is ironic that the analytic means that were expressly designed to mimic the architecture of that which powers human sensemaking end up being too complex for humans to consciously comprehend, even when highly scaled down (the human brain is made up of an estimated 100 billion neurons).

While there are sound reasons for why the mechanisms of automated decision-making are usually beyond simple description, it is nonetheless a formidable ADM acceptance obstacle. 'Trust, but verify' is a Russian proverb, popularized among English speakers by President Ronald Reagan who used it repeatedly in the context of nuclear disarmament discussions with the Soviet Union; it summarizes rather well the attitude of many users of data analysis-rooted recommendations. The 'trust, but verify' mindset can be accommodated by classical, i.e. distribution-based statistical analytic methods, such as the widely used regression analysis, because those

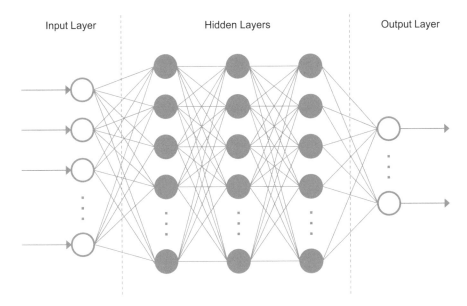

Fig. 8.1 A sketch of logical architecture of artificial neural networks

data analytic mechanisms produce results in ways that lend themselves to objective, rational decomposition and examination – the same, however, is not the case for computationally complex machine learning tools. Moreover, machine learning is also dynamic, which means that ostensibly alike outcomes continue to change in response to changing patterns in input data,[2] which also poses a challenge to assessing their efficacy.

In-use assessment offers what is commonly considered the most convincing evidence of predictive or explanatory efficacy; for instance, a model built to predict product choice is best evaluated by comparing the initial predictions to observed outcomes. That reasoning, however, implicitly assumes that a particular predictive algorithm is compositionally fixed, meaning that when seen as a mathematical formulation, once developed it does not change, unless wilfully amended. That assumption is important because it is a necessary prerequisite to meaningful efficacy attribution and the resultant interpretation – a positive in-use assessment evaluation suggests that the algorithm under consideration is worthy of deployment, but that conclusion is contingent on the algorithm remaining unchanged. It follows that an ever-evolving or compositionally dynamic algorithm is, in principle, untestable.

8.1.3 Accountability

For nearly a year, from November 20, 1945 to October 1, 1946, the Allied Forces held a series of 13 military tribunals that have come to be known as the Nuremberg trials, the purpose of which was to bring Nazi war criminals to justice.[3] Numerous high-ranking Nazi party officials and military officers, along with German industrialists, lawyers and doctors, were indicted on charges of crimes against humanity and peace – all but three were found guilty, even though many tried to disavow their responsibility by invoking the plea of due obedience (now commonly known as the Nuremberg defence), essentially claiming that their actions were merely manifestations of their duty to follow orders. The tribunals rejected that defence, instead finding that the 'fact that a person acted pursuant to order of his Government or of a superior does not relieve him from responsibility under international law, provided

---

[2] A typical non-machine learning-based system, such as the customer loyalty classification DSS algorithm discussed in Chap. 7, is compositionally fixed, which means that while input data may change, a scoring algorithm, expressed as a mathematical equation, remains unchanged (until it is redesigned or otherwise tweaked). In contrast to that, machine learning algorithms are compositionally dynamic, which means that the output producing mechanism (i.e. the composition of the hidden layers in Fig.8.1) continues to change with changing inputs.

[3] Though controversial at the time, primarily because of lack of clearly defined legal justification or precedent, the Nuremberg trials are now widely considered an important milestone towards establishment of a permanent international court for addressing crimes against humanity, such as genocide.

a moral choice was in fact possible to him.[4] In short, making choices entails responsibility.

One of the defining differences between decision support (DSS) and automated (ADM) systems discussed in the previous chapter is that the former offer advice to human decision-makers, while the latter altogether replace human decision-makers, which raises a host of legal and ethical questions, most of which are centred on accountability and liability. Within that context, DSS can be seen as just another tool controlled by its human user, who being capable of moral judgement bears full responsibility for how that tool is used. When that human, however, is taken out of the loop, the questions of accountability and liability no longer have clear answers. ADM systems lack free will, and without free will it is not possible to be capable of moral judgement; moreover, artificial systems have no intrinsic sense of worth and thus cannot be expected to 'understand' the sense of worth that humans attached not just to physical objects, but even more importantly to intangible ideals or beliefs. In view of that, it might be reasonable to expect that, for instance, life support system-monitoring automated control algorithms and human caretakers might evaluate the same set of medical facts quite differently, at least when it comes to some patients. If such a system made a decision to discontinue life support for a terminally ill patient and if that decision turned out to run counter to human ideals and beliefs, who would be accountable? The algorithm's original designers? The data scientists who trained it (in the sense of selecting and curating input data)? How about the healthcare organization that implemented it?

There are many questions, but most do not yet have meaningful answers because laws and regulations that are needed to answer those questions lag behind the pace of technological innovation, especially in the area of information technology. As noted in the previous chapter, the idea of applying the concept of personhood to non-humans is well established (it dates back to the early nineteenth century in the United States); given that, does that mean that the same general legal principles that apply to corporations and other non-person entities legally endowed with person-like status also extend onto ADM systems? After all, those systems are a lot more human-like (in terms of the functioning) than a business company; however, business firms have economic assets which adds an element of financial recourse. Or, should ADM systems be simply considered to be nothing more than property, with their owners and/or operators bearing full responsibility?

Turning to somewhat more pragmatic considerations, what, exactly, does the idea of algorithm accountability entail? That too is not a simple question; a well-considered potential answer has been offered by a group of industry practitioners and academics in the form of the following five distinct, general algorithm account-ability principles[5]:

1. Fairness: no discriminatory or unjust impact.

---

[4] Nuremberg Principle IV

[5] fatml.org/resources/principles-for-accountable-algorithms

2. Explainability: algorithm-produced decisions and the underlying data can both
   be validated.
3. Auditability: algorithmic logic can be probed and understood.
4. Responsibility: avenues of redress need to be available.
5. Accuracy: sources of error need to be identified, logged and articulated.

While intellectually compelling, the proposed principles are not operationally
clear. For instance, the notion of 'fairness' is surprisingly complex: it can entail
impartiality and honesty, freedom from self-interest or absence of prejudice or
favouritism, but at the same time it can also communicate conformance with the
established norms. Those conceptions are not only semantically distinct – they are
also value-laden and shifty, in the sense of evolving social understanding of fairness. Moreover, within the confines of statistical analysis and machine learning, a
'fair' classifier is a discrete-valued function capable of validly and reliably assigning class labels to individual data points; in other words, it is ultimately an expression of an algorithm's classificatory efficacy. Given its multifaceted essence,
assessment of the fairness dimension of algorithmic accountability is not only situational but also prone to subjective evaluation (to be fair, no pun intended, the
remaining four accountability principles are comparatively more operationally
straightforward).

While clearly important within the confines of organizational use of automated
technologies, the importance of accountability is particularly profound at the level
of societal governance, because here it touches on the most fundamental questions
of justice and fairness, which is an extension of the earlier raised idea of human
values. It is not only grasping the abstract and highly nuanced ideas of morality that
might be beyond the capabilities of artificial systems that view reality through the
prism of arrays of 0s and 1s – it could also be the more fundamental ability to correctly identify and group applicable fact patterns which should also not be taken for
granted. The vast majority of law and justice-related information is in the form of
text, yet computers are comparatively inept at the task of complete, valid and reliable analysis of text data. Even highly evolved text mining and sentiment analysis
systems are, at best, able to extract what amounts to reasonably close, though generally quite incomplete meaning out of a given body of text, and that is assuming that
those systems were trained on text that closely resembles the text to be analysed.
That alone is a considerable limitation in situations in which, quite often, small differences in specifics can give rise to materially different conclusions.

8.1.4 Concentration of Power

John Acton, a nineteenth-century English politician and writer, famously noted that
'power tends to corrupt, and absolute power corrupts absolutely'. Acton's observation almost immediately conjures up images of abuses of power by various political
dictators, but the idea applies equally well to those able to gain and exercise

outsized economic power. Here, the so-called robber barons of America's Gilded Age, best exemplified by legendary tycoons, such as John D. Rockefeller, Andrew Carnegie or Cornelius Vanderbilt, showed that the same holds true of non-ruling elites.[6] Is the combined effect of the rapid proliferation of algorithmic systems coupled with the concentration of the result economic, and by extension other types of power, giving rise to digital robber barons?

The behind-the-scenes automated algorithms already influence commercial (e.g. consumption habits), social (e.g. professional and personal networking) and societal (e.g. voting behaviours) dimensions of life; on a more personal level, those artificial decision-making systems have the power to influence individuals' state of mind, even the level of happiness, given how much of the latter now seems to be encapsulated in the growing attention to the number of 'likes' or social (media) followers. The operators of the vast majority of those systems are for-profit business enterprises who continue to amass growing power, as the ceaseless spread of the techno-informational infrastructure continues to reshape how work is done and how lives are lived. Though profound, the scale and the impact of those transformations cannot be said to be 'unparalleled' because the current evolutionary changes mirror the nature of discontinuities that characterized the transition from agrarian to industrial society (see Chap. 4 for a more in-depth discussion). Moreover, some of the defining characteristics that marked the onset of the Industrial Era, such as emergence of entirely new centres of wealth and power, are also among the defining traits of the now unfolding Digital Era. Hence, if the Industrial Era had its robber barons, does that mean that the Digital Era is destined to have its digital robber barons?

The monopolies of the Industrial Era were ultimately broken up, and legal and regulatory mechanisms were put into place (at least in the United States and other developed nations) to prevent their re-emergence; those mechanisms, however, may not adequately guard against the shadowy monopoly of algorithms. Looking to the future, it might be worthwhile to keep the past in the proverbial rear-view mirror, as a lot can be learned by re-examining of the social, political and economic history of the late nineteenth and the first part of the twentieth centuries. As succinctly but eloquently captured by British statesman Winston Churchill, 'those that fail to learn from history are doomed to repeat it'.

8.2 The Internet of Things

It seems that one of the favourite topics of apocalyptic storytellers is the rise of super intelligent machines, which, and that is the apocalyptic part, invariably turn against humanity. Given the many imperfect aspects of mankind, envisioning a machine that is governed by rational sensemaking mechanisms and which eventually

---

[6]The term 'robber baron' most likely originated in medieval Europe in reference to feudal lords who robbed merchant ships and travellers along the Rhine River; its modern use is attributed to a 1934 book by Matthew Josephson titled *The Robber Barons*.

comes to a conclusion that humanity is a menace warranting eradication does not seem far-fetched. And to some, the rapidly expanding Internet of things poses just that type of a threat.

The Internet of things, or IoT, is a network of sensor-, software- and other technology-enabled devices and systems, or 'things'; it is a catch-all term for the growing number of interconnected non-traditional[7] computing devices that are connected to the Internet to send data, receive instruction or both. It encompasses not only 'natively smart' devices such as Apple Siri or Amazon Alexa-like digital assistants, but also 'smart' versions of traditional appliances such as refrigerators or even light bulbs; moreover, it also encompasses systems within systems, such as Internet-connected sensors that automate factories, transportation and distribution systems and the ever-expanding aspects of healthcare. As of 2020, the IoT ecosystem encompasses more than 50 billion devices generating an estimated 4.4 zettabytes (or 4.4 trillion gigabytes) of data. How does it all work? Figure 8.2 offers a high-level schematic.

Individual devices, which can range from simple temperature sensors to wearable fitness trackers to autonomous vehicles operating on factory floors, transmit data to gathering points which include devices to aggregate data before sending it along – all of that is contained within the 'IoT Devices' section of the process depicted in Fig. 8.2. Once gathered and aggregated, data are sent, typically via Internet, to organizational data centres for processing; sometimes, however, that might not be feasible, as is typically the case with highly time-sensitive critical infrastructure, in which case data are processed in the cloud (i.e. on the Internet). In time-sensitive applications, *edge computing*, broadly defined as processing of data at or near the source of it with the help of distributed computing, is often utilized to

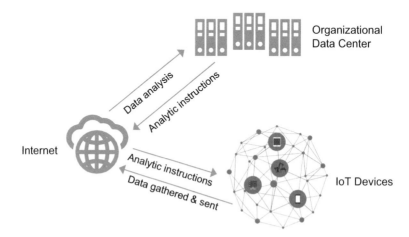

Fig. 8.2 High-level view of IoT mechanics

[7]Personal and larger computers are considered 'traditional' computing devices, while the more recent mobile, wearable and other forms of data processing digital electronics are framed here as 'non-traditional'.

minimize data transmission and processing-related delays (edge computing devices are also capable of sending data along for further processing and analysis; that functionality is referred to as upstream connectivity). What ultimately comes back to IoT devices are operational instructions, as in the case of autonomous vehicles roaming factory floors, or user feedback information, as in the case of wearable fitness devices.

The aspects of IoT's inner workings that are perhaps least visible are interconnectivity, interoperability and security – in order to work together, all those devices need to be properly configured, authenticated, monitored and updated, on as-needed basis. Here, the initial reliance on proprietary systems is slowly giving way to a standard-based device management model, which allows IoT devices to interoperate and ensures that devices are not left out or orphaned. That said, there is a wide variety of communications standards and protocols, which include household names such as Wi-Fi or Bluetooth, in addition to many others tailored to devices with limited processing capabilities or limited power supply. For example, ZigBee is a wireless protocol for low-power, short-distance communication, and message queuing telemetry transport (MQTT) is a messaging protocol for devices connected by low reliability networks.

8.2.1   Applications and Implications

While IBM's Jeopardy!-winning Watson platform may have fallen short of becoming the first AI medical doctor, it is still a powerful automated decision-making platform that can play an integral role in IoT applications. For example, Watson has proven itself to be quite capable in the area of predictive maintenance where it has been trained to analyse data from drones to distinguish, for example, between trivial damage to a bridge and cracks that need urgent attention. Not surprisingly, Watson-like systems are currently being trained for a host of purposes, including the following:

- Oil and gas exploration, where isolated drilling sites can be better monitored with IoT sensors than by human intervention
- Oil and gas transmission, where data gathered from pressure sensors are edge-analysed (i.e. at or near the source) looking for threats of pipeline rupture, detection of which can then trigger automatic shutdown procedure
- Agriculture, where IoT sensor-captured highly detailed data about crops' growth are used to increase yields
- Brick-and-mortar big box retail operations, where real-time data are used to micro-target customers in certain areas of a store
- In amusement parks, where phone apps that offer information about the park also send GPS pings back to parks' operators to help predict wait times in lines
- Smart home systems, to control lighting, climate and security
- Smart city systems, which, though still in early stages of development, show potential in helping to manage traffic, public transit, safety and more

Looking beyond those specific applications, the IoT infrastructure also enables a considerable situational agility in responding to unexpected developments, perhaps best illustrated by the global COVID-19 pandemic. An example of that is offered by Kinsa, a startup that is leveraging the IoT to deploy a network of connected thermometers; the idea is that being able to real-time track large amounts of anonymous, countrywide fever levels could offer a critical piece of information to support better, evidence-based public policy decisions. Data aggregates gleaned from the already deployed million-plus thermometers can be used to map out informative regional patterns and track changes over time, all in near real time. In fact, inspired by a similar idea of using the IoT functionality for tracking purposes, another company, X-Mode Social, a major data broker and a provider of location information, was able to develop a map tracking the location of Ft. Lauderdale March of 2020 spring break revellers (who chose to defy the rapidly spreading coronavirus pandemic), showing how they all spread out across the country.

While highly informative on the one hand, such applications illustrate the surveillance potential that is inherent in IoT connectivity. It also underscores the fact that since IoT is, for the most part, commercially owned infrastructure, much of the surveillance is commercially oriented, which gives rise to what has been labelled 'surveillance capitalism [1].' The essence of that largely invisible but insatiably nudging unspoken economic system is rooted in the offering of free online services/applications, which track their users in often shocking detail and quite frequently without their explicit consent; the resultant personal experience data are claimed as proprietary asset by service providers who then feel free monetize it in whatever way they deem it appropriate.

8.2.1.1 Privacy and the IoT

If, as noted throughout this book, the essence of how work is done and how lives are lived are both being transformed by the rapidly emerging automation, should the conception of what constitutes 'privacy' change as well? IoT technological infrastructure clearly gives rise to a conundrum: To know now increasingly requires being known. One aspect of how the proliferation of 'smart' technologies is effectively rendering the idea of privacy obsolete is what could be called *granular algorithmic identification*. IoT-enabled home appliances are capturing very granular behavioural details, as do the ubiquitous navigation systems, as do virtually all online applications and so on. Those rich details all go into vast collections of data known as data lakes, which when coupled with machine learning algorithms can compile surprisingly detailed pictures of individuals. The point is that the traditional conception of privacy was focused on non-disclosure of persons, but the modern detailed data-enabled technology can reconstruct persons from a sea of seemingly inconsequential details.

Moreover, the opaqueness of the underlying data capture, individual-level aggregation, analytic processing, group-level aggregation and, finally, deployment renders privacy enforcement practically meaningless. Also, in the commercial realm, individual-level identification is both necessary and inconsequential. It is necessary because it is the glue that holds together, in meaning-giving sense, the otherwise meaningless behavioural and attitudinal tidbits, and it is inconsequential because the value of big data is in its ability to provide meaningful information on multitudes, not individuals, because the ultimate goal is to predictively understand emerging needs, wants and other aggregate trends.

The owners and operators of data tracking systems are (usually) quick to offer assurances of anonymity, but a recent *The New York Times* report[8] showed how, for instance, the readily available cellphone data could be used to track the movements of individuals who could later be identified by name, in spite of those data having been manifestly anonymized. In contrast to what appears to be wilful breach of privacy, the embedded trackability of cellphone data can also lead to completely unintended consequences, as illustrated by a recent incident in which Strava heat map, which shows popular running routes for Fitbit users around the world, accidentally revealed several secret American military bases. Thus in many regards, the ever-expanding digitization of life effectively renders more and more aspects of the notion of privacy, as in being hidden from public view, operationally obsolete, because the mere participation in technology-enabled commercial and personal interactions is tantamount to de facto public disclosures of individual attitudes and behavioural choices and, ultimately, unmasking of self.

An almost impulsive conclusion here might be to put into place appropriate laws and regulations or at least push for more vigilant self-regulation. The latter seems unrealistic, at least insofar as for-profit organizations are concerned, because it would require wilful income stream reduction that would almost certainly follow any decrease in the precision of tracking data. The former might prove itself to be no less challenging as it would require political will, something that might be difficult to muster, considering that any meaningful digital privacy restrictions would cut at the core of data enterprises' business model, and as such it would pose an existential threat to those enterprises. Still, the Consumer Online Privacy Rights Act, or COPRA, has been introduced in the US Senate on December 3, 2019. The proposed legislation spells out limitations on the use of algorithmic decision-making systems; it broadly prohibits 'covered entities' from using 'covered data' to engage in discriminatory practices regarding eligibility for housing, education, employment or credit, to advertise or market for such purposes or to otherwise impose restrictions on public accommodations. It prohibits covered entities from processing or transferring covered data on the basis of an individual's actual or perceived demographic, biometric and other information to advertise, market, sell or engage in other commercial activities for housing, employment, credit or education. Still, a year after its initial introduction, no legislative action has been taken; thus it remains to be seen if COPRA will ever be enacted.

Given that, a more certain way forward is likely through concerted, social consciousness-oriented awareness building – simply put, societies need to demand

---

[8] https://www.nytimes.com/interactive/2019/12/19/opinion/location-tracking-cell-phone.html

that companies that capture data use those data only in the way that reflects broader social norms; an idea which is not unlike demands for social justice or environmental sustainability. There is early evidence suggesting that such efforts might prove themselves more fruitful than the often prolonged and uncertain political process: At the very end of 2020, the two leading tech giants, Apple and Google, banned X-Mode Social, the earlier mentioned data broker and a provider of location information, from collecting location details from users whose mobile devices run on iOS and Android operating systems; that decision appears to be a direct result of an outcry over how those data were used by commercial and governmental entities.

Given the transformational impact of automation on work and life in general, how the questions raised earlier in this chapter are answered will have a profound impact on the future of society. If history is to offer some lessons, it is that the road that is yet to be travelled might not be a smooth one. There are numerous parallels between the past transition from agrarian to industrial society and the current transition from industrial to information society – the former was triumphant to few, most notably the 'robber barons', but painful to many, if not most, as evidenced by well-known problems of injustice and abuse that were made possible by lack of adequate legal and social protections. Today, abuses take a different form – such as privacy-destroying behavioural and economic surveillance – but the underlying reason is the same: the lack of adequate legal and social protections. Still, should history once again repeat itself, there are reasons to be believe that the current challenges will also be overcome, but neither easily nor quickly.

8.3   Informational Automation and Organizational Learning

The MultiModal Organizational Learning framework, first introduced in Chap. 1, expressly differentiates between computational and simulational aspects of technology-based learning, while as discussed in the previous chapter, it also implies non-systemic but persistent association between observation- and simulation-inspired creativity. The Internet of Things plays a major role here, and its emergence and ceaseless expansive evolution have profound implications for organizational learning, graphically summarized in Fig. 8.3.

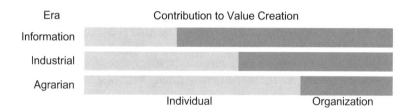

Fig. 8.3 The shifting locus of value creation

When looked at from the perspective of direct contribution (which sets aside the ideas of ownership and power), in the Agrarian Era, the skill and effort of individuals, most notably farmers and craftsmen, were the dominant driver of value creation; the organization, as in broadly conceived landowners, was a comparatively minor contributor to economic output. The onset of mechanization that gave rise to the Industrial Era more or less evened out the relatively individual vs. organization contributions, as organizations supplied factories and equipment, while individuals contributed equipment operating know-how, along with the necessary labour. In that context, Henry Ford's assembly line, a technological marvel of its day, was useless without skilled human operators. In the Information Age, however, the individual vs. organizational contribution ratio is beginning to tilt towards the latter as a direct consequence of operational and information automation and the corresponding decrease in production-related direct human involvement. The IoT-driven automation-originated data flows also expand the scope of organizational learning, as never before seen types of data give rise to never before available insights.

While the importance of data analytic skillset to making productive use of the rapidly emerging and expanding techno-informational infrastructure is largely self-evident, it is also important to have clear understanding of the essence of the closely related ideas of 'digitization' and 'digitalization'. Often used interchangeably, those two similarly sounding concepts are actually quite different: *Digitization* is a process of converting analog signals to digital (i.e. expressed as a series of 0 and 1) code, while *digitalization* is a process of utilizing digitized information to develop new systems, products and the like. In that definitional context, IoT data comprise broadly defined outcome of digitization, and those data or analysis-produced informational outcomes are among the key enablements of digitalization, which then leads to more data, giving rise to a largely self-perpetuating technological expansions, graphically summarized in Fig. 8.4.

Related to the concepts of digitization and digitalization is the notion of *digital transformation,* which captures the process of adapting to the emerging digital landscape, an idea that is well illustrated by the recently announced General Motors (GM) workforce transformation, an unprecedented move in the company's 100+ year history. An icon of the industrial age, GM has been experiencing a steady decline in the demand for the traditional mechanical skillset (e.g. machine operating, tool grinding), while also experiencing skyrocketing demand for the Information

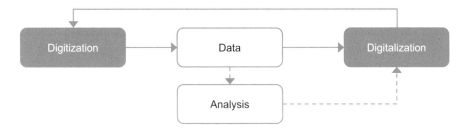

Fig. 8.4 Self-perpetuating techno-informational expansion

Age skills, such as computer network support specialists and data scientists, which culminated in a recently announced fundamental 'retooling' of the makeup of its roughly 164,000 strong (as of 2019) workforce. Clearly, what is true of GM is likely to also be true of numerous other, Industrial Era organizations, all of which points to the rapidly changing framing of the ideas of productivity and productive skills; as discussed earlier (Chap. 4), the transition from 'making' to 'conceiving' has profound implications for organizational learning.

Norm resetting decisions, such as the one just announced by GM, are rarely made without carefully and thoughtfully considering full spectrum of applicable evidence. One of the considerations (which may or may not have played a role in GM's decision) that continues to garner attention among Industrial Era organizations trying to adapt to the demands of the Information Age is of the idea of 'loose attribute coupling', an evolutionary step in the decade-old idea of mass customization made possible by less restrictive design schemas. While notionally compelling, operationalizing of the idea of loose attribute coupling poses considerable organizational learning challenges rooted in developing valid and reliable means of singling out meaningful attribute associations in torrents of digital details. Deceptively minor in its overt appearance, the idea of identifying material associations data, especially in vast quantities of data, has outsized profound impact on the overall efficacy of informational automation.

8.3.1 Variable Associations

The notion of association is one of the most all-encompassing tool concepts, used to describe enumerable embodiments of formal and abstract 'linkages', ranging from alliances, unions or coalitions to scientific properties and mental states. Within the realm of data and data analytics, the notion of association is used to communicate measurable and persistent relationships linking specific data elements and expressed in terms of concurrence of their variability.[9] In that very general sense, two data elements, or variables, can be considered to be associated when their cross-record change shows a meaningful pattern of similarity, so much so that knowing the value of one can yield some insight into the likely value of the other. And lastly, association can be either correlative or causal, which is an important distinction that will be addressed in more detail later in this section.

Implied in that coarse characterization of variable associations is their probabilistic and continuous nature, the combination of which has particularly profound

---

[9]This framing implicitly assumes a cross-sectional dataset in which an attribute, such as 'age', can take on different values across successive data records; in a longitudinal, or time-based dataset, however, a data record could represent time-based value, as illustrated by daily price of a particular stock. It is worth noting that a dataset can have a mixed, cross-sectional + longitudinal layout, as illustrated by a file containing daily stock prices for all New York Stock Exchange-listed companies for 2020.

implications for automated decision-making. In the measurement sense, an association can range from non-existent to strong, but in practice the focus tends to be on differentiating between spurious and material or, more generally, informative vs. non-informative associations. But doing so algorithmically and objectively is challenging because while numerous rules of thumb have been proposed over time (some discussed later in this section), there ultimately is no singular, objective threshold for distinguishing between trivial and important associations. At the same time, valid and reliable identification of meaningful associations is of central importance to learning from data, as it is at the core of the ability to extract decision-guiding insights out of data or to transform data directly into performance of specific tasks.

8.3.1.1   Informative Versus Non-informative Associations

IoT and the more broadly framed techno-informational infrastructure both produce and consume data. For instance, networked devices use data from sensors and other devices to perform specific actions, as illustrated by a self-driving car stopping or accelerating; moreover, IoT-linked devices also generate their action-tracking data, which together with operational details can also be are used as input into a wide range of exploratory or confirmatory[10] analyses. All told, the potential actuation (i.e. devices controlling other devices) and informational (input into human decision-making) potential of data captured by the modern electronic infrastructure is enormous to the point of being hard to grasp, but much of it is contingent on valid and reliable assessment of associations embedded in raw data.

Perhaps the best-known operationalization of the general idea of association is correlation,[11] which is a statistical measure of the degree to which a pair of variables are linearly related. As implied in that characterization, the strength of association is a continuum, which can range from no association (e.g. correlation = 0) to perfect association (e.g. correlation = 1.0); correlation-expressed associations can also be positive, which is when both variables move in the same direction (i.e. an increase in one is accompanied by an increase in the other one), or negative, which is when variables move in opposite or inverse directions. Of particular interest here, however, is the strength of association, abstractly summarized in Fig. 8.5.

---

[10] Broadly characterized, exploratory analyses seek to identify previously unknown patterns or associations, while confirmatory analyses are geared towards empirically testing of theoretically presumed associations (see the Conjectures and Refutations section in Chap. 3).

[11] The two most commonly used variants of correlation coefficients are Pearson's product moment and Spearman's rank. The former (which tends to be used more commonly, though sometimes incorrectly) is appropriate to use when both variables are continuous and normally distributed (which also means no outliers); the latter is appropriate when both variables are either skewed (i.e. not normally distributed) or ordinal (expressed as rank-ordered categories), and it is also robust with respect to the presence of outliers.

None   ⟶   Spurious   ⟶   Questionable   ⟶   Material   ⟶   Perfect

Fig. 8.5 The association continuum

None ⟶ Spurious ⟶ Questionable ⟶ Material ⟶ Perfect

In-the-Moment Association

Bayes Factor

Fig. 8.6 In-the-moment association

For interpretational convenience, the continuum-expressed association is commonly divided into several, more informationally meaningful categories, delineated in Fig. 8.5. In practice, 'none' and 'perfect' (e.g. correlation = 0 and 1.0, respectively) are exceedingly rare; in an overwhelming majority of cases, the association of interest will fall somewhere in between those two extremes. 'Spurious' associations are generally considered non-informative, in contrast to 'material', which are typically deemed informative; 'questionable' are just that – of uncertain meaning. While interpretationally straightforward, the boundaries of those categories are inescapably arbitrary because there are no universally accepted thresholds for what constitutes, strength of association-wise (e.g. the magnitude of a correlation coefficient), material association. Different situations and different analytic perspectives may render the same association estimate, e.g. a correction of 0.4, material, questionable and possibly even spurious. Statistical significance tests offer some help in differentiating between spurious and non-spurious associations, but those tests are highly susceptible to sample size (as the sample size used to calculate test statistics increases, so does the likelihood of the association being deemed statistically significant). To be more explicit, concluding that a particular association is 'statistically significant' may have little-to-no practical meaning, given that even a moderately sized (e.g. a few thousand records) analyses are likely to find all but the nearly microscopically sized expressions of association statistically significant.

The progression summarized in Fig. 8.5 hints of the recommended approach, which is effect size-based *in-the-moment association*, an idea which roughly parallels the logic of Bayesian probability estimation (which was also the conceptual foundation of the cognitive diversity quotient discussed in Chaps. 3 and 7). In essence, it amounts to framing of the magnitude of an association of interest as the current state of knowledge, the magnitude of which, along with the resultant qualitative or directional interpretation, can be expressed using the Bayes factor discussed in Chap. 5 (it is a ratio of the likelihood that the observed association is material, and the likelihood that the observed association is spurious). Consider Fig. 8.6.

The ideas outlined above appear to dismiss the importance of the strength of association, which to some extent is correct, because in practice the analog-like expression of the strength of association is not always informationally clear. For instance, what is the difference, in the sense of implied decision, between a correlation of 0.4 and 0.6? It is not necessarily clear, due to several factors. Firstly, as summarized in the well-known 'correlation is not causation' adage, an association between action X and outcome Y should not be confused with action X causing outcome Y (because estimates of association do not account for the presence of other factors which may influence both X and Y, as expressly discussed in the next section). That means that the numerically stronger correlation of 0.6 does not convey stronger impact of X on Y, which would imply greater desirability of investing effort or resources into action X in hopes of increasing outcome Y. Secondly, there are no universally accepted benchmarks for what constitutes 'weak' and 'strong' associations. For instance, in psychological research, correlations greater than 0.4 are often considered relatively strong, and those in the range of 0.2–0.4 are considered to be moderate, whereas in medical or demographic research, to be considered strong, correlations need to be at least 0.7, and correlations weaker than 0.3 are often considered negligible or spurious. And thirdly, correlation coefficient-expressed association between action X and outcome Y is assumed to be linear, which is an often overlooked, yet a substantially limiting assumption, graphically illustrated in Fig. 8.7.

To be considered a linear association, the relationship between the rate of change in X and the rate of change in Y has to remain constant across all possible values of X and Y, as graphically illustrated by the straight line (labelled 'linear') in Fig. 8.7. Thus, if 1 unit increase in X is associated with 1 unit increase in Y, that rate of change has to remain constant, in principle infinitely, but in practice at least across all plausible values of X and Y. While it is relatively easy to find examples of linear

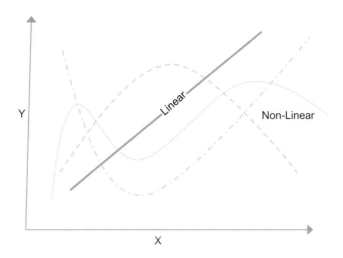

Fig. 8.7 Linear and non-linear associations

relationships in everyday life – an obvious one being the association between turning of the steering wheel and the change in the vehicle's direction – comparatively
few organizational decisions related phenomena are linear (and estimation of non-
linear effects is analytically taxing and interpretationally abstract, thus rarely undertaken in applied settings). As a result, in most applied estimation contexts, linear
associations are merely computationally manageable and interpretationally inviting
approximations. It is usually understood that because of micro factors such as
fatigue or waning interest, or macro effects such as need saturation or diminishing
returns to scale, relationships between actions and outcomes and other pairings of
interest tend to be non-linear and thus can take on many different forms (with a few
sample ones depicted in Fig. 8.7). It follows that it is important to be mindful of the
difference between weak linear associations, as in the case of magnitudinally small
correlation coefficient, and weak associations – a magnitudinally small correlation
coefficient, for instance, may warrant a conclusion of weak linear association, but
not week association.

8.3.1.2   Causation: A Special Case of Association

Merely asserting that there is an association between X and Y has some but ultimately limited value in the context of applied decision-making, because it leaves the
key consideration of *causation* (or causality, the two terms are used interchangeably
here) unresolved. Will investing more resources or effort into action X lead to measurable increase in the outcome Y? Many organizational decisions are predicated
upon that type of general cause-effect relationship, but the earlier discussed notion
of association is geared towards ascertaining the existence of any type of a relationship, which could be causal or non-causal.

   Given its central role in knowledge creation, the notion of causation has been a
subject of centuries-long debate among scientists and philosophers alike. At the
core of the debate has been the line of demarcation separating cause and effect from
simple concurrence-based relationships. From the standpoint of philosophy of science, to be deemed causal, an association has to exhibit temporal sequentiality (X
has to precede Y) and associative variation (changes in X must be systematically
associated with changes in Y), the association between X and Y has to be non-
spurious (it should not disappear upon introduction of a third, moderating factor or,
as discussed earlier, have to be material) and lastly, it also has to have sound theoretical rationale (it has to 'make sense' or exhibit basic face validity). Temporal
sequentiality and associative variation have fairly simple meanings and applications
(in spite of their somewhat foreboding names), as the conditions of occurring in
sequence and doing so persistently are both intuitively obvious and relatively easy
to demonstrate. For instance, a customer has to be exposed to a promotion, such as
a television or online ad, prior to making a purchase in order for that promotion to
have causal impact on the subsequent purchase (temporal sequentiality) and that
sequence has to be recurring (associative variation). However, the rationale and the
proof of the remaining two causality prerequisites – non-spurious association and

theoretical support – are often more elusive. The commonly used measures of association, such as correlation analyses, are only capable of ascertaining the extent to which X-Y association is linear; moreover, what constitutes non-spurious association is inescapably interpretation laden and situational. The main challenge associated with showing evidence of 'sound theoretical rationale' is that, absent a well-established scientific theory, any such 'evidence' is not only highly subjective, but it is also a function of the state of knowledge at a point in time, not to mention being perspective or value laden.

Until Copernicus offered evidence to the contrary, the geocentric model was the accepted theory of the cosmos for many centuries; more recently; Newtonian mechanics were believed to be universal, until Einstein's relativity constrained it to just low velocities. There are practically countless examples of the evolving character of what constitutes sound theoretical rationale, and that problem is particularly acute in social sciences which lack objectivity and precision of measurement of hard sciences such as chemistry, biology or physics. In fact, the notoriously relativistic framing of many socio-behavioural explanations lends themselves to competing theoretical frameworks. For instance, an analyst trying to understand reasons behind economic contractions can employ the Keynesian, the Friedmanite or the Fisherian theory, each of which will produce materially different explanations, ultimately lending 'evidence' to different economic policy decisions. Stated differently, the same fact patterns may be interpreted differently depending on the often arbitrary choice of theory; hence, it is important to make well-thought-out and deliberate choices.

8.3.2   Data Reimagined

The relentless march of digitization and automation is not only reshaping how work is done and how lives are lived – but it is also remoulding the long-standing conception of data. Traditionally, data were viewed as collections of essentially standalone facts, and discerning patterns of interconnections among them embodied the essence of extracting insights from data. In a more abstract sense, data were seen as a source of auxiliary or additional insights. However, that somewhat simplistic view is beginning to change, partly because of technological advances which are enabling more complex and informationally richer data structures, partly because of necessity stemming from the continuously expanding volumes and varieties of data and partly because of the emergence of data-powered infrastructure in the form of the IoT. Stated differently, while at the onset of the Information Age data were seen as worthwhile but auxiliary to social and commercial functioning, the ongoing maturation of the Information Age is ushering in the era of data centricity, where torrents of sensor readings and related digital flows can be seen as the 'other' source of energy needed to power modern society. To be sure, as summarized in the MultiModal Organizational Learning framework discussed in Chap. 1, the utility of data did change as much as it expanded; still, the sheer data volumes coupled with high levels of granularity are nonetheless rendering detailed data 'tinkering' difficult, if not outright implausible. Moreover, combining greater level of measurement detail (i.e. more granular data) with vastly greater volumes is also amplifying the earlier discussed methodological challenges associated with identifying cross-variable associations, which are of foundational importance to extracting insights out of data.

Core to the idea of reimagining data is the interplay between the 'what' and 'how' of data capture and the nuanced data usage demands. Much has been written about the staggering volumes and varieties of data, and though less often expressly addressed, the volume of highly disaggregate metrics, best exemplified by point-in-time sensor readings, is also exploding, especially when IoT device-generated data are taken into account. And to top it all, data are also increasing real time – in fact, according to IDC, a consultancy, by 2025 about a third of all data captured will be real time. Setting aside instances where data are used to control automated devices such as thermostats or autopilot systems (in which situations specific sensor readings trigger predetermined responses, without the need for sensemaking oriented analyses), exploratory or confirmatory analytic usage of 'as-captured' highly disaggregate data is becoming increasingly incommensurate with the mechanics of established data analytic means and modalities. Considering that it is common for IoT device-generated data files to combine both cross-sectional (different attributes) and longitudinal (point-in-time measurements of a single attribute) elements, even sophisticated machine learning algorithms may not be able to correctly differentiate between, for instance, different measurements of a given attribute and informationally distinct attributes, further diminishing the efficacy of already probabilistic insights. All considered, within the confines of organizational learning, even if the actual data analytic tools are highly automated, input data may still need to be adequately feature-engineered by human analysts, a daunting task when dealing with highly detailed data (some of those considerations were discussed in Chap. 5; see Fig. 5.4 for a graphical illustration of a simple data feature re-engineering example). It follows that while, on the one hand, the highly detailed, 'as-captured' IoT data are essential to the very functioning of automated, interconnected systems, on the other hand, that level of granularity may be counterproductive to meaningful discovery of new knowledge. And while it may not be possible to completely alleviate that problem of overabundance, the extent of it can be materially diminished through recasting of as-captured data into bespoke aggregates, metadata and graph datasets.

8.3.2.1 Bespoke Aggregates

The traditional way of thinking about data is through the prism of either cross-sectional or longitudinal measurements, where the former represents point-in-time values of attributes for a set of entities, while the latter encapsulates repeated, as in different points in time, measurements of attributes of interest. In a more tangible context of the commonly used two-dimensional data layout structures framed by rows and columns (e.g. an Excel spreadsheet), in a cross-sectional dataset, rows are used to delimit entities, such as distinct customers or companies, and columns are used to delimit variables, which are usually attributes of entities. In a longitudinal dataset, on the other hand, rows are used to delimit distinct time periods, while columns are used to delimit variables or attributes being (repeatedly) measured. The distinction between cross-sectional and longitudinal data layouts is important because it goes to the heart of what constitutes an individual *data record*, defined here as a collection of related fields treated as a unit, which plays a pivotal role in how data are processed and analysed. In both cross-sectional and longitudinal datasets, a single row of data constitutes an individual data record, but as implied by their contents, the informational meaning of cross-sectional and longitudinal data records are considerably different: The former encapsulates standalone entities (e.g. customers, companies) described in terms of specific attributes, while the latter encapsulates time-dependent measurements of a single entity. While the specifics of cross-sectional and longitudinal analyses fall outside the scope of this discussion, let it suffice to say that the either-or structure is highly suggestive of plausible data utilization and data analytic approaches, but when a single dataset encapsulates both cross-sectional and longitudinal data layouts, the possibility of misuse increases sharply, and potentially substantial amounts of data pre-processing may likely become necessary. Those considerations are graphically summarized in Fig. 8.8.

Nowadays, IoT device-generated data commonly combine cross-sectional and longitudinal layouts, which necessitates at times considerable amounts of data pre-processing. To avoid or at least significantly reduce the amount of that time- and resource-consuming work, the originally captured data can be structured into *bespoke aggregates,* which are groupings of disaggregate variables linked by 'natural' closeness of informational content. Perhaps the best illustration of that general idea is offered by psychometric measurement. Most psychological constructs, such as attitude or commitment, are latent, meaning not directly observable; their presence or their magnitude is typically assessed using proxies, usually in the form of multiple directly measurable correlates. Due to the fact that, in the vast majority of situations, no single proxy can be considered a perfect indicator of the latent construct of interest (primarily because of the somewhat interpretation-prone character of latent constructs), multiple proxies are usually grouped together to form a singular composite factor,[12] which is then used as a singularly distinct variable. And because the resultant factors are typically developed with a particular purpose in

Longitudinal Layout					Cross-Sectional Layout					Mixed Cross-Sectional & Longitudinal Layout				
Entity	Date	Variable 1	Variable 2	...	Entity	Date	Variable 1	Variable 2	...	Entity	Date	Variable 1	Variable 2	...
Patient A	Week 1				Patient A	Week 1				Patient A	Week 1			
Patient A	Week 2				Patient B	Week 1				Patient A	Week 2			
Patient A	Week 3				Patient C	Week 1				Patient B	Week 1			
:	:				:	:				Patient B	Week 2			
Patient A	Week n				Patient n	Week 1				Patient C	Week 1			

Fig. 8.8 Data layouts

---

[12] Those are commonly referred to as 'factors' because factor analysis, a variable grouping statistical technique, is often used to construct statistically valid and reliable indicator composites.

mind, such as enabling quantification of the aforementioned latent constructs of attitude or commitment, they can be considered bespoke aggregates.

In a more general sense, bespoke aggregates are composites of naturally coupling variables that offer greater (then their individual components) analytic utility, are designed to manage linked data effectively and are also able to support fast extraction of aggregates, all while retaining rich relationship information. In addition, bespoke aggregates hold probabilistic variable linkage data in the form of all material pairwise associations (see Figs. 8.5 and 8.6, and the associated discussion), allowing dynamic aggregate variable recompositions. Lastly, bespoke aggregates can accommodate mixed data layouts, illustrated in Fig. 8.8, which greatly reduces manual data pre-processing.

Enriching the original disaggregate data with those general types of amendments will not only make the increasingly more untenable task of quickly analytically 'consuming' the ever-expanding arrays of data more manageable – but it will also support more meaningful real-time analytics. By embedding rich heterogeneous information about data features' relationships, real-time data queries will be capable of producing more refined and thus more informative results. To make the amended data even more usable, the association-related statistics can also be graphically expressed, which will effectively enable data visualization at a more rudimentary data level.

8.3.2.2 Graph Metavariables

A logical next step in pre-analysis refinement of disaggregate data is to join bespoke aggregates into graphically expressed *metavariables*.[13] In the computer science sense, graph structures can be used to store natural representations of linked records; thus, in contrast to originally captured data where the focus is on data elements (i.e. variables and records), in graph data structures the focus is on associations between data elements, where 'nodes' are used to store data elements and 'edges', or paths, are used to store associations between them. An edge always has a start node, end node, type and direction, and it can describe parent-child relationships (e.g. customer-purchase), actions (e.g. purchase-repurchase), membership (e.g. person-household) and just about any other type of association; in fact, there is no limit to the number and kind of associations. Figure 8.9 shows a simple schematic of the graphical A-B-C-D metavariable.

The A-B-C-D metavariable in Fig. 8.9 can be traversed along specific edge types (e.g. C to A) or across the entire graph. Traversing the individual joins or associations is very fast because the relationships between nodes are not calculated at query times but are persisted or permanently stored in the data structure, which translates into significant gains in performance. In fact, in the traditional, data element-centric

---

[13] Formally defined, metavariable is an idea used in the study of semantics where it is framed as an element of a metalanguage (language used to describe expressions of another language) that can be used to refer to expressions in a logical language.

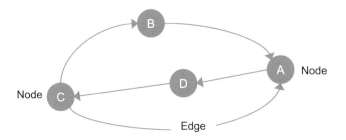

Fig. 8.9 Graphical representation of A-B-C-D metavariable

structures as the number of data elements increases, the number and the depth of potential relationships grows exponentially, and since relationships are calculated at the time of a query, as the size of data grows the query time increases – in contrast to that, the query time of graph data structures does not increase with the size of the dataset. Not surprisingly, graph data structures are already used extensively in social networking, recommendation engines and fraud detection, all of which are characterized by immense volumes of associations and quick turnaround demands.

A somewhat esoteric but informationally important aspect of graph metavariables is *recursivity*, which arises in the context of causal relationships. It addresses the sequencing of objects encapsulated in a particular graph structure, such as the A-B-C-D metavariable shown in Fig. 8.9, expressed in terms of the direction of causal relationships, which can be either recursive or non-recursive, as graphically illustrated in Fig. 8.10.

The core difference between recursive and non-recursive relationships is the nature of directionality: The former are unidirectional, while the latter are bidirectional. From the perspective of organizational decision-making, recursive relationships are more informationally meaningful simply because their implications are unambiguous, as demonstrated by the recursive C $\rightarrow$ A relationship in Fig. 8.10 (assuming, of course, that both C and A are meaningful variables). The non-recursive C $\leftrightarrow$ A relationship, on the other hand, is comparatively more informationally baffling, suggesting lower decision-making utility. Still, non-recursive relationship can ultimately offer a more empirically correct assessment of a relationship of interest. For example, intuitively, advertising should have causal impact on sales, but it would be empirically unsound to express that belief recursively the requisite temporal sequentiality cannot be assumed in view of the common practice of tying advertising spending to sales (thus changes in ad spending cannot be said to precede sales level changes[14]).

---

[14] An additional complication stems from advertising being a 'slow moving' variable, meaning that its level (as measure either in terms of spending or exposure) changes slowly, while sales levels fluctuate constantly.

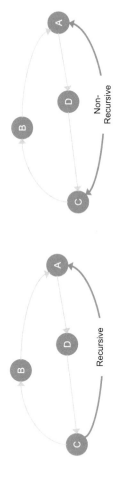

Fig. 8.10 Recursive versus non-recursive causal relationships

8.3.2.3   Metadata

A yet another logical data enhancement step is to create *metadata,* somewhat tauto-logically defined as data about data. Whereas the earlier discussed bespoke aggre-gates and graph metavariables attempt to remedy record and variable count-related access and usability challenges, the goal of metadata is to address validity and reliability-related considerations. Simply put, the idea behind metadata is to answer a seemingly simple question: Are the available data robust enough to give rise to dependable insights?

Operationally, metadata offers a summary view of data contained in datasets, expressed in terms of key statistical descriptors such as mean and median as expres-sions of the average, minimum and maximum values as indicators of possible outli-ers and the proportion of missing values as a high-level assessment of usability. Although historically more familiar to academicians than to practitioners, the con-cept of metadata is gaining popularity among the latter as the amount and diversity of data contained in organizational repositories continue to grow and the task of keeping track of what data are available and usable is becoming ever more daunting. The need for metadata typically grows out of the realization that organizational data warehouses, even the well-annotated ones, i.e. those accompanied by clear data model descriptions and comprehensive data dictionaries, tend to be a hodgepodge of robust as well as sparsely populated, discontinued and definitionally amended (and thus informationally discontinuous) variables. As could be expected, the pri-mary value of metadata is geared towards enhanced data analytic efficiency and efficacy: Without metadata information, data utilization efforts can become riddled with time-consuming and confidence-shaking corrective rework, precipitated by the fact that various potential data deficiencies are often hard to spot without readily available guidance. Moreover, the addition of bespoke aggregates heightens the importance of metadata, as those computationally embedded enhancements need to be expressly delineated and defined to serve their intended purpose.

Although the type and the number of specific variables can differ considerably across datasets, the informational foundation of dataset-specific metadata is rela-tively constant. A general outline of a commonly used template is shown in Fig. 8.11.

The scope and the exact informational content of metadata can vary across types of data; in a general sense, structured numeric datasets lend themselves particularly well to metadata summaries (the sample template shown in Fig. 8.11 was in fact

Metavariable	Variable	Properties						
		Type	Mean	Std. Dev.	Median	Min	Max	Missing
X	X1	scale type	value	value	value	value	value	%
	X2							
	X3							
Y	Y1							
	Y2							
	Y3							

Fig. 8.11   Sample metadata template

developed with that general data type in mind); templates developed for unstructured texts data need to emphasize objectively definable characteristics of those datasets, which could include descriptors such as source (e.g. online product reviews, insurance adjustor notes, etc.), origin (e.g. Yelp, Manta, Foursquare, etc.), time period, unique record count, etc. Overall, the main focus of metadata is on the description of the basic properties of the focal dataset using a combination of actual (e.g. continuous vs. categorical variable 'type'), computed (e.g. mean, median, % missing) and appended (e.g. bespoke aggregates designated as 'metavariable' groupings in the sample template shown in Fig. 8.11). The overall goal is to provide as informationally complete as possible, explicit and unambiguous overview of potential informational value of individual datasets.

8.4 Algorithmic Decision-Making and Organizational Value

Value creation is a broad idea, even if its scope is narrowed to just organizational context; in a very general sense, it can be framed as a series of actions that produce net positive outcomes. To become more operationally meaningful, the process of value creation needs to be considered in the qualifying context of commercial and non-commercial entities. The former, which are largely for-profit, business enterprises, usually frame the notion of 'value' in the economic context of maximization of their owners' wealth, while the latter tend to see it through the prism of attainment of social (relationships of groups with society) or societal (functioning of organizational systems) goals. Of the two, economic value creation is conceptually more straightforward as it centres on the efficiency with which a firm can transform inputs, such as financial and intellectual capital, raw materials and effort into economic output; social value creation, on the other hand, tends to be comparatively more nuanced as it is built around harder to tangibly express ideas of social impact and social return on investment. While in many regards economic value is considered to be more quantifiable than social value, ultimately both are subjective and contextual to the extent that both are judged by their core constituents, and both are evaluated in the context of alternatives. More specifically, a new product or service that yields, for instance, 10% profit will be viewed as a success by someone expecting a 5% return, but the same 10% profit will be viewed as a failure by an investor expecting a 20% return; similarly, incremental improvements in currently available social services tend to be seen as less impactful than introduction of previously non-existent benefits. What does all that have to do with automated decision-making (ADM)? Quite a lot, actually. While technologically marvellous, ADM systems are ultimately a yet another source of value, which is to say that their appeal cannot be fully understood outside of the context of perceived user benefits; moreover, that evaluation also needs to take into account plausible ADM alternatives, which suggests that it is possible for algorithmic decision-making (ADM) to not just create but also to destroy organizational value.

In a more operational sense, ADM-related value creation (or destruction) is not just a function of the appropriateness and the efficacy of decision automation solutions. Just as important is organizational readiness, an umbrella notion that encompasses a wide array of factors ranging from the availability of critical resources, such as the supporting data infrastructure and the appropriate mix of skills and competencies, to the ability and commitment to change, which typically means organizational structures, behaviours and culture. It is common for organizations to embrace the idea of ADM while lacking the institutional will or discipline to translate that idea into an operational reality. Deeply ingrained ways of organizational functioning, inertia and resistance to change are some of the more common ADM implementation impediments; moreover, decision-making automation may evoke manifestly different reactions from distinct groups of organizational constituents. For example, what might be seen as a positive development by shareholders might be, and quite often is, seen as a threat by employees.

An even more operationally granular set of ADM value defining considerations that warrant close and careful attention can be grouped together under the general heading of human-computer interaction (HCI) design choices, discussed in more detail in the previous chapter. One particular aspect of HCI that has not yet been expressly addressed is not just the impact, but the very essence of what constitutes 'analytics' in the context of automated decision-making. While the idea of analytic automation is not new, it is typically approached from the perspective of 'how' data are analysed, with comparatively less attention paid to 'where' data are analysed. Supervised and unsupervised machine learning algorithms are nearly synonymous with big data analytics, but within the confines of IoT, the less talked about edge, gateway and cloud analytics are all important contributors to the overall digital value creation process, warranting closer examination.

8.4.1 Analytics: From Edge to Augmented

The previous chapter examined the idea of data-enabled creativity, encapsulated in the MultiModal Organizational Learning (MMOL) typology. Embedded in that expanded conception of creativity is the theme of fostering higher-level creative problem-solving: As informational automation is replacing many of the functions traditionally performed by humans while also adding new capabilities, new opportunities to advance the level of human sensemaking arise, with data as the springboard to new ideas. To be operationally meaningful, those abstract ideas need the qualifying context of the earlier mentioned 'where' of data analytics, as graphically summarized in Fig. 8.12.

It is now commonly accepted that human sensemaking operates at two, somewhat independent levels: the subconscious, instinctive 'snap judgement' and the

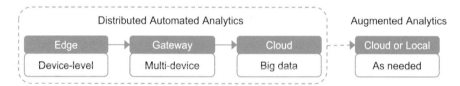

Fig. 8.12   The continuum of IoT analytics

conscious, deliberate rational thinking.[15] With that as the frame of reference, the
now rapidly automated sensemaking capabilities can be described as being comprised of notionally similar duality of (1) distributed automated analytics and (2)
augmented analytics, as shown in Fig. 8.12. Vast volumes of real- or near-real-time
data, often coupled with quick turnaround time response requirements that are common in numerous IoT control systems, necessitate automated processing of incoming data streams using distributed analytic model, which replaces the traditional
centralized collection and analysis methods with a decentralized model, where data
are analysed at a point of capture, such as a connected device or a sensor. The distributed analytic model is based upon the premise that the continuum of data analytic tasks can be spread across the continuum of data flows, starting with a singular
point of capture (edge), typically in the form of a device or a sensor, followed by a
natural grouping of sensors or devices (gateway) and lastly the centralized repository (cloud). More specifically, data first captured on the edge are analysed and
acted on (device-level edge analytics); oftentimes, individual devices or sensors
function as parts of a larger system which creates the next distributed data analytic
opportunity in the form of multi-device gateway analytics. Like a river, as data flow
and accumulate, the large, centralized, cloud-based data repositories provide the
most aggregate level of distributed analytics, which also supports 'as-needed' augmented data analytic capabilities.

Interestingly, while it seems that the overtly passive control-oriented edge·ana-
lytical applications are geared largely towards in-the-background infrastructure
support, such as intelligent heating and cooling systems, increasingly more complex
decision-requiring tasks are being handled by those point-of-capture data analytic
technologies, with numerous advanced edge analytical applications currently under
development. One such system entails the use of closed-circuit television (CCTV)
camera systems for emergency management and traffic control purposes. Numerous
CCTV cameras positioned at different points throughout a city street can be leveraged to quickly 'notice' accidents; upon spotting an accident, the nearby camera can
send emergency notification to local authorities and also enact emergency patterns
for nearby traffic lights (edge analytics). When shared with the larger network, the
accident information can then be used to automatically update traffic controls in the
accident-impact section of the city (gateway analytics). And lastly, when combined

---

[15] Formally, this phenomenon falls under a general umbrella of dual-processing theories of cognition, which recognize two types of information processing: conscious and subconscious (or
non-conscious).

with ever a broader set of data in the cloud, the accident-response and related details can be a source of invaluable planning insights. In the more general sense, using real-time information and working together, multiple CCTV cameras can react to the said accident by adjusting multiple patterns and durations of multiple traffic lights in an effort to counteract congestion buildup while also contributing to a larger knowledge base.

As evident in the above examples, the edge-gateway-cloud distributed model is operationally focused, meaning that pre-planned, automatically executed analyses are geared towards near-time, functional responses. However, as discussed in earlier chapters, a considerable part of organizational learning entails creative problem-solving, which encompasses not just reactive responses to emerging environmental developments but also proactive independent ideation. With that in mind, it is important to note that IoT-generated data will likely play an increasingly more star-ring role in both. To continue with the traffic monitoring example, looking beyond the immediately obvious emergency response and traffic flow re-routing benefits, the CCTV network-captured traffic pattern data can be invaluable for city planners working on long-term infrastructure planning projects. Moreover, as conveyed by Fig. 8.12, the informationally rich details can also be leveraged to support flexible, question-and-answer types of augmented analyses, graphically summarized in Fig. 8.13 (first shown in Chap. 1 and adapted to the traffic planning example here).

The one characteristic of augmented analytics that becomes immediately obvious upon closer examination of Fig. 8.13 is that it is an open-ended undertaking, where, in principle, any data can be analysed using any applicable techniques to answer any question. That over flexibility is also a manifestation of another key feature, which is informational parsimony. To be a source of organizational value, the pursuit of analyses-derived knowledge needs to be guided by well-defined informational needs to avoid falling into the trap of what is commonly known as the 'analysis paralysis', an endless cycle of analyses producing little-to-no action.

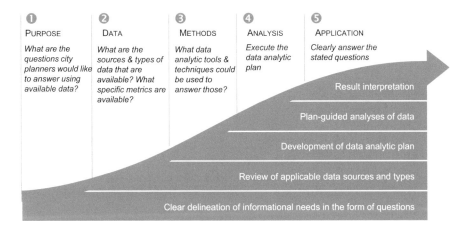

Fig. 8.13 Question-and-answer augmented analytic process

As summarized in Fig. 8.13, the first step in the open-ended augmented analytic process is to clearly delineate business questions, or more generically, questions that emanate from decisions or choices facing the organizational. In a general sense, those can be exploratory in nature, such as 'what is the optimal length of traffic signal during rush hours?', or confirmatory, such as 'is there an association between the length of traffic signals and the frequency of accidents?'. Once informational needs have been identified, an examination of what data are available and can be used needs to be undertaken, followed by a careful review of applicable methodologies that could be used to address the stated informational needs. Next comes the development of an explicit data analytic plan, the overt purpose of which is not only to guide the ensuing data analytic work but also to facilitate a holistic assessment of the stated knowledge creation initiatives from the standpoint of informational goals, data inputs, data processing and analysis throughputs and decision-guiding outcomes. Using the analytic plan as a guide, the next step of the process is to conduct the appropriate analyses and produce the initial data analytic results. Quite commonly, those initial data analytic findings may need considerable amounts of post-analytic refinement focused on reshaping of the often-esoteric statistical details into practically meaningful insights, all while being mindful of addressing validity and reliability considerations. The last step of the augmented data analytic process outlined in Fig. 8.13 is to expressly answer the stated questions. Depending on the type of questions that were asked and the particulars of data analytic means that were utilized, the analytic results-to-decision-guiding insight translation could range from a simple restating of results to drawing probabilistic inferences.

Reference

1. Zuboff, S. (2019). *The age of surveillance capitalism: The fight for a human future at the new frontier of power*. New York: Public Affairs.

Chapter 9
Letting Go: Knowing When and How

9.1 Brain as a Computer

It is almost intuitively obvious to most that even the slowest electronic computers are incomparably faster than human in-the-mind computing. While that is undeniably the case when comparing the speed with which even a computationally gifted human can, for instance, find the product of two large numbers, it is not at all so in a more general sense or when all, meaning conscious and subconscious, brain processing prowess is taken into account. When considered from the all-in perspective, the opposite is actually true, as it turns out that even today's fastest supercomputers cannot keep up with human brain. It thus follows that understanding of human brain's functioning is critical to building and sustaining robust organizational learning capabilities.

In a physical sense, human brain can be described as approximately three pounds of soft and highly fatty (at least 60% – the most of any human organ) tissue, made up of some 100 billion nerve cells known as neurons, the totality of which comprises what scientists refer to as 'grey matter'. Individual neurons are networked together via exons, which are wirelike connectors numbering in trillions (it is believed that each neuron can form several thousand connections, which in aggregate translates into a staggering 160+ trillion synaptic connections) and jointly referred to as 'white matter'. Function-wise, grey matter performs the brain's computational work, while white matter enables communication among different regions of the brain which are responsible for distinct functions and where different types of information are stored; together, this exon-connected network of neurons forms a single-functioning storage, analysis and command centre, which can be thought of as the biological human computer.

Although billions of cells linked by trillions of connections make for a super large network, if each neuron was only capable of storing a single piece of information, the entire human brain network would only be capable of a few gigabytes of

A. Banasiewicz, *Organizational Learning in the Age of Data*, EAI/Springer Innovations in Communication and Computing, https://doi.org/10.1007/978-3-030-74866-1_9

storage space, or roughly the size of a small flash drive. However, research suggests that individual neurons 'collaborate' with each other so that each individual cell helps with many memories at a time, effectively exponentially increasing the brain's storage capacity from just a few gigabytes (one neuron = 1 piece of information) to around a million gigabytes or 2.5 petabytes (1 neuron = multiple pieces of information). Mechanics of that collaboration are complex and not yet fully understood, but they appear to be rooted in neurons' geometrically complex structure characterized by multiple receptive mechanisms, known as dendrites, and a single, though highly branched outflow (called axon) that can extend over relatively long distances. To put all of that in less abstract terms, the effective amount of human brain's raw storage makes it possible for the brain to pack enough 'footage' to record roughly 300 years of nonstop TV programming, or about three million individual shows, which is more than enough space to retain every second of one's life. Moreover, according to the emerging neuroscientific research consensus, while the brain's storage capacity can grow, it does not decrease. What drops off, at times precipitously, is the retrieval strength, especially when memories – including semantic and procedural knowledge critical to abstract thinking – are not reinforced. It is all a long way of saying that while electronic computers have, in principle and in practice, infinite amount of storage as more and more external storage can be added, human brain's storage capacity, though limited, is nonetheless sufficient.

When expressed in terms of raw computing power, as measured in terms of the number of calculations performed in a unit of time, electronic computing devices appear to be orders of magnitude faster than humans. However, as noted earlier, that conclusion is based on a biased comparison of the speed of rudimentary machine operations compared to the speed of higher-order, or largely conscious, human thinking. Abstracting away from deep learning-related applications (which entail layered, multidimensional machine learning discussed earlier), machine-executed computational steps can be characterized as one-dimensional and explicit, which is to say they tend to follow a specific step-by-step logic built around sequential input-output processes. In contrast to that, human brain's computation is dominantly non-explicit and multidimensional, which means that the bulk of computations performed by the brain are running subconsciously in the 'mental background'. Stated differently, while it is tempting to compare the speed with which a human can consciously execute a specific computational task to the speed with which the same task can be accomplished by a machine is tantamount to comparing the speed of high-order human reasoning with rudimentary machine-based computation. And while such comparisons are tempting, it is nonetheless a bit like comparing the amount of time required for an author to write a captivating novel to the amount of time required by a skilled typist to re-type the content of that novel.[1]

---

[1] Interestingly, in a 2014 experiment a group of clever Japanese and German researchers managed to simulate a single second of human brain activity using what was then the fourth fastest super-computer in the world (the K Computer powered to nearly 83,000 processors) – it took that super-computer 40 minutes to complete the requisite calculations. In other words, that which was accomplished in just a single second by the roughly 3 pounds of grey and white matter took the 100

It all comes down to efficiency. The fundamental computer design separates storage and processing, which is not the case with human brain where the same interconnected areas can simultaneously store, analyse and redistribute information. As a result, a process that might take a computer a few million steps to complete can be achieved by a few hundred nearly instantaneous neuron transmissions in human brain. The amazing processing efficiency of human brain is perhaps most vividly illustrated by the amount of required energy: when performed by the world's fastest supercomputers, a computational task, such as the simulation of 1 second of brain's activity outlined in footnote #1, consumed an amount of energy about equal to what was required to power a sizable building; the same task performed by the brain would only consume about what it takes to light a single light bulb.

9.1.1 The Decision-Making Brain

Before delving into finer details of decision-making, it is important to expressly frame the core notion of *decision*. From the perspective of cognitive functioning, decision can be characterized as a commitment to course of action, intended to serve a specific purpose or accomplish a specific objective. It can be seen as a product of implicit, or subconscious, and explicit, or conscious, cognitive processes that are broadly characterized as boundedly rational[2] or, simply put, imperfect. Implied in that characterization is an element of indeterminacy of the mental heuristic processes and proximal mechanisms that produce judgements. Human sensemaking is not just paradoxically complex (how does one make sense of one's sensemaking?), but it is also shaped not only by cognitive but also by emotional, cultural and social factors discussed in Chap. 3.

And yet, making decisions is an inescapable aspect of not just organizational functioning but living in general; interestingly, it has been estimated that a person makes on average about 35,000 decisions a day. If that estimate seems exceedingly large, that is probably because many choices are so routine or unordinary that they simply do not seem like 'real' decisions, as illustrated by mundane choices required to, for instance, cross a street. Thus in a more general sense, decisions can vary significantly in terms of frequency, importance and difficulty. Many everyday choices, such as street crossing, are routine and important (given the potentially dire consequences), but typically fairly easy because, under most circumstances, the available information is nearly perfect, the amount of time is usually adequate and

---

or so ton device made up of roughly 83,000 processors 40 minutes. In a more general sense, the current estimates suggest that if the human brain were a computer, it could perform about a billion billions (a quantity known as quintillion, or 10^{18}) of calculations per second, which is on the next order higher than the fastest supercomputer (which, as of the end of 2020, is Fugaku at the RIKEN Center for Computational Science in Kobe, Japan).

[2]Bounded rationality is a notion suggesting that rational decision-making is constrained by the tractability of decision problems, cognitive limitations and time constraints.

requisite cognitive processing is straightforward. In a sense, those do not seem like 'real' decisions, in contrast to which choosing a college, for example, can be characterized as non-routine, important and complex and thus more in line with the intuitive connotation of the notion of decision.

An aspect of cognitive functioning, decisions are ultimately a product of brain's computations, and as astounding as human brain's computational capabilities are, they are ultimately finite, which at times necessitates information processing trade-offs. As a result, when facing prohibitively large amounts of stimuli, the brain will often make use of *decision heuristics*, or shortcuts, to quickly reach the needed conclusions with a minimal commitment of cognitive resources. While, as discussed earlier, human brain has, in principle, enormous computational capacity, the amount of information that can be cognitively processed at any given time is quite limited, a phenomenon known as channel capacity.[3] More specifically, as originally suggested in 1956 by G. A. Miller, an average person can simultaneously process only about 7 ± 2 of discrete pieces of information in his/her working memory, which is due to limitations of short-term and working memory, a phenomenon not unlike limits of random access memory (RAM) in computers, or physical limits of a simple container.

There are numerous situations in which heuristic-produced snap judgements can work quite well, but there are also situations when cognitive shortcuts can to lead to what may turn out to be unwarranted conclusions or actions. According to cognitive psychologists, the difference maker is cognitive bias, defined earlier as systematic pattern of deviations from norm or rationality in judgement. Not only are there many manifestations of cognitive bias (likely more than a hundred), biased-expressed warping of rationality takes place at a subconscious level, which makes rectifying the effects of cognitive distortions all that much more difficult. That is not all, however. Unlike computers that 'remember' all information stored in them equally well at all times, the brain does not. Another inner information processing characteristic know as brain plasticity, or the brain's persistent self-rewiring, renders older, not sufficiently reinforced memories progressively fuzzier and more difficult to retrieve. One of the direct consequences of that is that human recall tends to be incomplete and selective – for instance, positive events are generally remembered in more detail than negative events, and more recent events appear to be more significant and thus likely to recur than older ones. Moreover, details of individual events are often filled in after the fact, leading to self-deception, a situation which is further exacerbated by the almost instinctual inclination towards generalizing from (often) non-representative anecdotal evidence. Further chipping away at the representativeness of one's self-captured evidence is that, even under the best of circumstances, the totality of one's individual, subjective experiences typically constitutes only a small and likely non-representative sample of such experiences. In other words, one person's slice of reality is just that – a set of experiences encountered

---

[3]As used in computer science and information theory, it refers to the upper limit of the rate at which discreet amount of information can be reliably transmitted over a noisy communication channel. Originally proposed by C. Shannon in 1948, as a part of his information theory.

and encoded (into memory) by a particular person, which may differ substantially from experiences of others. And if all of those information processing hindrances were not enough, cognitive psychologists have still more bad news.

9.1.1.1 Food for Thought

Two similarly credentialled business consultants, Louise and Todd, were hired to independently determine the most likely causes of a brand's customer attrition problem. After a few weeks of pouring over customer and sales data, conducting customer interviews and reviewing the brand's current customer relationship management practices, the consultants presented their findings to their client. Interestingly, though both manifestly used the same informational inputs, the results of their analyses were quite different: Louise's findings largely paralleled the beliefs currently held by the brand's management team, while Todd put forth a new explanation, which effectively disputed the core aspects of the current management's beliefs. Which of the two competing sets of conclusions is the organization more likely to embrace?

According to a psychological theory of self-affirmation, information that is congruent with one's pre-existing beliefs is more likely to be accepted than information that contradicts those beliefs. The theory postulates that the core reason for that self-serving choice is instinctual self-defence, particularly in the form of protection of self-integrity, which compels managers (and others) to shield their beliefs simply because they see themselves as competent professionals. Though hypothetical, the case of Louise and Todd is emblematic of those types of situations: explanations that affirm whatever happens to be the accepted view are more likely to be positively received and, ultimately, believed; those that contradict the currently held beliefs are likely to face heightened scrutiny aimed at anything from the underlying data to analytic reasoning. It is important to note, however, that while such scrutiny can be overtly directed towards informational contents of the accepted view-bucking explanation, the theory of self-affirmation suggests that it is in fact the disconfirmation of existing beliefs that triggers an almost automatic 'defend the accepted view' response.

Often characterized as confirmation bias, that cognitive instinct compels, for example, more than 90% of college professors to believe themselves to be better at their jobs than their colleagues and 25% of students to believe themselves to be in the top 1% in terms of their ability to get along with others [1]; it is also the reason gamblers swear by their 'hot' and 'cold' streaks and alter their bets accordingly, or the reason sky gazers see faces of God in cloud formations, etc. And then there are subjective prior conceptions known as mental models (discussed at length in Chap. 3), or abstract representations of the real word that aid the sensemaking aspects of cognitive functioning, particularly reasoning about problems or situations not directly encountered. All considered, self-affirmational tendencies compel individuals to look for familiar clues, which in turn predisposes decision-makers to affirm the familiar, rather than to embrace the new.

When considered in a broader organizational setting, additional sensemaking-influencing effects come into view. Groupthink, social cognition and organizational structure and culture are just some of the common group dynamics that further shape individual-level sensemaking mechanisms. Not to be overlooked, so are all manner of management miracle cures ranging from one-size-fits-all frameworks promising to solve even the most intractable problems to unsubstantiated axioms and anecdotes, many of which pervasive enough to unjustifiably achieve the status of immutable facts. The now largely discredited 'emotional intelligence', 'business process re-engineering', 'employee engagement' and numerous other management frameworks once exerted profound impact on organizational sensemaking, in spite of being nothing more than transitory collective beliefs built on analytically flawed leaps of faith. The same holds true for some of the best-known business truisms: Walt Disney's famous axiom 'if you can dream it, you can do it', the widely repeated expression 'sky is the limit' or the enduring belief that 'thinking big' leads to big outcomes are just a few of the nearly axiomatic maxims that still permeate organizational psyche, even if empirical evidence suggests that, for instance, dreaming big can actually be detrimental to individuals' performance, even to their career success. (That is because adopting unrealistic fantasies of success can lead to becoming accustomed to the fantasy as it is was already a reality, which in turn can lead to overconfidence and failure to commit the effort required to actually achieve big goals.)

Another important (in the sense of being commonly encountered) contributor to cognitive reasoning is the use of anecdotal evidence – a good story is a verbal equivalent of a picture that is worth the proverbial thousand words. Anecdotes, which are brief, often amusing or humorously revealing accounts of a person or an event, have the power to capture attention and deliver a clear message. Widely used to convey abstract ideas underpinning many core human values such as morality or justice, anecdotes have been used by teachers, preachers and countless others to deliver important messages, as it has been long known that listeners are more likely to remember notable rather than typical examples. But that, however, can also be a problem.

Any informal account or a testimonial can attain the status of anecdotal evidence. In most situations, such evidence is limited to either a single case (an event or an individual) or a small set of cases, and it is intended to communicate a specific story, often in the form of a personal account or testimony. For instance, a person may argue against compulsory use of seatbelts by citing an example of an acquaintance who was not wearing a seatbelt while driving and got in a head-on collision resulting in being ejected from the car, which shortly thereafter burst into flames. Using that example as anecdotal evidence, the individual may then argue that seatbelt laws should be abolished. Logically, that would appear to be a sound inference because anecdotal evidence does not need to meet any specific standards of appropriateness, or to use more scientifically clear language, it does not need to exhibit satisfactory degree of validity and reliability. In most instances, to be appealing, anecdotal evidence is expected to be on-point, which is to say it needs to depict a topically related person or situation, in addition to also being interesting, which is the crux of the problem. It is the notable or atypical cases or situations that are interesting and memorable, but being 'atypical' renders such examples non-representative, which can be problematic from sensemaking point of view. Unorthodox business strategies that led to successful outcomes are detailed in case studies, and a good part of business education is focused on dissecting those atypical cases, ultimately further conditioning future business managers to trust anecdotal evidence; at the same time, failed unorthodox strategies receive comparatively little attention and tend to be quickly forgotten. Unique events, maverick leaders and norm-bucking choices are all so much more 'colourful' and thus captivating than steady, systematic and persistent learning, just as the overnight sensation, rags-to-riches success stories are so much more captivating than a slow and predictable building of a successful career. And yet, it is the persistent and dependable knowledge that forms the core of sound organizational sensemaking.

9.1.2 Objective Evidence Versus Subjective Beliefs

The conjectures and refutations (also referred to as hypothesis testing) model of learning discussed in Chap. 3 is a widely used mechanism for advancing the state of scientific knowledge. Within that framework, scientific knowledge is built by means of empirical, meaning using objective evidence, testing of prior beliefs, which could encompass anything ranging from formal scientific theories to a subjective 'hunches'. However, as discussed in the previous section, in applied organizational settings objectively derived insights often compete with subjective beliefs, especially those in the form of deep-rooted mental models, cherished axioms or compelling anecdotes. A nearly instinctive reaction might be to assert that objective evidence should simply override subjective beliefs, but there are reasons to believe that mechanics of the objective-subjective duality are more complex than just picking the 'more correct' of the two. Recent neuroscientific studies suggest that cognitive processing of information requires the brain to hold some beliefs about reality, as absent those we would be unable to learn from available informational inputs, which suggests that blending of subjective prior beliefs and objective evidence is more plausible than simply swapping out the former for the latter. It thus follows that, as illustrated earlier by the hypothetical case of the two consultants (Louise and Todd), even when situational factors and demand effects are controlled for, the two similarly qualified individuals utilizing essentially the same objective information may arrive at materially different conclusions, primarily because of the impossible to see or examine differences in their subjective mental models.

In its purest form, management by the numbers (see Chap. 6 for a more in-depth discussion) assumes no underlying prior beliefs; instead, it approaches the pursuit of decision-guiding insights as a purely exploratory endeavour. The resultant knowledge is built 'from scratch', cobbled together from individual insights extracted from available and applicable data in a largely dispassionate manner, which is a compelling idea, but one that is at odds with the aforementioned neuroscientific evidence. And just as machine learning algorithms are not truly objective, as their inner workings reflect perspectives and choices of their designers, management by the numbers is similarly coloured by perspectives of individual data users. When looked from that perspective, the very idea of objectivity becomes a little questionable.

Or perhaps the idea of objectivity is overstated? After all, the goal of organizational management is to make effective choices, ones that will yield the greatest benefit to organizational stakeholders; if information-producing mechanisms are not truly objective, yet are effective, does that matter?

Still, the 'it is all about the data' mindset is certainly in vogue. Organizations are awash in data, but many are quipping about being data-rich but knowledge-poor. An organization with lots and lots of data is a bit like a farmer with lots of usable land – it is certainly a good start, but not a guarantee of a good harvest. As clearly illustrated by successful organizations, the biggest single source of informational advantage is the superior knowledge creation know-how. Overall, the most significant factor that consistently explains why some data-rich organizations are also knowledge-rich while other, equally data-rich and technologically enabled firms are comparatively knowledge-poorer is the advanced data analytical skillset of the former. At a time when data are ubiquitous and standard data processing tools yield informationally generic outcomes, it is the ability to go beyond the basic data crunching functionality that is the key determinant of the ultimate value of data.

And that, for many organizations, is a problem. Though manifestly important, the knowledge creation know-how is arguably the least developed and certainly the least formalized aspect of the new, digital world. That may be surprising as, after all, the domain of quantitative analysis is a well-established, long-standing field of study. And indeed it is, but mostly in the academic, rather than applied sense. Similarly to a number of other fields of study, quantitative methods tend to be inwardly oriented and primarily focused on methods rather than outcomes. Those trained in that domain tend to acquire substantial amounts of abstract knowledge, but comparatively little understanding of application-related knowledge. The emergence and rapid proliferation of 'business analytics' and 'data science' curriculums, coupled with growing emphasis on hands-on experiential learning, is beginning to change that, but the process is slow while needs are urgent. And even if and when more and better trained analysts are mining the vast quantities of organizational data, the problem of informational overload will not disappear – in fact, it may grow as more analysts doing more analyses will translate into higher volume and wider diversity of information. It is already happening: the avalanche of data is triggering the avalanche of information, which implies that the earlier discussed analytic literacy (see Fig. 1.6) and digital literacy (see Fig. 5.1) should be seen as building blocks of the broader domain of informational literacy.

9.2   Informational Literacy

The long-standing notion of *literacy* encompasses the ability to read and write and the ability to use numeracy, while the more recent concept of *analytic literacy* entails the ability to comprehend and construct rational arguments. A still more recent notion of *digital literacy* emerged to capture the ability to use information and communication technologies to find, evaluate, create and communicate information, and the concept *data literacy* has been suggested as a way of capturing the ability to utilize data; lastly, the notion of *data analytic literacy* was put forth in this book with the goal of expressly accounting for the ability to extract meaningful insights out of raw data. While potentially confusing, the expanding conception of what it means to be 'literate' can be seen as a reflection of scientific and technological progress – in fact, the term *multiliteracies* emerged in the 1990s as an attempt to capture the expanding expressions of the basic idea of literacy. Still, the inherent overlap among the distinct but not exclusive variants of the general idea of literacy suggests a need for some definitional simplification.

There are two themes that cut across the various aspects of the general idea of literacy: information and communication. Whether it is the most rudimentary reading and writing skills or the ability to conduct advanced analyses of data, the goal is to extract information contained in whatever form it was encoded, which could be text produced by a writer or data captured at point of sales. Before the advent of modern electronic infrastructure, data were synonymous with manual recordings of facts; thus the ability to read, write and understand numbers was all that was required to be literate; the emergence of electronic data vastly expanded the scope of what constitutes data, which in turn broadened the literacy-related skillset; the still more recent emergence of new media and machine learning systems broadened the scope of literacy even further. Throughout it all, however, the goal remained the same: to extract, comprehend and communicate information, from which it follows that analytic literacy, digital literacy, data literacy and data analytic literacy can all be seen as manifestations of a single metaconcept of *informational literacy,* graphically summarized in Fig. 9.1.

Comprised of four distinct dimensions of digital, data, analytic and communication facets, the high-level conceptualization graphically summarized in Fig. 9.1 is meant to encompass all core predicates of informational literacy. Building on the earlier overview of analytic and digital literacies, the contribution of each of the four distinct sub-domains to the broad notion of informational literacy is discussed next.

9.2.1   The Digital Dimension

While the notion of digital literacy traces its roots to the 1980s' emergence and subsequent rapid proliferation of personal computing, it was the birth of the World Wide Web in the early 1990s that truly began to shape that aspect of informational

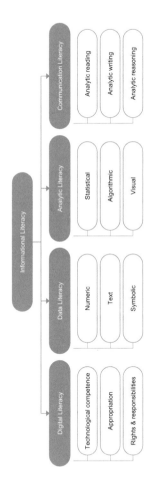

Fig. 9.1 Informational literacy metamodel

literacy. It is commonly defined as the ability to use information and communication technologies to find, evaluate, create and communicate information, when considered within the confines of organizational learning-focused Informational Literacy Metamodel graphically summarized in Fig. 9.1; digital literacy can be framed as a set of technical, procedural and cognitive skills and competencies that manifest themselves in technological competence, appropriation and rights and responsibilities.

Technological competence entails functional understanding of man-made systems designed to facilitate communication- and transaction-oriented interactions, including the manner in which those systems capture, store and synthesize data. At a more granular level, it can be seen as the ability to evaluate and use technology as a means of accessing and interacting with data or information; moreover, it entails understanding of planned and unplanned consequences of technology and the interrelationships between technology and individuals and groups.

Appropriation is the ability to meaningfully sample and mix media content in a way that makes positive and material contribution to stated informational objectives. Though overtly reminiscent of the notion of digital nativity (discussed in Chap. 5), which is a relatively recent label used to describe the general information technology leanings of generations born into the Digital Age, the notion of appropriation is phenomenologically distinct and different from the idea of digital nativity: It is focused on cognitive discernment, in the sense of rational usage, rather than the mere acceptance of digital tools.

The third dimension of digital literacy encapsulates understanding of rights and responsibilities that are connected to using and contributing to broadly defined digital content. On the one hand, it is somewhat more intuitively obvious than the other two digital literacy dimensions as it stipulates awareness of applicable laws and regulations, perhaps best exemplified by privacy laws, as well as the more technically involved proactive data protection and usage considerations, as illustrated by the earlier discussed differential privacy. On the other hand, however, it is a lot more interpretation prone to the extent to which it also encompasses the comparatively more nuanced and individually interpreted ethical and moral considerations. Moreover, in the always important context of organizational structure- and culture-shaped group dynamics, the broad notion of rights and responsibilities also encapsulates the important-but-elusive idea of social consciousness, broadly defined as 'doing the right thing'. All considered, the rights and responsibilities dimension of digital literacy is ultimately highly situational and nuanced.

9.2.2 The Data Dimension

Fuel of the Digital Age, data are nowadays so varied and so ubiquitous as to almost defy meaningful categorization. Given the central importance of data to modern organizational learning, as described in this book, a number of data-related considerations have been discussed in the previous chapters including data types (structured vs. unstructured, numeric vs. text, transactional vs. social vs. online search vs. crowdsourced vs. machine to machine), data due diligence, data feature engineering and data security – when considered from the standpoint of informational literacy, it is instructive to look a data from the perspective of core data forms: numeric, text and symbolic. It should be noted that the ability to read and write, which is the oldest and still the most fundamental aspect of literacy, clearly implies the ability to communicate meaning using text, numbers and symbols; the emphasis here is on familiarity with how those distinct means of conveying meaning are used as inputs into computerized analyses.

Perhaps the most intuitively obvious type of data is *numeric*, which, when considered from the standpoint of communications theory, are signals encoded using numbers and representing a measurement of an underlying quantity. The first part of the definition is nearly tautological, but the second part contributes an important qualifier: Only numbers that represent a measurement can be considered 'numeric' (data). That is an important distinction because while numbers may not make the most informative labels, there is no reason why manifestly non-quantitative values, such as 'gender', could not be encoded using numbers (e.g. '1' to denote 'female' and '2' to denote 'male'); in other words, while all numeric data are encoded using numbers, not all value encoded using numbers are numeric data. For decades, numeric data, and particularly the computer analysis-friendly structured numeric data, have been the staple of organizational analyses so, at least notionally, it is the most familiar or at least familiar-sounding of the three general data forms.

Text data, or stated differently, *text* seen as data, seen again from the perspective of communications theory are signals encoded using letters. In the vast majority of cases, those letters are organized into words which communicate meaning, but from the point of view of computer processing, any string of letters or letters and other characters such as digits or special characters constitutes text.[4] As the dominant form of non-verbal human communication, text data are as easily understandable to humans as binary strings are to computers and vice versa – are as difficult to 'understand' for computers as binary strings are to humans, which is the essence of that aspect of data literacy. Text data are explosively high-dimensional,[5] semantics and syntax are both critical to informational meaning-making and highly nuanced, and factors such as punctuation, misspellings, abbreviations, acronyms, etc., in addition to any implied, tacit meaning lacking any type of tangible encoding, all render text data enormously challenging to computer-analyse (and yet, by some estimates, upwards of 90% of the world's data are text). And lastly, a growing share of text data are verbal recordings, which adds a yet another level of data analytic difficulty. In

---

[4] In practice, non-word text, meaning a combination of letters or letters and other characters that lack semantic structure (meaning), is referred to as strings or alphanumeric values.

[5] To put that into perspective, in a sample of documents, where each document is k words long and each word comes from a vocabulary of m possible words, the resultant dimensionality will be k^m. It is easy to see that even when the said sample is modest, the resultant dimensionality will be staggeringly high.

short, understanding the key informational aspects of text data is the essence of that aspect of data literacy.

The third and final broad form of data, *symbolic*, is also the broadest category as, in principle, it encompasses any form of data that is not either numeric or text. Admittedly, such a broad characterization overlooks some otherwise meaningful differences, such as the distinction between a symbol used as an indication of an object and an image, which offers a direct visual depiction of that object. It thus follows that from a strictly definitional perspective, symbolic data category is highly heterogeneous as it encompasses distinct subtypes that reflect not only differences in the types of expression (e.g., an abstractly rendered symbol vs. directly interpretable image) but also in the types of media, which could be a static image, a voice recording or multimedia. Echoing some of the challenges associated with computer processing of text data, the essence of that aspect of data literacy is familiarity with the core facets of symbolic data and their key informational aspects.

9.2.3 The Analytic Dimension

Commonly seen as the centrepiece of informational literacy, analytic literacy gets at the core of what it takes to extract informative insights out of the ever-expanding arrays of data, ranging from the easy to work with structured numeric data to analytically challenging diverse arrays of unstructured data. In a very abstract sense, the essence of analytics is to find meaning in arrays of transactional, communication and other details. As graphically summarized in Fig. 9.1, there are three distinct, broad avenues for doing so: statistical, algorithmic and visual.

The oldest (dating back to at least the latter part of the eighteenth century, but with much earlier origins, as summarized earlier in Fig. 4.4) and thus the most established of the three aspects of analytic literacy, statistics is the '…science of learning from data, and of measuring, controlling, and communicating uncertainty[6]'. And as noted by John Tukey, the father of exploratory data analysis, 'statistics is a science… [that] uses mathematical models as essential tools'. It encompasses a wide array of approaches, methodologies and techniques ranging from relatively simple univariate descriptive analyses to comparatively complex multivariate predictive modelling techniques. When considered from the broader perspective of analytic literacy, and as discussed in Chap. 5, statistical proficiency entails, at a minimum, adequately effective competency with basic conceptual and computational tools of exploratory, confirmatory and predictive analyses. That said, it should be noted within the confines of organizational learning what constitutes 'adequately effective' statistical knowledge is also partly shaped by factors such as the organizational level of analytic maturity. All considered, the statistical facet of the analytic dimension of informational literacy is rooted in carefully considered use of

---

[6]The definition given in the April 6, 2012, editorial in journal *Science* by the then-president of the American Statistical Association M. Davidian and a former president of the International Biometric Society T. Louis.

well-defined mathematical models by sufficiently knowledgeable analysts seeking to extract probabilistic meaning out of data.

However, not all data and data utilization contexts lend themselves to statistical analyses. For example, statistical models are of little help in extracting general sentiments out of online product reviews because, as discussed in Chap. 5, unstructured text data (a typical online product review data format) simply do not lend themselves to such methods of analysis. That is where the more computationally intensive, mathematical optimization-based machine learning techniques, which comprise the algorithmic facet of the analytic dimension of informational literacy (see Fig. 9.1), are particularly beneficial. In contrast to statistical methods of extracting insights out of data which, at their core, can be seen as mechanisms designed to contribute to human judgement, modern-day machine learning algorithms are best seen as means of effectively replacing human judgement, in situations in which either the nature of data (e.g. extracting key insights out of millions of documents used as evidence in a large class action lawsuit) or the nature of decision-making (e.g. self-driving vehicles) renders that desirable or even necessary. It follows that when compared to statistical literacy, algorithmic/machine learning literacy is structurally noticeably different. When using statistics, human analyst makes the key outcome-shaping decisions, whereas when using machine learning algorithms, the bulk of those decisions are a product of (largely) hidden algorithmic logic, which has direct impact on required competencies. Hence the thrust of algorithmic competency implied knowledge is on understanding algorithm-specific input requirements, algorithm choice and result validation and interpretation. That said, similarly to statistical knowledge, environmental and organizational factors play an important role in determining what constitutes adequately effective algorithmic know-how.

The third and final facet of analytic literacy is visual data learning, discussed in more detail in Chap. 5. It can be seen as complementary to statistical and algorithmic modalities in two different ways: firstly, as an alternative where, for instance, a bar graph and numerically expressed frequency counts offer two different ways of communicating the same information, characterized earlier as informational visualization and secondly, as an enhancement where, for instance, mapping applications can greatly enhance the utility and the appeal of information (see Fig. 5.14), more generally known as scientific visualization. Not surprisingly, the visual facet of analytic literacy shares a number of overt similarities with statistical and algorithmic modalities: Like statistics, the choice of visualization types and many visualization design considerations, such as colours or features, are heavily analyst judgement dependent. On the other hand, similarly to machine learning algorithms, visualization-focused processing of data is highly automated, making heavy use of specialized software applications. In view of that, the core determinant of visual analytic literacy is tied to the degree of understanding of interdependencies between data types and visualization outcomes, manifesting itself in the degree of appropriateness of selected graphs, charts and other forms of visual data analyses.

9.2.4 The Communication Dimension

In recent years, the idea of *data analytic storytelling* has been gaining popularity, primarily as a reaction to growing recognition of how often potentially impactful data analytic outcomes are poorly communicated. To be fair, however, effective conveyance of data analytic outcomes can be a surprisingly challenging task. Imperfect data analysed using tools designed to produce approximately correct outcomes ultimately can only support nuanced conclusions, and validly capturing those nuances in a way that is easily understandable is full of not immediately obvious challenges. One of the most commonly encountered examples is the use of statistical significance tests in business analytics, as illustrated by the need to determine if observed differences are significant at a chosen level of statistical significance, such as 95% (or $\alpha = 0.05$). For instance, a marketing manager may want to know if the observed difference between two sets of response rates is significant at 95% level. Doing so calls for ascribing statistical significance to a point estimate (in the form of the difference between the two rates), which is something that, as discussed in the opening chapter, is methodologically invalid. What the manager could do, to be methodologically correct, would be to express the said response rate differential as a 95% confidence interval, but doing so would recast an informationally unambiguous value, such as 1.5%, into a comparative ambiguous range, such as 0.5–2.5%. So, the manager has a choice of either reporting the observed 1.5% differential without ascribing any statistical efficacy assurances or reporting the generally less practically palatable 0.5–2.5% range, but supported by a widely recognized measure of statistical efficacy assurance. That common dilemma captures the essence of the challenge associated with methodologically correct and informationally meaningful communicating of data analytic outcomes.

Embodied in the above reasoning are the three distinct aspects of the communication facet of analytic literacy: analytic reasoning, analytic writing and analytic reading. Analytic reasoning refers to the ability to evaluate available information in a consciously systematic and rational manner; it is often seen as the analog to intuition, which relies on automatic, often subconscious associations. In more explicit, information processing sense, analytic reasoning is purposeful and effortful, thus comparatively slow, in part due to being impeded by a myriad of factors ranging from individual, mostly cognitive limitations such as the size of working memory, to organizational and situational influences, such as resource or time constraints. And precisely because it is slow and effortful, analytic reasoning needs to be carefully developed in terms of focus (i.e. not getting sidetracked by tangential pursuits) and evaluation logic or mechanics. The overall process of developing disciplined and robust analytic reasoning capabilities is quite extensive and thus beyond the scope of this summary; that said, the 3E (empirical and experiential evidence) framework detailed in *Evidence-Based Decision-Making* [2] offers a comprehensive overview of the recommended approach and its operational mechanics.

Analytic writing, the second of the three aspects of the communication facet of analytic literacy, can be seen as a variant of the broader domain of technical writing used in various occupational fields such as computer and software engineering, robotics, finance or medicine, to name just a few. It is characterized by a combination of technical understanding (of the subject at hand) and a clear, direct and neutral style. In keeping with those general ideas, effective analytic communications need to reflect the earlier discussed nuanced character of data analytic outcomes, presented in a clear and dispassionate manner, so as to allow readers to correctly interpret available information, without any intended or unintended analyst bias.

Analytic reading, the third and final aspect of the analytic communication, reflects the ability to understand, summarize and paraphrase the often highly technical language of analytics. There is perhaps no better example than academic research studies, which often encapsulate pages and pages of methodological and related details; while beneficial from the perspective of peer review and replicability, the depth of methodological and related details has been a substantial obstacle to those studies' wider utilization. It follows that the ability to correctly and completely summarize the most informationally and usage pertinent details is an important manifestation of analytic reading capabilities. A yet another dimension of that competency is the ability to soundly summarize manifestly non-technical text. For example, each year, 300–400 companies traded on public US stock exchanges are sued by their shareholders because of incomplete, incorrect, misleading or untimely disclosures of pertinent financial information, a phenomenon known as shareholder litigation. The details of individual allegations are readily publicly available, which makes possible distilling of the often-voluminous legal arguments into informative summaries of the most often seen triggers – with that in mind, the ability to compress many pages of legal, financial and other details into a comprehensive and complete set of underlying litigation root causes exemplifies a yet another manifestation of analytic reading capability.

9.3 In Data We Trust

There are some hard to believe facts associated with folding an ordinary sheet of notebook paper. First, regardless of the size of a sheet, no one has been able to fold a sheet of paper more than 12 times.[7] However, what is even more extraordinary about that exercise is the height of the resultant stack. Starting with an appropriately sized sheet of ordinary notebook paper, folding it seven times (the number of folds once believed to constitute the upper limit) will result in a stack approximately equal in height to the thickness of an average notebook. Extra three folds will result

---

[7]The 12-fold threshold was reached using very long stretch of thin paper, resembling toilet paper, using what is known as 'single direction fold'. A number of people, however, questioned the validity of that outcome, believing that a proper folding approach entailed folding a sheet in half, turning it 90° and then folding it again – under those constraints, a single sheet of thin paper has been physically folded 11 times, with the first eight folds accomplished manually and the remaining three with the help of mechanical equipment (a steam roller and a forklift).

in the stack height about the width of a hand (thumb included), and additional four (for a total of 14 folds) would push the height of our stack to be roughly that of an average person. If one were able to continue to fold, the resultant heights would quickly enter the realm of unimaginable: Seventeen folds would produce a stack the height of an average two-storey house; extra three, for a total of 20 folds, would yield a stack reaching approximately a quarter of the way up the Sears (currently known as Willis) Tower, the landmark Chicago skyscraper. If folded over the total of 30 times, the resultant stack would reach past the outer limits of earth's atmosphere, and lastly, if folded over 50 times, an ordinarily thin (albeit extraordinarily large in terms of area to allow such a large number of folds) sheet would produce a stack of paper reaching all the way to the sun, which is roughly 94 million miles away. Yet that is not all – a little more than doubling the number of folds, for a total of 103 total folds, would result in a stack of paper thicker or taller than the observable universe, which measures 93 billion light-years (and given the speed of light of approximately 186,000 miles per second amounts to an absurd height of 5.4671216×10^{23} miles). A standard 1/10 millimetre-thick sheet of paper folded over a seemingly small number of times…even though the underlying math can be shown to be correct,[8] are those conclusions believable?

Years of education tend to imprint human minds with a variety of abstract notions while also conditioning human psyche to accept as true a wide range of ideas. So long as those scientific and other truths do not contradict one's experience-based sense of reasonableness, most educated people are willing to accept even somewhat far-fetched ideas. However, when that is not the case, that is, when a particular claim violates what one deems to be reasonable, the result is cognitive dissonance. In those situations, many, perhaps most people will find it difficult to accept intuitively unreasonable assertions, even if the underlying rationale and empirical data and methods both seem correct. It is simply difficult to embrace a conclusion that does not make sense, and that is precisely the case with the paper folding exercise. That particular example is an application of a well-known mathematical concept of exponential growth, which is a phenomenon where the rate of growth rapidly increases as the quantity (e.g. the above stack of paper) gets larger. Since it is a well-defined mathematical notion, its value can be computed without the need for physical measurements, which is the reason the paper folding example is estimable, though clearly practically impossible to execute. It is likely that most of the readers of this book have been introduced to the concept of exponential growth at one or more points in their educational journey, and it is probably not a stretch to assume that while learning that concept, most found the idea of exponential growth to be intuitively clear and reasonable. Finally, once properly explained, the computational logic and mechanics also likely made sense. Yet when put to a bit of an extreme test, the otherwise acceptable and reasonable idea now yields conclusions that many – the writer of this book included – find unimaginable and impossible to grasp. It

---

[8]This is an example of exponential growth; the underlying equation can be expressed as $W = \pi t 2^{\frac{3}{2}(n-1)}$, where W is the width, t is the thickness and n is the number of folds.

simply does not make sense that folding a thin sheet of papers a relatively small number of times could produce such a staggeringly high stack.

The above example underscores both the value and the challenge associated with using data analysis-derived knowledge as evidence. It is generally easy to accept findings that fall in line with one's existing beliefs, though quite often little incremental value comes out of such 'discoveries'. It is altogether a different story when data analytic findings contradict one's prior beliefs, which frequently prompts almost instinctive, doubt-ridden reactions. Is there a problem with the data? Is the approach flawed? Are there any computational errors? To be fair, data can be corrupted, an approach can be flawed and logical, methodological and other mistakes are always possible. But what if data, methods and logic are all sufficiently robust? Doubts can linger, and truly provocative findings that could pave a new avenue for organizational thinking can instead join that dusty repository of insights that were just too far-fetched to be believable. What, if anything, can be done to remedy such situations?

Awareness-building reminders of the earlier discussed reason-warping cognitive bias, recall-altering brain plasticity and cognitive attention-limiting channel capacity need to be in full view of decision-makers, because those sensemaking obstructions operate at a cognitively imperceptible level. In the United States, business and other organizations post reminders of anti-discrimination or anti-harassment laws – why not do the same with factors that can subvert the basic sensemaking abilities? As is well-known, the first step towards addressing a problem is recognizing and acknowledging that there is a problem.

Making the potentially adverse impact of reason-warping factors more visible, in the manner of speaking, would also likely have a positive spillover effect of indirectly making the case for data-driven decision-making. One of the common bases for rejecting predictive models, for instance, is that at least some of the predictions 'do not make sense'; although certainly not always but still often enough, the problem with that line of reasoning is that the implied sensemaking benchmark in the form of one's intuition may itself not be informationally robust. Moreover, many well-established scientific principles, such as quantum superposition,[9] can be maddingly difficult to make sense of and thus to accept as being intuitively correct. Still, notions such as quantum superposition have not only been experimentally confirmed but are also in fact driving advanced technologies, such as quantum computing. One of the core barriers to truly unbound creative thinking and problem-solving is the inability, or the unwillingness, to accept that to be true or correct, an inference or an idea does not necessarily need to make sense.

---

[9] One of the key postulates of quantum mechanics, which is the study of sub-atomic structures, that posits that in a quantum system, particles can exist in several separate states at the same time, and each state can be expressed in terms of probabilities. If that idea were to be applied to the human realm, it would be akin to saying that a person can be simultaneously dead and alive, which would imply that human existence can be expressed as a sum of non-zero probabilities of being simultaneously dead and alive.

9.4 Probabilistic Thinking

Organizational decision-making commonly entails making choices under conditions of uncertainty, defined as a general state of ambiguity regarding the future, which typically connotes non-specific opacity or insecurity regarding the future. When painted with such broad strokes, the notion of uncertainty is quite vague; thus it might be beneficial, at least within the confines of organizational sensemaking, to reframe the idea of uncertainty in the context of the degree of knowability. When considered in that context, uncertainty can be seen as a product of lack of information or the inability or unwillingness to make use of available information. For example, an insurance claim manager trying to assess the potential future cost of a particular claim may not have pertinent benchmarking data, or s/he may be unwilling to use available information in the form of probabilistic projections. While the former is often a possibility as low information decision contexts are relatively common, there are also numerous examples of reluctance to trust available information, particularly in the form of probabilistic data analytic conclusions. The latter tends to be particularly pronounced in organizational sensemaking situations that historically relied heavily on tacit knowledge, such as insurance claim management, where to be trusted, evidence needs to make (intuitive) sense.

And therein lies the problem: the oft-implied choice between intuition and data. Ironically, many, perhaps even most of those mental impulses that manifest themselves as intuition are themselves products of inescapably probabilistic subconscious information processing heuristics. To continue with the insurance claim processing example, claim handlers who rely on their experience to estimate the potential future cost of individual claims are effectively using a small handful of cues, in the form of select claim details, to draw probabilistic inferences; just because it all happens in that fuzzy, informal mental processing realm does not mean it is notionally different from formal statistical estimation. In the end, the difference is not so much the underlying mechanics, but rather the thoroughness with which the bundle of attributes that comprise a particular insurance claim is evaluated.

Becoming a thoughtful and confident user of probabilistic evidence calls for rudimentary competency in three, somewhat distinct dimensions of probabilistic thinking: computational, inferential and evidentiary. *Computational thinking* reflects the understanding of general data sourcing, data feature engineering and curation processes that produce what ultimately becomes the informational foundation from which probabilistic estimates are derived. *Inferential thinking* is a manifestation of familiarity with scientific means of extracting valid and reliable insights from data, which encompass the overall logic of drawing inferences from data, as well as the core tools used in that process, such as hypothesis testing. And lastly, *evidentiary thinking* encapsulates the ability to use analysis-derived insights as evidence in the decision-making process.

While it is generally obvious why a user of inferences drawn from data should have at least a general understanding of the logic of statistical inference (inferential thinking), and the manner in which such information should be used (evidentiary thinking), it might be less obvious why that individual should also be familiar with data sourcing and curation processes (computational thinking). The reason is the earlier discussed believability of evidence. Data analytic results that confirm one's prior beliefs do not pose a threat to one's self-concept and thus tend to be easily assimilated; in fact, it is rare to see instances where the validity of belief-confirming information is questioned. However, situations in which new information contradicts one's prior beliefs often give rise to cognitive dissonance, which then tends to lead to questioning the efficacy of data and/or data manipulation processes; it follows that the most direct and effective dissonance mitigation strategy is computational thinking proficiency.

9.4.1 Understanding Probability

Prior to the seventeenth century, the term 'probable' was synonymous with 'approvable' and was used in reference to opinions or actions reasonable individuals would hold or undertake. The roots of what today is known as a probability theory, or mathematically explicit treatment of chance, can be traced back to the sixteenth century and the first mathematical analysis of games of chance undertaken by a noted mathematician (and a gambler) G. Cardano. Initially focused on the more mathematically manageable discrete events, Cardano's ideas were later expanded to also include the more analytically complex continuous variables, largely through work of Pierre-Simon Laplace; still, the current conception (and computations) of the probability theory is based on the seminal work of A. Kolmogorov [3].

There are two distinct approaches to probability estimation: Bayesians and frequentist. The former is named after Thomas Bayes, a clergyman and a statistician who developed what is now known as Bayes' Theorem, which expresses probability as a measure of the state of knowledge at a point in time. More specifically, it frames probability to be a function of subjective prior beliefs (of the decision-maker and/or decision influencers) and of past outcomes encoded in objective data; as such, it is particularly well suited to problems characterized by sparse or otherwise undependable historical outcome data, such as instances of civil unrest. The frequentist approach, on the other hand, treats probability strictly as a product of past events. Hence, to a frequentist, if a particular event occurred 2% of the time in the past, that event has a roughly 2% chance of occurring in the future. It follows that the frequentist approach is particularly well suited to problems characterized by abundant historical outcome data and relatively stable trends, as exemplified by automotive accidents.

In the everyday language of business, probability estimation is simply an attempt at predicting unknown outcomes using estimable parameters. For instance, weather forecasters strive to predict future conditions such as air temperature and the amount of precipitation using historical trends and weather element interdependencies. Using statistical likelihood estimation techniques, forecasters can estimate the probability of different weather scenarios using historical weather data. That said, the users of the resultant estimates are rarely cognizant of which of the above two broad approaches was utilized, even though the estimates can differ substantially based on the choice of an approach. The reason for that is that while the frequentist approach only leverages historical outcomes data, the Bayesian method combines objective historical outcome data with subjective judgement, which can significantly alter forward-looking projections. As suggested earlier, when historical data are robust and trends fairly stable, the frequentist approach is likely to produce more dependable estimates because it bars the often biased subjective beliefs; on the other hand, Bayesian approach might be preferred when data are sparse and/or the phenomenon in question is highly variable, provided that care is taken to minimize the intrusion of cognitive bias and other evaluation warping effects.[10] Understanding those fairly general considerations plays an important role in sound utilization of the probability estimates.

9.4.2 Probabilistic Evidence

'Chance favors the prepared mind' wrote Louis Pasteur, nineteenth-century biologist and chemist renowned for discovering the principles of vaccination, fermentation and pasteurization (the last named after him). His words are timeless, although what constitutes the 'prepared mind' continues to change. Early in the evolution of human civilizations, prepared mind was informed by description of natural phenomena, as exemplified by writings of Aristotle, which eventually (circa seventeenth century) gave way to scientific theoretical and experimental knowledge, as exemplified by Newton's laws; nowadays, prepared mind looks to data-derived insights, derived from mining of vast quantities of multisourced data. Interestingly, whether it was Aristotle musing over the mysteries of the natural world some 2500 years ago or a modern-day physicist pouring over the vast volumes of data generated by the Large Hadron Collider,[11] one of the elementary challenges faced by both is to differentiate between informative 'signal' and non-informative 'noise', as a necessary step of discerning sensemaking evidence.

What is *evidence*? Broadly defined, it is facts or other organized information presented as justification or support of beliefs or inferences; the term itself is derived from the Latin term 'evident', or 'obvious to the eye or mind'. The notion of evidence has been the bedrock of modern science for more than two centuries, where it has been used to lend substantiation to theoretical postulates. More recently, starting in the later 1990s with medical practice, the notion of evidence has also been

---

[10] Discussed in more detail in Banasiewicz [2].

[11] Opened in 2008 and operated by CERN (the European Organization for Nuclear Research), it is the world's largest and highest-energy particle accelerator and the largest machine in the world, consisting of a 27-kilometre (nearly 17 miles) ring of superconducting magnets with a number of accelerating structures to boost the energy of the particles along the way; its goal is to enable experimental research into fundamental properties of matter.

used to guide professional practice; evidence-based management, for instance, is built around the idea of systematic identification, evaluation, synthesis and dissemination of effective organizational management practices. Given that 'facts and other organized information' framing of evidence can yield a confusing mishmash of disparate, even potentially contradictory insights, it is ultimately the 'weight of evidence' rather than individual pieces of information that form the informational basis of evidence-based practice (see footnote #13).

An aspect of evidence that cuts to the core of its validity and reliability is source. Tacitly implied in the above outlined conception of evidence is the notion's external-to-self aspect – in order to offer defensible corroboration of an assertion, supporting facts or information need to be derived from sources that are external to that assertion. In other words, offering one subjective belief as a 'proof' of another subjective belief is, in philosophical sense, tautological and in practical sense simply not credible. Scientific theories are validated with the help of independently generated data, not other abstract ideas; similarly, in legal disputes, a party's claims are substantiated by independent (to that party) facts, and in medical practice treatment strategies are compared to best established practices. In short, to be of substantive value, evidence needs to be independent of the claim it is to support.

9.5 Organizational Change

For most established organizations, making fuller use of organizational informational resources entails amending of often long-established structural and cultural norms. Implementation of decision support systems and automated decision-making discussed in Chap. 7 requires re-engineering of aspects of organizational functioning which has not only process but also cultural implications. In short, becoming more data-driven requires change.

The idea of change, however, is impossibly broad because it can be seen as an inescapable consequence of the dynamism of natural and socioeconomic ecosystems. In a sense, it might be easier to describe what change is not, rather than what it is – that ubiquity can render change imperceptible, especially if the underlying mechanism is a continuous series of small steps, a bit like ageing. Not all change, however, takes the form of continuous ad hoc adaptations – organizations also undertake discrete and deliberate alterations that can happen comparatively abruptly and can have profound consequences. For example, a business firm may opt to amend its governance structure, it can re-engineer some or all of its operational processes to better suit its evolving needs or it can implement a newly developed automated decision-making engine. Each of those purposeful changes can have a direct and a dramatic impact on at least some of the organizational stakeholders and can upend long-standing organizational processes or practices; hence it follows that planned organizational change needs to be thoughtfully and carefully managed. Even though self-initiated change is usually motivated by organizational betterment, it can nonetheless disrupt organizational equilibrium; hence a closer look seems warranted.

9.5.1   Understanding Organizational Change

Heraclitus, a pre-Socratic Greek philosopher, is believed to be one of the first thinkers to formally acknowledge the ubiquity of change, as captured in his doctrine of universal flux and its derivative, and now quite popular claim that 'change is the only constant in life'. However, it was the late nineteenth century's wave of great sociocultural changes, coupled with the rise of rationality and the embrace of the scientific method, that spurred wider interest in the nature and the process of change. It was within that context that the leading social thinkers of the era, including Herbert Spencer, began to conceptualize social change as an evolutionary process, much like the biological ones described by Charles Darwin (in fact, it was Spencer who coined the term 'natural selection', often erroneously attributed to Darwin).

Building on the conceptual foundation of the concept of social change, Kurt Lewin introduced the notion of planned change, which since became one of the foundations of the field of organizational development. Seeking deeper understanding of the mechanics of social transformations, Lewin developed his field theory, which posits that mapping out the totality and complexity of the 'field' (the environment) is key to making sense of the manner in which an 'entity' (an individual or an organization) interacts with its environment. Ultimately, he was able to capture the essence of change in a simple three-step model of 'unfreeze-change-refreeze'. Although researchers today tend to dismiss his model as overly simplistic and outdated, Lewin's conceptualization should be credited with drawing attention to the two distinct aspects of self-imposed organizational change: interactions between individuals and organizations and interactions between organizations and their ecosystems. Recognizing and expressly addressing that distinctiveness is critical, as the former captures the process of change, while the latter sheds light on factors that precipitate episodic self-imposed organizational transformations.

An important aspect of intentional organizational transformations is that organizational change is ultimately personal and as such cannot be fully understood without expressly taking into account the human dimension. Empirical evidence from social psychology and organizational behaviour lends credence to that assertion by pointing out that when one's self-interest is threatened (as discussed earlier in the Probabilistic Thinking section) by organizational change, that individual is likely to resist change. That is particularly so when change entails replacing human judgement with automated decision-making mechanisms.

9.5.2   Managing Organizational Change

No organizational transformation is truly value-free. No matter the type of system or organizational context, informational automation usually has its proponents and opponents; those who see it as benefiting their self-interest are likely to applaud it, while those who perceive it as a detriment are likely to oppose it. Overt behaviour-wise, as posited by the cognitive-emotional model, those nearly instinctive responses can be expected to influence not only how individuals handle change related tasks but also how they perceive the change itself.

Understanding and carefully evaluating and managing those realities is one of the keys to successful organizational transformations. For example, when contemplating a large-scale rollout of an automated insurance claim evaluation and management system, the overall strategy ought to be clearly mapped out, and change-enabling tactics should be controlled and carefully directed. That entails anticipating requirements and obstacles, and designing means to fulfil needs and overcome obstacles, all while staying focused on successful transition from the current to future state. Given the earlier discussed complexities stemming from the combination of personality traits and group dynamics, the use of an appropriate conceptual framework can be highly beneficial. Over the past several decades, multiple planned organizational change conceptualizations have been proposed, some quite similar and others offering drastically different perspectives.

The earlier noted Kurt Lewin's conceptualization dating back to the 1940s is commonly viewed as the first formal attempt to capture the individual and group mechanics governing the process of organizational change. Viewing organizations as highly resistant to change, Lewin theorized that to be successful – i.e. to reach the desired state and to endure in that state – organization-initiated change must begin by first disrupting the current equilibrium, to be then followed by the introduction of a new state and lastly the creation of a new equilibrium, as captured in his unfreeze-change-refreeze process (which itself was modelled on the thermodynamic process of breaking down, changing and then re-constituting of matter, as exemplified by changing the shape of a block of ice). Intuitive and simple, the three-step process has been the staple of theoretical research and practice for several decades following its introduction, but its use waned as researchers began to view it as overly simplistic and only applicable to relatively small-scale change initiatives.

More recent change management frameworks aimed to provide more process-guiding granularity, perhaps best illustrated by Judson's 1991 conceptualization [4] which views self-imposed organizational change as being comprised of five distinct phases: analysing and planning the change, communicating the change, gaining acceptance of new behaviours, changing from the status quo to a desired state, and consolidating and institutionalizing the new state. Although manifestly more granular than Lewin's three-step model, Judson's framework was also criticized as too simplistic, vague and difficult to operationalize. Not surprisingly, change management conceptualizations that followed offered considerably more operational details – among the better known of those are the Kotter's 1995 eight-steps [5] and

Galpin's 1996 nine-wedges [6] (so named due to being graphically depicted as nine wedges forming a wheel) model. An altogether different view of self-imposed organizational change management process has been put forth by Armenakis, Harris and Field in 1999 [7], who proposed an even more involved two-phase, multi-consideration model.

References

1. Gilovich, T. (1991). *How we know what isn't so: The fallibility of human reason in everyday life*. New York, NY: The Free Press.
2. Banasiewicz, A. D. (2019). *Evidence-based decision-making: How to leverage available data and avoid cognitive biases* (1st ed.). New York, NY: Routledge.
3. Kolmogorov, A. N. (1950). *Foundations of the Theory of Probability* (Originally Published in German in 1933)
4. Judson, A. (1991). *Changing behavior in organizations: Minimizing resistance to Change*. Cambridge, MA: Basil Blackwell.
5. Kotter, J. (1995). Leading change: Why transformation efforts fail. *Harvard Business Review, 73*(2), 59–67.
6. Galpin, T. (1996). *The human side of change: A practical guide to organizational redesign*. San Francisco, CA: Jossey-Bass.
7. Armenakis, A., Harris, S., & Field, H. (1999). Paradigms in organizational change: Change agents and change target perspectives. In R. Golembiewski (Ed.), *Handbook of organizational behavior*. New York: Marcel Dekker.

Chapter 10
The Future?

10.1 Human-Machine Interweaving

Aviation offers one of the best examples of human-machine integration. A few short years after Wilbur and Orville Wright famously completed the first powered flight of a heavier-than-air aircraft near Kitty Hawk, North Carolina, another, though generally lesser-known aviation pioneer, Lawrence Sperry,[1] created the first successful autopilot. Nowadays, it is hard to think of an area of aviation, from design to pilot training to piloting itself, that does not entail a significant degree of human-machine interweaving; the now ubiquitous drone flying devices even take that idea a step further, towards human-machine cooperation, discussed in more detail in Chap. 7. In general, systems consisting of hardware and software that allow user inputs to be translated into signals for machines that, in turn, provide the required result to the user are now commonplace in the ever more automated manufacturing sector; moreover, the ongoing growth of the Internet of Things functionality is quickly rendering dependence on 'intelligent' systems ubiquitous.

However, the coming together of human and machine 'physicalities' is already decades in the making; the next logical step is to begin to merge machine capabilities with human consciousness. The modern data infrastructure with it rich real- or near-real-time detail manifesting itself in maddening varieties and volume simply can no longer be analytically consumed using the long-established, tinkering-like conjectures and refutations model of knowledge discovery discussed earlier in Chap. 3. That is not to say that the time-honoured hypothesis testing-based analytics are obsolete, for that is not the case; it does, however, mean that the bulk of big data value creation opportunities are in systems that combine innate human imagination

---

[1] It was known as the known as 'gyroscopic automatic pilot'; Sperry is also credited with developing the artificial horizon, which is still in use today.

© The Author(s), under exclusive license to Springer Nature
Switzerland AG 2021
A. Banasiewicz, *Organizational Learning in the Age of Data*, EAI/Springer
Innovations in Communication and Computing,
https://doi.org/10.1007/978-3-030-74866-1_10

with virtually unlimited computing power of electronic systems. Since the bulk of what could be called 'creative cognitive computing' that powers imagination occurs subconsciously, and thus it is lightning-fast, the bulk of 'rational cognitive computing' is infinitesimally slow since it happens in the conscious sensemaking realm, and combining human imagination and machine data crunching seems like a natural step towards the earlier discussed data-enabled creativity. Maybe, after all, Walt Disney was right; maybe if we can dream it, we can indeed do it. The challenge of separating signal from noise (in data) is still there, but advanced systems that combined adaptive signal processing with neural interfaces while also leveraging artificial reasoning functionality are already beginning to integrate the immense computational power of digital systems with the human capacity to experience insights and apply intuition.

For example, a recent DARPA, the US defence research agency, project produced a revolutionary prosthetic hand that connects directly to the brain and allows paralyzed individuals (who are unable to experience physical sensation) to actually 'feel' physical sensation. The device is sensitive enough to enable its users to identify exactly which mechanical finger has been touched, even if more than one have been touched at once. To do that, DARPA scientists had to first figure out how to route electrical signals from touch receptors in the prosthesis back to the sensory cortex in the brain. Once that challenge was surmounted, by wiring a sense of touch from the mechanical hand directly into the brain, the researchers demonstrated the possibility of seamlessly integrating human cognitive functions (in this case, thoughts) and biotechnological devices to bring about near-natural functioning. Long a staple of futuristic sci-fi movies, cyborgs are no longer just a cinematic fantasy.

10.1.1   Emulation, Simulation and Virtualization

The original Sperry autopilot *emulated* some of the human pilot functions, which is to say it duplicated basic flight functionality allowing the pilot to perform other tasks or to just rest. Flight *simulators*, widely used to train pilots, replicate the actual flying experience in a classroom-type setting, which allows trainees to safely learn how to operate complex flight systems. When designing new aircraft or new flight systems, *virtualization* is commonly used, which is the art and science of emulating or simulating the function of an object digitally in a way that is identical to that of the corresponding physical reality. While on the one hand emulation, simulation and virtualization all serve a different purpose thus are ultimately complementary, those distinct bundles of functionalities can nonetheless be seen as manifestations of the underlying technological progress. While considerably more complex today, the original Sperry autopilot was a simple mechanical device, flight simulators are effectively special purpose electronic computers, and virtualizations are complex systems that encompass high-power computers, advanced machine learning algorithms and appropriately curated and sufficiently rich data. If the technological advancement that underpins the emulation-simulation-virtualization progression was to be extrapolated into the future, what can be expected next?

10.1.1.1 Singularity?

The technological singularity hypothesis argues that humanity is headed towards the creation of artificial superintelligence in the form of accelerating-self-improvement-capable computers running software-based general artificial intelligence, destined to, ultimately, far surpass human intelligence. Though the idea may seem far-fetched at first, it is actually quite rational: Since one of the ultimate manifestations of human intelligence is the creation of artificial intelligence, an ultra-intelligent system initially designed by mankind can be expected to continue to improve upon its initial design leading to what is often referred to as 'intelligence explosion', ultimately leaving its creator – the mankind – far behind. This is just an idea for now, though one that encapsulates a seemingly inescapable conclusion to the clearly evident accelerating technological progress, especially in the area of information technology. In that context, some futurists are promoting the view that within the next couple of decades we, the humans, will connect our neocortex, which is the part of brain responsible for thinking, to the cloud, which is effectively a metaphor for the Internet. That line of reasoning may also seem far-fetched, but would not the now ubiquitous addiction to the Internet as a source of information and the primary mean of communication seem far-fetched just a couple of decades ago?

Central to the technological singularity hypothesis is the idea of 'intelligence', which is seemingly obvious, at least until an attempt is made to frame it in terms of what it is and it is not. To some (primarily academic researchers, but also behavioural health practitioners), intelligence is an 'entity' which is fixed and unchanging; thus it is present or not; to others, it is an 'incremental property' that can be developed through effort (most notably, learning). When approached as an empirically measurable scientific concept, the most common, taxonomy-wise, representation of intelligence is a hierarchical structure with one or more broad abilities at the apex of the hierarchy and one or more levels of narrower abilities arranged below the broad abilities, as graphically summarized in Fig. 10.1. Here, the latent dimensions of abilities (shown by the dark-shaded circles) are arranged in a subordinated fashion, where a more general ability, i.e. the higher-order factor, is comprised of several more narrowly defined abilities, i.e. Factor 1, Factor 2, etc. Only the lower-order abilities (Factors 1, 2 and 3 in Fig. 10.1) are directly measurable or observable (as depicted by the Indicator 1, 2, etc. boxes); the high-order ability (higher-order factor) is analytically discerned, meaning it is inferred from the lower-order abilities. (An alternative to the hierarchical model is the so-called bi-factor model, so named because it assumes that each measurable indicator points towards two or more factors; it is different primarily insofar as it lacks the higher- vs. lower-order subordination).

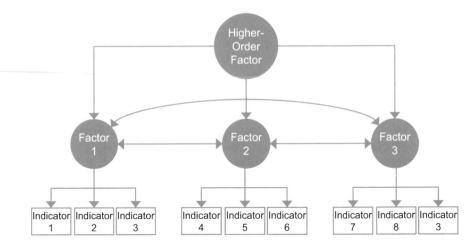

Fig. 10.1 The hierarchical structure of intelligence

Looking beyond the formalities of scientific conception of intelligence, the general framing of that idea points towards the ability to adapt to and to shape one's environment; as a sensemaking construct, it is both discovered and invented. It is discovered, because as illustrated by Fig. 10.1 it is a target of ongoing, dedicated research efforts; it is invented because it is a latent construct that is understood differently in different situations, cultures and other settings. Moreover, the consensus view among academicians and practitioners alike is that while intelligence can be broadly characterized as an aspect of individuals' information processing prowess, there is no universally agreed-on mechanism enabling full and objective assessment its true essence. In short, intelligence is, at best, partly understood as an idea and partly measurable as a property; given that, while the earlier mentioned notion of 'ultra-intelligent' systems is compelling, it is not yet clear what, beyond the ability to rapidly process vast quantities of data, that would entail. It is entirely possible that the truly intertwined human and artificial sensemaking and information processing capabilities will give rise to a whole new realm of intelligence.

An informative illustration is offered by a yet another forward-looking DARPA initiative: the Big Mechanism project. Here, researchers are developing systems that use advanced natural language processing algorithms to read scientific journal articles about a particular medical problem, such as coronavirus, and feeding the machine-generated insights (no human in the loop) into a continuously evolving model of coronavirus-related genetic factors. As could be expected, the text mining system can process thousands of scientific articles each day, which is orders of magnitude more than a team of scientists could manage. Then, working in collaboration with human experts who study those results, the uber-machine learning system is generating promising new avenues of detection, treatment and prevention.

Will such technologies render the good old-fashioned human reviews obsolete? Not long ago, *wetware,* or technologies modelled after biochemical processes of biological life forms, were in the realm of science fiction, but are now an operational reality. Armed with high-resolution neural devices, neuroscientists are able to not only decode the electrochemical signals in the brain but also to encode instructions directing neurons to respond in the desired manner (which is how, for instance, the earlier mentioned robotic prosthetic hand is capable of creating physical touch-like sensation). Setting aside the formidable technological complexities, it is nonetheless easy to imagine machine-augmented human cognition, leading to much faster assimilation of knowledge, improved retention and development of new cognitive skills. When framed in that context, the idea of technological singularity seems a lot less otherworldly.

10.1.2 *From Doing to Thinking to Being*

When considered from a more abstract perspective, technological advancement can be seen as a manifestation of a progressively higher-order enablement: Prosthetics and self-driving cars enable their users to *do* more, while big data and machine learning algorithm-powered systems help their users *experience* more. Does that mean that digital technologies can ultimately remould the essence of what it means to be human, in the sense of how one experiences reality? And going a step further, will the very essence of what constitutes 'reality' change?

The idea of axial periods, first discussed in Chap. 1 and later expanded on in Chap. 4, is rooted in the biological notion of punctuated equilibria, which posits that human development is marked by isolated episodes or rapid change followed by long periods of little or no change. The idea behind axial periods is that in contrast to the slow, smooth and steady march of evolution suggested by Darwinism (and its many derivatives), historical evidence clearly suggests that human cognitive development[2] is fuelled by yet-to-be-explained developmental jolts, or discontinuous leaps in sensemaking capabilities, which propel human development to the proverbial 'next level'. There are no good explanations for what lit (and then dimmed) the fire of reason close to 3000 years ago, nor is there a compelling explanation for what, exactly, precipitated the incredible burst of science after nearly 2000 years of the cognitive lull known as the Dark Ages; there might, however, be an explanation for the currently underway transformation to the Age of Transcendence. Consider Fig. 10.2.

The Age of Reason was not cumulative. Confucius in China, Buddha in India and Plato and Aristotle in Greece all contributed to lighting of the fire of reason, but by and large, their contributions were not incremental, in the sense that their ideas did not build on one another, but instead offered different perspectives (perhaps best illustrated by the fact that while Aristotle was Plato's student, their philosophies

---

[2] The same applies to biological development; in fact, the idea of punctuated equilibria, as developed by Gould and Eldredge, was meant to address numerous inconsistencies (and the resultant leaps of faith often embodied by the famous 'missing link' notion) between the slow and steady evolution postulated by Darwin's theory and fossil-based evidence.

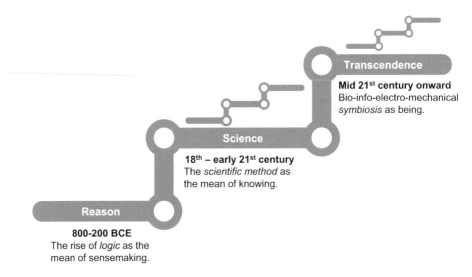

Fig. 10.2 Axial Ages' mechanics

were starkly different: Plato's was abstract and utopian, whereas Aristotle's was empirical and practical). Walled in by cultural and language divides, Confucius, Buddha, Plato and other great thinkers of antiquity each contributed powerfully to human sensemaking, but their wisdom did not trigger wider socio-economic transformations, as judging by the many centuries of no meaningful intellectual and technological progress separating the Axial Ages of Reason and Science.[3]

Science, however, is by its very nature cumulative; moreover, the core aspects of scientific thought are communicated using universally understood language of mathematics. Consequently, as illustrated in Fig. 10.2, in contrast to the 'flash in the pan' emergence of reason, once ignited (by means that are, once again, not entirely clear), science begot more science as one set of brilliant insights led to another and then another. And so while for the most part science is pursued one project at a time, when considered from an aggregate perspective, it is the ultimate team endeavour.

Moreover, the now-emerging Age of Transcendence is a direct result of the growth and maturation of science; in a sense, it could even be seen as being another stage of scientific and technological development. However, the Age of Science was about 'thinking and doing', whereas the Age of Transcendence is about 'being'. The information technology-enabled cognitive blossoming is gradually enabling mankind to peer into previously inaccessible realms through transformative

---

[3] The very abbreviated argument presented here is not meant to suggest that the wisdom of the great thinkers of antiquity had no impact at all, but merely that it had no immediate advancement-related impact. For instance, the ideas of Plato and Aristotle were rediscovered centuries later, as for a long time, especially during the Middle Ages period in Europe, they were painted as pagan heresy (even if two of the greatest Christian theologians, Augustine of Hippo, also known as Saint Augustine, and Thomas Aquinas, quietly 'repurposed' Plato's ideas of abstract forms into core parts of Christian doctrine, most notably the Holy Trinity).

bio-info-electro-mechanical symbiosis discussed earlier and, quite literally, other-worldly advancements such as quantum simulations. It is no longer about better thinking or being able to do more – it is about being more. It is about unleashing completely new ways to learn, made possible by combining the most human aspects of mankind – the ability to reason, imagine and feel, in a metaphysical sense – with data-powered technologies that are beginning to enable us to transcend our current physical, three-dimensional limitations in search of deeper understanding of the nature of reality, and the essence of our own existence, perhaps even allowing us to redefine our existence.

10.2 Transcendent Creativity

The idea of 'data-enabled creativity' has been a recurring theme throughout the preceding chapters, primarily because it encapsulates the ultimate goal of the organizational learning perspective discussed in this book: to go beyond the established 'what-is'. Chapter 7, in particular, explored that notion in the context of the MultiModal Organizational Learning (MMOL) typology, which argued that human sensemaking and machine learning ought to be seen as a singular, overarching mechanism for transforming data into insights. The focus of the current chapter has thus far been on informational automation, or more specifically, on emerging bio-info-electro-mechanical symbiosis, and the impact it is beginning to exert on the very essence of human experience. But, when the earlier discussed data-enabled creativity ideas are placed in the context of artificial augmentation of the brain's natural (i.e. biological) functioning, an even broader conception of the general

Fig. 10.3 Data-enabled creativity: the transformative power of interacting modalities

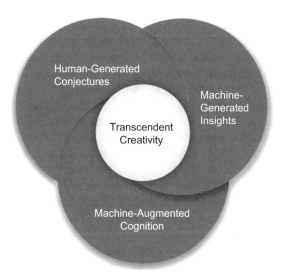

notion of 'creativity' emerges, one characterized here as *transcendent creativity*, graphically summarized in Fig. 10.3.

The 'transcendent' label is meant to communicate a couple of the key defining characteristics of the changing conception of creativity. First, the long-standing understanding of the idea of creativity equates that notion with the use of imagination, which in turn is (typically implicitly) framed as forming of new ideas, not already present to or in the senses. But, is creativity an act of conjuring up something completely otherworldly, meaning something that has not been previously experienced, or is it creating new arrangements or forms using known elements? Evidence from the field of child psychology, which focuses on cognitive and behavioural human development from birth through adolescence, points to the latter as being a more plausible framing of creativity, as it suggests that the bulk of what ultimately comprises creativity is a product of ongoing learning. And what is learning? It is purposeful and structured immersion in a domain, situation or a problem, aimed at developing lasting understanding. As so eloquently phrased by W.B. Yeats, a poet, 'education [and thus learning] is not the filling of a pail, but the lighting of a fire'. Those words ring particularly true in the Age of Data, where immersion in rich data patterns and relationships, in addition to computer-supported and/or rendered 'reality', in the form of simulations, could spark ideas that might not otherwise arise.

While the total storage and processing power of the brain is formidable (as noted earlier, more than enough to record every moment of one's life), human ability to engage in cognitive tasks is nonetheless constrained by numerous individual-level recall and evaluation limitations in the form of memory decay, limited channel capacity, brain plasticity and cognitive bias, in addition to also being adversely impacted by social effects such as groupthink. In view of that, the idea of transcendent creativity is also meant to communicate the shifting focus of what it means to 'know' and to 'learn'. In a simplistic sense, the dominant conception of learning and knowing implies a twofold task of (1) committing specific information to long- or short-term memory and (2) developing the ability to recall that information on as-needed basis. Going forward, however, when 'making' and 'doing' are going to become progressively more dominated by automated and autonomous systems, the value of 'knowing stuff' will continue to decline, while the value of 'knowing how to know' will continue to grow.

Turning back to the idea of transcendent creativity, as depicted in Fig. 10.3, it is comprised of three distinct elements: human-generated conjectures, machine-generated insights and machine-augmented (human) cognition. Human-generated conjectures represent a continuation of the scientific method, which frames knowledge creation in the context of empirical tests of theoretical ideas, meaning those derived from established theories, general principles or some other assumptions. The logic of this approach to knowledge creation implicitly assumes that the informational essence of conjectures that are to be tested, be it in the form of formal theories or informal beliefs, holds the greatest knowledge potential, or stated differently, the conjectures to be tested are the right questions to ask. And as could be expected, it is a mixed bag, as some questions lead to meaningful advancements of knowledge, while other ones perpetuate the established, and perhaps not entirely sound, beliefs. Influences such as researchers' mental models (discussed at length in Chap. 3) are among the important factors that shape not just what questions are asked but also how they are evaluated; in short, human conjecture-derived knowledge is never truly value-free and thus never truly objective.

Lacking consciousness, machine learning systems are unaffected by perception-shaping mental models; moreover, their practically unlimited data processing capabilities make those systems ideally suited for finding previously unknown patterns and relationships in vast pools of data. Thus in contrast to human-generated conjectures, which are typically rooted in prior beliefs,[4] machine-generated insights are truly emergent, in the sense of being value-free (although, as touched on earlier chapters, it could be argued that there are no truly value-free man-made algorithms). And while at least some of machine-generated insights might represent already known patterns or relationships, the possibility of finding previously unknown, unnoticed or perhaps even dismissed associations is nonetheless ever present.

Whereas the human-generated conjectures and machine-generated insights can be seen as supplying informational inputs in the broadly defined creative process, machine-augmented cognition supplies more capable sensemaking mechanisms. Setting aside the rich and colourful science fiction implications, to make better use of the vast amounts of information and knowledge that are already available and being added to at a dizzying rate, human cognition needs to transcend its biological limits. For instance, only being able to cognitively consider, on average, between five and nine distinct pieces at a time (human channel capacity) necessitates the type of prioritization that favours prior beliefs and thus effectively short-changes new ideas. At least notionally, machine-augmenting human cognition is not different than building machines that enable humans to fly high up in the air or dive deep into the oceans – it is just another manifestation of human ability to go beyond human biological limits. Ultimately, it is the power of the human mind, not the confines of the human body that will map out the course of human evolution.

A distinct but related idea is that of educational preparation. In order for organizations, both commercial and nonprofit, to be able to learn from data, the learners need to know how to learn, which is an idea that goes to the heart of the current educational mindset. Given its importance, that topic is discussed in more detail next.

10.3 Liberal and Digital Arts

The core of Western higher education tradition is commonly known as *liberal arts*, an umbrella term that encompasses natural sciences, social sciences, arts and humanities. In the United States, numerous institutions of higher learning describe

---

[4] As clearly exemplified by Copernicus, who drew radically different conclusions from what was commonly available data, all such generalizations are subject to some notable exceptions, which is why human-generated conjectures are characterized as being 'typically', meaning most of the time, rooted in prior beliefs.

themselves as liberal arts colleges, and so it is rather surprising that the question of 'why' is not asked more often. Why a grouping of subjects as diverse as philosophy, history, mathematics, biology, psychology, sociology and fine arts, to name just a few constituent academic areas, are labelled 'liberal arts'? Natural and social sciences are clearly not 'arts', and neither are the humanities; moreover, why are those learning pursuits characterized as 'liberal'? Given that liberal connotes being open to new ideas, aren't all academic domains endemically liberal?

The notion of liberal arts education first emerged in ancient Greek and Roman societies as a manifestation of different educational needs, and the underlying rights, of the ruling and serving classes. The term itself traces its roots to the Latin word 'liberalis', which in the context of education was meant to spell out knowledge that was 'appropriate for free men'; more specifically, it was an expression of the belief that the study of grammar, rhetoric and logic was an essential enabler of free citizens' of ancient Greece and Rome ability to participate in civic life. That idea was carried forward into the feudal societies that followed in their wake, where clergy and nobility (who replaced the free citizens of Greece and Rome as the ruling class) carried forth the core ideals of liberal arts education, which was also gradually expanded to included mathematics, music and astronomy, to comprise what was considered 'proper' education for the ruling classes. During the same time periods, education of first slaves and later commoners was limited to specific, typically servitude-related skills, now broadly characterized as vocational training (it is worth noting that what could be framed as advanced vocational training, such as in medicine, law or engineering, simply did not exist in those days). If the spirit of the idea behind liberal arts education was to be applied to the modern context of the Information Age, what would be its modern equivalent?

The idea of education as an enabler of full societal participation is undeniably powerful, even timeless; at an individual level, it offers a pathway towards self-actualization, which is the fulfilment of one's talents and potentialities.[5] In Greek and Roman societies, logic and communication (i.e. the study of rhetoric, grammar and logic) were essential to reaching one's potential as a (free) member of those societies; relying on the same line of reasoning, it seems reasonable to suggest that informational literacy, as described in Chap. 9 (see Fig. 9.1), is essential to becoming a self-actualized participant in the current, digital society.

It is, however, important to note that informational literacy is a complement, not a substitute for liberal arts, because the idea of self-actualization implies the capability to 'know' that encompasses understanding of past contributions to human understanding, as well as the ability to peer into the future. With that in mind and given that the essence of liberal arts education is to immerse oneself in the cumulative pool of human knowledge as seen from the perspective of enlightening philosophical thought, great scientific achievements and imagination-awakening artistic

---

[5] Commonly attributed to Abraham Maslow and his hierarchy of needs, first proposed in his seminal 1943 paper 'A Theory of Human Motivation', the term and the idea of 'self-actualization' can actually both be traced to Kurt Goldstein's 1934 book titled *The Organism: A Holistic Approach to Biology Derived from Pathological Data in Man*.

contributions, robust grounding in liberal arts is essential. It is so simply because, today, the role of that broad and varied body of knowledge is to instil a sense of greater purpose and understanding of who we are, as a specie, and where we are going. And that is where the goals of liberal arts education and the benefits of informational literacy intersect: Firstly, as more and more informational content is digitized, meaningful utilization and ultimately absorption of that content calls for digital and analytic literacy; without robust foundation of information utilization know-how, one's ability to immerse oneself in human heritage studies will effectively be limited to just the traditional media and learning modalities. Secondly, as more aspects of work and life are also digitized, becoming a contributing participant of modern society is also contingent on what could be characterized as digital realm participation skills; moreover, the earlier discussed goal of fostering data-enabled creativity is unachievable without the foundation of informational literacy.

Implied in that line of reasoning is a growing entanglement of the traditional (i.e. liberal arts) and the emergent (i.e. informational literacy) learning dimensions, which can have profound implications for the very ideas of learning and education, especially in application-oriented domains of knowledge. Perhaps the most obvious of those is the growing importance of discovery-based learning, which can bring about re-interpretations and re-examinations of past insights, in addition to new contributions. In fact, the idea of a person being 'qualified' or 'competent' might need to be conditioned on demonstrated ability to engage in ongoing learning, something along the lines of continuing professional education that is commonplace in the practice of law or financial services.

10.3.1 Managing Organizational Know-How

It is intuitively obvious that in the Information Age organizational decision-making is inextricably tied to the ability to translate abundant varieties of data into meaningful information. The resultant surge of interest in data analytics as a core organizational competency is perhaps best illustrated by the rise of data science as a new field of study and practice and even more so by that profession's rapid ascendance to becoming one of the most talked about employment areas. Interestingly, the focus seems to almost entirely on creation, rather than utilization of data analytic insights, even though, it is not those insights as such, but rather the manner in which those insights are utilized that creates value; a simple scenario might help to illuminate the difference.

A cohort of engineering students immerses itself in their chosen field of study; upon graduating, most accept employment with established companies, while one chooses to pursue the development of a breakthrough technology, which ultimately spawns a new company. This not so hypothetical a scenario underscores the need to extend the current conception of what constitutes knowledge – in the era defined by volatility and change, it is myopic for organizations to frame knowledge as just a static collection of information and skills; it is the capability to innovatively use information and skills that ultimately gives rise to organizational value. Overtly, the hypothetical engineering students acquired essentially the same domain knowledge, yet the innovator among them somehow managed to get more out of that initially shared knowledge, most likely because of that individual's ability to mentally manipulate it in a unique way. Hence from the organizational perspective, it is not 'having' but 'using' knowledge that matters; similarly, it is not the ability to analyse data, but the capability to use data analytic outcomes productively that matters.

The preceding scenario highlights the importance of *thinking* as a constituent element of knowledge: Though undeniably one of the more elusive and hard to objectify aspects of cognitive functioning, thinking is the actuating force that turns informational potential into realized value. Within the realm of organizational decision-making, it is that something that transforms generic knowledge into value competitively advantageous insights; oftentimes, it is the difference maker that helps to explain why organizations that have access to essentially the same types of information end up on markedly different economic trajectories (Walmart vs. K-Mart quickly come to mind). Together with existing informational assets (knowledge) and the ongoing process of knowledge acquisition (learning), innovative organizational thinking comprises the organizational know-how, and as such it is at the core of organizations' competitive advantage in the Information Age.

But here comes the critical part: Since the onset of modern Industrial Age, business organizations entrusted the explicit aspects of their workforce training to educational institutions, while the tacit or experiential side of learning was essentially left to chance. By and large, colleges and universities' liberal arts education offers robust 'learn-to-think' intellectual training, but some aspects of professional preparation offered by some of the same educational institutions, most notably business education, have been criticized as being overly theoretical, divorced from practical problems, jargon-laden and generally numb to the needs of practitioners. At the same time, it is rarely a stretch to characterize experiential knowledge prized by practitioners as catch-as-catch-can, in view of the often-considerable professional competency variability across individuals with overtly comparable professional experience. And just as was the case in the earlier example of engineering students, overtly similar qualifications can produce vastly different outcomes. All considered, in the era in which creative use of available information and thoughtful fostering of new knowledge are of paramount importance, organizations simply must take a more active interest in their employees' 'learning to know' and perhaps to a somewhat lesser extent, 'learning to think' processes and outcomes.

To maximize the economic value of their value-producing assets, business and other organizations need to more thoughtfully and purposefully address creation, dissemination and cataloguing of organizational know-how, map out the foundational skills and competencies expected of their value creation-contributing constituents and develop and institute sound organizational learning strategies and tactics. While devising those comprehensive knowledge creation, dissemination and usage mechanisms, organizations also need to take affirmative steps to mitigate the potentially adverse influences of various individual-level effects, such as cognitive bias or informational literacy, as well as broader structural factors, which commonly take the form of groupthink, and restrictive organizational structure and culture. Not a trivial undertaking, to be sure, but being competitive in the Information Age demands no less.

10.3.2 The Evolution of Learning

Recently, scientists have demonstrated that organisms devoid of brains (and nervous systems), such as simple unicellular bacteria, are capable of what is known as 'habituation learning'. It is a form of rudimentary learning that has long been known to exist in all animals, but there is now clear empirical evidence pointing to its existence in far more basic, non-neural life forms; it entails a decrease in response to a stimulus after repeated exposures, such as becoming accustomed to initially bothersome sounds (hence a frequently used term of becoming 'habituated' to a condition). That finding suggests that learning, in all its many manifestations, can be considered one of the most fundamental characteristics of living organisms, which is in a sense surprising given that, intuitively, learning connotes some degree of intelligent behaviour.

Organizations, of course, do not behave like cognition-lacking bacteria. While the reasoning behind some organizational decisions might at times be perplexing, organizational choices are ultimately products of reasoned action, and thus the scope of organizational learning is ultimately limited to conscious, purposeful behaviours. The MultiModal Organizational Learning (MMOL) conceptualization, first introduced in Chap. 1 and further expanded on in Chap. 6, offers a broad framing of the core learning modalities and learning-related tasks that encompass the full scope of organizational learning means and modalities. The one aspect of learning that has not yet been expressly addressed is expected outcomes, which can take one of two general forms: assimilation of enduring truths or construction of new knowledge. Figure 10.4 offers a graphical summary of the expanded MMOL model.

At the most aggregate level, the MMOL conceptualization expressly differentiates between learning *modes* and learning *tasks*. The former is framed as two meta-categories of reason- and technology-based learning, each comprised of two distinct but interdependent mechanisms of knowledge creation (experiential and theoretical and computational and simulational), while the latter differentiate between creative and choice and negotiation tasks. The interactions of different learning modes and learning tasks produce two distinct types of learning *outcomes*: assimilation of enduring truths and construction of new knowledge. The first of the two types of learning outcomes, enduring truths, are framed here as a broad category of generally agreed on, foundational, objective body of knowledge, perhaps best exemplified by rules of mathematics, fundamental concepts of physics or biology or principles of social justice. As implied in their characterization, enduring truths are largely unchanging, across time and contexts, which means they transcend individual-level experience, which implies that those truths need to be understood the same way by all learners. Those 'tried and true' elements of general knowledge play an important

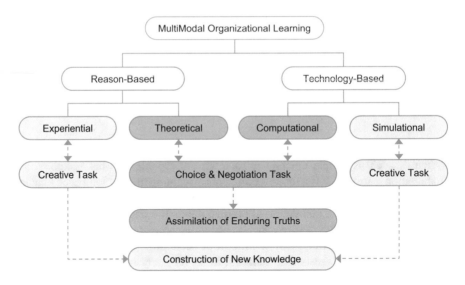

Fig. 10.4 The expanded MultiModal Organizational Learning model

role as tools of everyday rational reasoning; thus within the confines of organizational learning, as well as learning in general, enduring truths are the foundation on which pursuit of new knowledge is built. As famously noted by one of the greatest scientific minds, Isaac Newton, 'if I have seen further, it is by standing on the shoulders of giants'.

New knowledge, on the other hand, is precisely just that: newly found or developed insights; in a very general sense, new knowledge can be framed as the desired outcome of the earlier discussed data-enabled creativity. It follows that in contrast to enduring truths, which by their nature are universally accepted and, at least in the near term largely unchanging, new knowledge can be situational and individualized and thus ultimately speculative in a broader sense. The implied enduring truths – new knowledge duality – notionally parallels the logic of dual-aspect monism, more commonly known as body-mind duality. Assimilation of enduring truths is body-like in the sense that it has a clearly defined essence and structure, and its mechanics are predictable; on the other hand, discovery or creation[6] of new knowledge can be seen as mind-like because the inner workings of the underlying processes often defy

---

[6] While a deeper investigation of the difference between 'discovery' and 'creation' of knowledge falls outside of the scope of this book, it is at least worth mentioning that the said distinction hides some rather weighty philosophical considerations, not the least of which manifests itself in the fundamental incommensurability of the positivistic (i.e. there are absolute truths waiting to be discovered) and relativistic (i.e. no knowledge claim is absolute as all claims are shaped by individual perspectives) viewpoints. The implied philosophical mindset that guides the reasoning outlined throughout this book aims to bridge the gap between the 'fixed' and 'relative' perspectives on knowledge by asserting that some knowledge is absolute, as evidenced by mathematics being the de facto universal language of science (e.g. the value of π, which is the ratio of the circumference of a circle to the diameter of that circle, is about 3.14 for any circle, anywhere and every-

objective characterization, which is why new insights can be quite unexpected. While, on the one hand, what is initially seen as new knowledge can ultimately be shown to be untrue, on the other hand, this is also the realm in which great ideas are born, ones that go beyond long-held beliefs or altogether break with those beliefs, as exemplified by scientific leaps of Copernicus or Newton or the creative genius of da Vinci or Mozart. And that is precisely the challenge of organizational learning: how to validly and reliably differentiate between flashes of genius and, well, flights of fantasy.

There is little doubt that the ability to go beyond what is currently known or believed cannot be taught, because teaching is largely a means of transferring the knowledge of 'what-is', whereas new, norm-breaking ideas spring from the vaguely defined – and poorly understood – art of imagining of 'what could be'. When freed from limits imposed by currently in place knowledge and beliefs, human brain can become infinitely more productive, both in terms of volume and diversity of newly conjured up ideas. That is because of the brain's computational – and thus creative – potential is enormous, which is due to its inner workings combining analog as well as digital properties. Very briefly, in contrast to artificial computational units, such as those used in machine learning algorithms, which are digital, the building blocks of human brain, neurons, are analog. Using an ever-changing continuum of signals, analog computational units have vastly greater computing power than discrete value-based digital units. At the same time, however, the manner in which neurons communicate is binary, and thus digital-like (because they either fire or not), which implies the ability to construct distinct, standalone ideas. But there is more: Information stored in the brain is represented using quantum-state-like, non-deterministic statistical approximations, whereas digital computers rely on deterministic binary logic, a difference that further accentuates the enormous ideation potential of the brain. All considered, when provided with richer sets of inputs, as, for instance, those derived from data-driven simulations, the brain's capacity to conceive can grow exponentially. And that is before the earlier discussed bio-info-electro-mechanical cognitive enhancements are taken into account. Harvesting that potential is both the opportunity and the challenge of organizational learning.

10.4 Up to the Challenge?

It has been argued earlier that the ongoing sensemaking evolution of the roughly 6000-year-old human civilization has been propelled forward by the three discontinuous advancement-inducing axial periods of Reason, Science and the currently underway, Transcendence. The Age of Reason gave rise to formal, even somewhat institutionalized pursuit of knowledge as embodied by the Great Library of

---

where), while other knowledge is relative in the sense of being at least somewhat shaped by cultural, historical or other perspectives.

Alexandria in Egypt or Plato's Academy in Athens, Greece, and it can be seen as the first chapter in an ongoing quest to untangle the mysteries of nature and of the essence of human existence. The Age of Science gave birth to the modern conception and tools of science and technology, and it also gave rise to the modern conception of occupational profession. Individually and jointly, both of those axial periods broadened and deepened the idea of what it means to learn and to know, which ultimately spurred creation of new institutions and new ways of living and working, all while spelling doom for many of the 'old ways', which begs a number of questions, such as: Will the currently underway transition to the Age of Transcendence give rise to similar discontinuities? Will it precipitate previously unimagined or even unimaginable opportunities for individual growth and organizational development?

The current twenty-first century is a crucible moment for the mankind. While 'far-out' ideas such as the earlier discussed singularity hypothesis are, at the very least, speculative, imagination-jarring bio-info-electro-mechanical symbiosis, perhaps best exemplified by DARPA's 'feeling prosthesis' work, offers a sneak peek at a hard to dispute evolutionary path, which can be seen as a natural extension of the last two centuries of scientific progress. While on the one hand future is known to no one, on the other hand it seems more than probable that the mankind's problem-solving and creative genius will enable us to rise above our biologically rooted cognitive limitations, just as the technological advances that triggered the first Industrial Revolution propelled mankind to rise above our physical limitations. And so rather than being left behind by some form of super-intelligent artificial system, it seems more likely that mankind will ascend, through bio-info-electro-mechanical symbiosis, to that uber level of intelligence.

What do those obtuse ideas have to do with the everyday reality of organizational learning? To the degree to which those trends are suggestive of general societal evolutionary patterns, they underscore a sense of urgency that should mark organizations' skills and competencies focused on 'retooling' efforts. This book is rooted in the idea that broadly defined utilization of available data is the core part of organizational value creation processes, but extraction of value out of data demands robust data analytic capabilities. Those capabilities span a continuum of know-why, know-what and know-how, as encapsulated by the summary notion of informational literacy, which expressly differentiates between digital, data, analytic and communication literacy. In a sense, there is nothing novel about emphasizing the importance of those skills, the importance of which is now almost taken for granted, but that does not mean that all organizations are taking meaningful strides towards retooling of their organizational skillsets. There are always seemingly more urgent near-term priorities, and so it is easy to put informational literacy development efforts on the proverbial 'back burner'. Doing so, however, can be a monumental mistake as it effectively curtails long-term organizational success, perhaps even the very organizational viability; it is akin to a sports team choosing to not invest in developing younger players.

It is well-known that the rate of technological change, which is already rapid, is accelerating; competition, both in the commercial and non-commercial realms, is in hyperdrive, and socio-politico-economic volatility is the new normal, all of which underscores the need for organizational problem-solving agility and creativity. To meet that impossibly broad challenge, organizations, which at their core are collections of individual contributors, need to transform themselves into collections of informationally literate individual contributors. That is not to say that teamwork is no longer important – it means that to be a meaningful contributor to a team, an individual needs to be endowed with meaningful skills and competencies. To once again use a sport parallel, what is the value to a soccer team of a player who lacks the basic ball handling skills? Yes, the general personality traits that are indicative of being able to work in groups are important, but the ability to make materially meaningful contributions is paramount.

In every transition there are winners, typically those who correctly anticipated the oncoming changes and took steps to ready themselves accordingly, and there are losers, or those who simply did not prepare for what is next. It matters little, really none outside of academic case studies, why those who did not see and prepare themselves for the unfolding future did not do that, because there are many individual reasons that all lead to the same outcome. The Eastman Kodak Company invented digital photography in the 1970s, long before it became the staple it is now, but decided to kill their own innovation out of fear that it would cannibalize their highly profitable film business, to which they clung all the way to their 2012 bankruptcy. It is perhaps one of the most jarring examples of the difference between managing 'to the numbers' and managing 'by the numbers'; interestingly, another photography focused company, Polaroid Corporation, offers an equally illuminating example of that very distinction. In a more general sense, the cases of Kodak and Polaroid underscore the perils of informational inefficacy, which, in those two cases, manifested itself in value-laden evaluation of the available information. This book argues that organizations, big and small, commercial and non-commercial, can greatly enhance their value creation capabilities by means of systematically and thoughtfully learning from available data. However, it also argues that doing so is contingent on the development of broad-based, meaning cutting across all contributors to organizational value creation, skills and competencies needed to transform massive volumes of messy data into decision-guiding insights. It goes far beyond a handful or data science and analytic experts charged with 'infusing' of data analytic discipline into various organizational functions – it calls for each and every organizational function to become informationally literate, which entails a fundamental reimagining of what constitutes a productive organizational skillset. It is a formidable challenge, but that is what it will take to flourish in the rapidly emerging digital reality of the future.

Index

© The Author(s), under exclusive license to Springer Nature
Switzerland AG 2021
A. Banasiewicz, *Organizational Learning in the Age of Data*, EAI/Springer
Innovations in Communication and Computing,
https://doi.org/10.1007/978-3-030-74866-1

Printed in the United States
by Baker & Taylor Publisher Services